Anjum Sayed

Handbook of Petroleum Exp

Handbook of Petroleum Exploration and Production

2

Series Editor

JOHN CUBITT

Cover painting 'Solitude' (detail) is from the 'Wilderness' series of paintings by British artist Jenny Ryrie, inspired by the landscape of the American Southwest.
Artist contact details:
Tel: +44-1244-880781
E-mail: jryrie@outsource.freeserve.co.uk

ELSEVIER

Amsterdam – Boston – London – New York – Oxford – Paris
San Diego - San Francisco – Singapore – Sydney – Tokyo

STATISTICS FOR PETROLEUM ENGINEERS AND GEOSCIENTISTS

2nd Edition

Jerry L. Jensen
Department of Petroleum Engineering
Texas A&M University
College Station, TX 77843-3116, USA

Patrick W. M. Corbett
Department of Petroleum Engineering
Heriot-Watt University
Edinburgh, Scotland EH14 4AS, UK

Larry W. Lake
Department of Petroleum &
Geosystems Engineering
University of Texas at Austin
Austin, TX 78712, USA

David J. Goggin
Chevron Overseas C-1132
6001 Bollinger Canyon Road
San Ramon, CA 94583, USA

ELSEVIER

Amsterdam – Boston – London – New York – Oxford – Paris
San Diego - San Francisco – Singapore – Sydney – Tokyo

Elsevier
Radarweg 29, PO Box 211, 1000 AE Amsterdam, The Netherlands
Linacre House, Jordan Hill, Oxford, OX2 8DP, UK

First edition 2000
Reprinted 2003, 2004, 2005, 2007

British Library Cataloguing in Publication Data
A catalogue record for this book is available from the British Library

Library of Congress Cataloging-in-Publication Data
A catalog record for this book is available from the Library of Congress

ISBN–13: 978-0-444-56395-8

Transferred to Digital Printing 2008

DEDICATIONS

To the glory of God and to my parents, Jim and Jean . . . J. L. J.

In memory of my mother, Ina B. Lake, 1923-1994 . . . L. W. L.

To Kate for her continuing support . . . P. W. M. C.

To Holly, Amanda, and Janet – My support and inspiration . . . D. J. G.

PREFACE to the FIRST EDITION

Upon seeing this book, the reader's first reaction might be, "Another book on geostatistics?" This reaction is natural, for there are excellent texts on the subject, ranging from the seminal works of Matheron (1967) and Journel and Huijbregts (1978) through the more recent works of Agterberg (1974), Isaaks and Srivastava (1989), Cressie (1991), Hardy and Beier (1994), and Carr (1995). Some books come complete with software (Davis, 1973; Deutsch and Journel, 1992). The more general statistical literature also contains numerous excellent works (e.g., Hald, 1952; Kendall and Stuart, 1977; Box *et al.*, 1978).

Yet there are several important aspects that are not treated together in the currently available material. We write for academic engineering audiences, specifically for upper-level or graduate classes, and for practicing engineers and geoscientists. This means that most of the material and examples are based on engineering quantities, specifically involving flow through permeable media and, even more specifically, permeability. Our experience suggests that, while most engineers are adept at the mathematics of statistics, they are less adept at understanding its potential and its limitations. Consequently, we write from the basics of statistics but we cover only those topics that are needed for the two goals of the text: to exhibit the diagnostic potential of statistics and to introduce the important features of statistical modeling. The reader must look elsewhere for broad coverage on probability, hypothesis testing, properties of distributions, and estimation theory. Likewise, we will direct the reader to other papers or books that treat issues in more detail than we think necessary here.

The role of geology is emphasized in our presentation. For many engineers, statistics is the method of last resort, when no deterministic method can be found to make sense of the geological complexities. In our experience, however, statistics and geology form a very powerful partnership that aids geosystem diagnosis and modeling. We aim to show that the data and the geology often have a story to tell and analysis of one informs us about the other. When heard, this story will provide information about further sampling and the model formulation needed to emulate the important features. The alternative to reconciling the geology and petrophysical properties is to grasp any model on the shelf and let the statistical realizations cover our ignorance. Unfortunately, this latter approach only partially meets engineering needs.

About half of the book is devoted to the diagnostic, or "listening" topics. This includes the usual tools such as histograms and measures of variability, along with some newer concepts, such as using geology to guide sampling and the role of variable additivity. The other half is then aimed at methods for model development, once the important aspects of the geosystem behavior have been detected and quantified. Here, we present a variety of modeling methods, including linear regression. We devote two chapters to linear regression because it is so common and has considerable utility; we think even experienced users will find a few surprises here. The final chapter centers on several field modeling studies that range from the highly deterministic to the strongly random. In all cases, the statistical diagnosis and geology were essential to the choice of the modeling method.

The term geostatistics was intended to reflect the quantification of geologic principles and the reconciliation of the disciplines of geology and statistics. Matheron (Journel and Huijbregts, 1978, p. v) points out that "geo" was intended to reflect the structure while "statistics" signified the random aspect of assessing and modeling ore deposits. However, geostatistics has more commonly come to refer to a small part of statistical analysis and modeling. We have abandoned the term "geostatistics" in this text in favor of the older, and more appropriate, usage of Matheron (1967) of regionalized or autocorrelated variables. This frees us to cover all types of statistics that might be of benefit to the geopractitioner. Pedagogically, this allows us to show the progression of statistics from probability through correlation to autocorrelation using the same terminology and notation.

As a collaborative effort, this book has afforded all the authors many opportunities to learn from each other and expand their interests. A book written by one or two authors would have been easier to produce, keeping the logistical problems to a minimum. The result, however, would have suffered by being either much more geological or statistical. We hope the reader will also benefit from this interplay of views and agree that "life on the rocks" is not so painful as it sounds.

No work of this kind succeeds without help from others. We are especially indebted to Joanna Castillo, Mary Pettengill, Samiha Ragab, and Manmath Panda for their help. Heidi Epp and Dr. I. H. Silberberg are to be specially commended for their copy-editing and technical diligence in dealing with the multiple revisions and random errors that are the way of life in producing camera-ready books.

LWL is pleased to acknowledge the patience of Carole and the kids, Leslie and Jeffrey, for allowing him to hog the computer for so long. JLJ would like to express thanks for the patience and support of his wife, Jane, and his daughters Cathy and Leanne; the latter two have waited for that tree-house long enough! PWMC expresses thanks to Kate, William, Jessica and Hugo for their patience over the years and the Elf Geoscience Research Centre for financial support. PWMC and JLJ both wish to acknowledge the support of colleagues in the Reservoir Description Research Group. JLJ particularly thanks Heriot-Watt University for a sabbatical leave to work on this book and his colleagues in the Department of Petroleum Engineering for covering for him during that absence. PWMC would also like to acknowledge Kingston University for offering the opportunity for practicing geologists to get an appreciation of statistical methods -- an opportunity that led, many years later, to his involvement in this book. DJG expresses his love and thanks to Janet, the "mom", and Amanda and Holly, the "soccer dudettes", for their patience during the never-ending crunch times. DJG also thanks Chevron Petroleum Technology Co. (formerly Chevron Oil Field Research Co.) for support of reservoir characterization research, and his many friends and colleagues on the Geostatistics Team for spirited discussions on the broad application of this technology.

PREFACE to the SECOND EDITION

Since the first edition appeared, we have had the opportunity to discuss the book with readers and teach courses using the text. While the general consensus is that we met our two objectives, namely to exhibit the diagnostic potential of statistics and to introduce the important features of statistical modelling, there is undoubtedly more work to be done. With this in mind, we have set out to offer this revised edition, aimed as a teaching text to support university courses and industrial training.

We have made amendments in three areas. First, the discussions of some material have been expanded. In particular, conditional probabilities and Bayes Theorem (Chapter 2), correlation (Chapter 11), and Kriging (Chapter 12) have been addressed. Second, the coverage of estimation (Chapter 5), variability (Chapter 6), and modeling applications (Chapter 13) has been updated. Third, a number of errors have been corrected and the figures improved.

Since going to print there have been several new geostatistical texts - Kitanidis (1997), Dubrule (1998), Hohn (1998), Chiles and Delfiner (1999), and Olea (1999), to mention a few. As in the first edition, we will refer the reader to these texts where supplementary discussion beyond the needs of the text is required. There have also been a number of published examples, which we have drawn on to illustrate some of the concepts introduced.

Two reviews of the first edition have been helpful and encouraging for revising the text. Brian Russell in his review to the first edition of this book (The Leading Edge, July 1997), said that the book was "..definitely aimed at the engineering (rather than geological or geophysical) audience..." Brian's comments stem from our emphasis on the statistics of permeability. This focus was intentional since permeability, because of its variability in reservoirs and importance in fluid flow prediction, has to be the most important end objective of many analyses of reservoir data. Our aim was, and continues to be, to stimulate geoscientists and engineers alike to use statistics to explore the relationships between geology and petrophysical properties for the engineering end-user.

Brian Moss' review in the Log Analyst (November-December, 1998) suggests that we were able to meet our more explicit goals: to exhibit the diagnostic potential of statistics and to introduce the broad features of statistical modeling. He also suggested the work would benefit from some repackaging. We expect that this version will be a great improvement in that regard.

With the new edition there are also new people to acknowledge. First, we thank all those who took the time to discuss the first edition and related issues with us. We have benefited greatly by your ideas. Second, the authors wish to thank John Cubitt, Friso Veenstra, and Elsevier for the opportunity to make this new edition. Third, we would like to thank our students and co-workers, particularly Leon Barens, Joanna Castillo, Natalya Hunter-Williams, Dirk Seifert, and Simon Thomas, who have put forward their best and helped us in many ways. Finally, we thank Jim Jennings and I. H. Silberberg for their time and suggestions for improvements.

PWMC would like to thank colleagues at Heriot-Watt University for covering for him during a working visit to College Station, and the Jensen family for hosting him, during the revisions to the book. JLJ is pleased to acknowledge the support of Jane, Cathy, and Leanne, and extends special thanks to his colleagues at Texas A&M, University of Alaska, and Heriot-Watt for their help. Of course, any errors are the responsibility of the other authors and an errata can be found at http://pumpjack.tamu.edu/~jensen. DJG would like to thank colleagues and friends at Chevron, in both the technology and international subdivisions, for the many "white board" discussions on characterization and flow modeling. LWL acknowledges the help of the students in the geostatistics class at The University of Texas and several helpful discussions with colleagues.

ACKNOWLEDGEMENTS

We wish to thank the following for figures used here.

Elsevier Science NL for Figs 1-3 and 6-13

American Association of Petroleum Geologists for Fig. 1-4

Society of Petroleum Engineers for Figs. 3-4; 6-5; 6-10 to 12; 12-8 to 10; and 13-13

American Geophysical Union for Fig. 11-13

Geological Society for Figs. 13-14 to 18

Academic Press for Figs. 13-8 to 11.

CONTENTS

CHAPTER 1

INTRODUCTION

The modern study of reservoir characterization began in the early 1980's, driven by the realization that deficiencies in advanced oil recovery techniques frequently had their origin in an inadequate reservoir description. The litany of problems became commonplace: wells drilled between existing wells did not have the interpolated characteristics, chemicals injected in wells appeared (when they appeared at all) in unforeseen locations, displacing agents broke through to producing wells too early, and, above all, oil recovery was disappointing. Each of these problems can be traced to a lack of understanding of the nature of the distribution of properties between wells.

It is surprising that inferring the nature of interwell property distributions is not firmly rooted in petroleum technology, despite the maturity of the discipline. Engineers generally constructed a too-simple picture of the interwell distribution, and geologists generally worked on a much larger scale. Consequently, both disciplines were unfamiliar with working in the interwell region at a level of detail sufficient for making predictions. Into this gap has come the discipline of geological statistics, the subject of this book.

Geological statistics has many desirable properties. As we shall see, it is extremely flexible, fairly easy to use, and can be made to assess and explore geological properties easily. But it flies in the face of many geologic traditions in its reliance on quantitative descriptions, and its statistical background is unfamiliar to many engineers. This statement, then, epitomizes the overall objectives of this book: to familiarize you with the basics of statistics as applied to subsurface problems and to provide, as much as is possible, a connection between geological phenomena and statistical description.

1-1 THE ROLE OF STATISTICS IN ENGINEERING AND GEOLOGY

The complexity of natural phenomena forces us to rely on statistics. This reliance brings most of us, both engineers and geologists, into a realm of definitions and points of view that are unfamiliar and frequently uncomfortable. Therefore, the basic idea of this section is to lay some general groundwork by explaining why we need statistics.

We take *statistics* to be the study of summaries of numbers; a *statistic*, therefore, is any such summary. There are many summaries: the mean, standard deviation, and variance, to name only a few. A statistic is quite analogous to an abstract or executive summary to a written report that exists to convey the main features of a document without giving all the details. A statistic gives the main feature(s) of a set of numbers without actually citing the numbers. Unlike abstracts, however, the specific statistic needed depends on the application.

For example, the mean and standard deviation of a card deck have no relevance compared to the ordering, a measure of which is also a statistic.

In a broader sense, statistics is the process of analyzing and exploring data. This process leads to the determination of certain summary values (statistics) but, more importantly, it also leads to an understanding of how the data relate to the geological character of the measured rock. This last point is quite important but often overlooked. Without it, the data are "boiled down" to a few numbers, and we have no idea if these numbers are representative or how significant they are. As we shall see below, the essence of statistics is to infer effects from a set of numbers without knowing why the numbers are like they are. With exploration, the data can "tell their story," and we are in a position to hear, understand, and relate it to the geological character. The geology then allows us to understand how to interpret the statistics and how significant the results are.

As we will see numerous times in this book, statistical analysis on its own may produce misleading or nonsensical results. For example, the statistical procedures can predict a negative porosity at a location. This is often because the statistical analysis and procedures do not incorporate information about the geological and physical laws governing the properties we are assessing and their measurement. It is up to the analyst to understand these laws and incorporate them into the study. Statistics is a powerful method to augment and inform geological assessment. It is a poor substitute for physics and geological assessment.

1-1.1 Definitions

Let us begin by considering two extreme cases of determinism and randomness. Borrowing from systems language, a system is *deterministic* if it yields the same output when stimulated several times with the same input. It is *random* if it yields a different output (unrelated to each other) from the same input. *Stochastic* describes a system that is part deterministic and part random, a hybrid system. The essential feature here is that the descriptors apply to the system, although, since knowledge about the system is limited, we frequently use them to apply to the output. Stochastic implies that there is *some* unpredictability in the description of a set of numbers or in a statistic; random implies complete unpredictability.

On this basis, reservoirs can be placed into one of three categories. The first is strictly deterministic, having a recognizable, correlateable element at the interwell (km) scale with a well-understood internal architecture. The second is mixed deterministic and random (i.e., stochastic), having a recognizable, correlateable element at the interwell (km) scale with a well-understood gross internal architecture but local random variability, noise, or uncertainty. Finally, there is the purely random, with no readily identifiable geological control on the distribution of properties in a heterogeneous flow unit. We will see examples of all three types.

Some types of statistical operations are quite deterministic because repeated application of the same input to the same problem will give the same estimate. These procedures include both regression and Kriging. By saying that a variable is stochastic, we are not implying that there is no determinism in it at all—merely that there is at least some unpredictability associated with it. In later chapters, we shall produce several sets of numbers that are all different but that have the same statistic; each set is called a *stochastic realization*.

1-1.2 Deterministic Versus Stochastic Systems

In general, deterministic predictions are superior to stochastic predictions. After all, deterministic predictions have no uncertainty and usually contain significant information about the nature of the system. Consequently, if the system under consideration is perfectly understood (in our case, if the nature of interwell property distribution is perfectly known), then determinism is the preferred prediction method. But if we are honest, we recognize that there rarely is such a thing as a perfectly deterministic system—certainly no system that can predict permeable media property distributions. In fact, exactly the opposite is true; many properties seem to be random, even when their underlying physical cause is understood.

Randomness has always been an unsettling concept for physical scientists who are used to solving precise, well-defined equations for well-defined answers. Einstein himself said that "God does not play dice" with the universe (quoted by Hoffman, 1972). We know that all physical measurements entail some random error; nevertheless, only a cursory look at the real world reveals an incredible amount of unexplained complexity and *apparent* randomness—randomness that usually exceeds any conceivable measurement error.

1-1.3 Causes for Randomness

How can there be randomness when nature follows well-established, highly deterministic physical laws such as the conservation of mass, energy, and momentum? Curiously, this dilemma, which is the origin of a breach between physical scientists and statisticians, has only recently been resolved. We try to illustrate this resolution here.

Consider the epitome of a random event: the flipping of a two-sided, unbiased coin. The rise and fall of the coin are governed by Newton's second law and the rate of spin by the conservation of angular momentum—both well-defined physical laws. The first requires as input (initial condition) the initial velocity at which the coin is tossed. This is equivalent to the height of the toss. The second law requires the initial angular momentum or the rate of spin. With both the height and rate of spin specified, we can plot the results of the deterministic solution as in Fig. 1-1, where the shading indicates regions having a common outcome, heads or tails.

At the origin, the regions are large and it should be easy to predict the outcome of a flip (MacKean, 1987). But as the height and rate of spin increase, the regions become small. If we imagine that there is some small error in the height and/or rate of spin (such uncertainty always exists), then it is apparent that we can no longer predict the outcome of the flip and the coin toss is random.

Randomness, then, is the manifestation of extreme sensitivity of the solution of an equation (or a physical law) to small uncertainties in initial conditions. From this point of view, then, few systems are truly random but a great many are *apparently* random.

There are two requirements for this sensitivity to exist: a nonlinear physical law and a recursive process. Both are present in nature in abundant quantities (e.g., Middleton, 1991). However, it's surprising just how simple an equation can be and still exhibit apparent randomness.

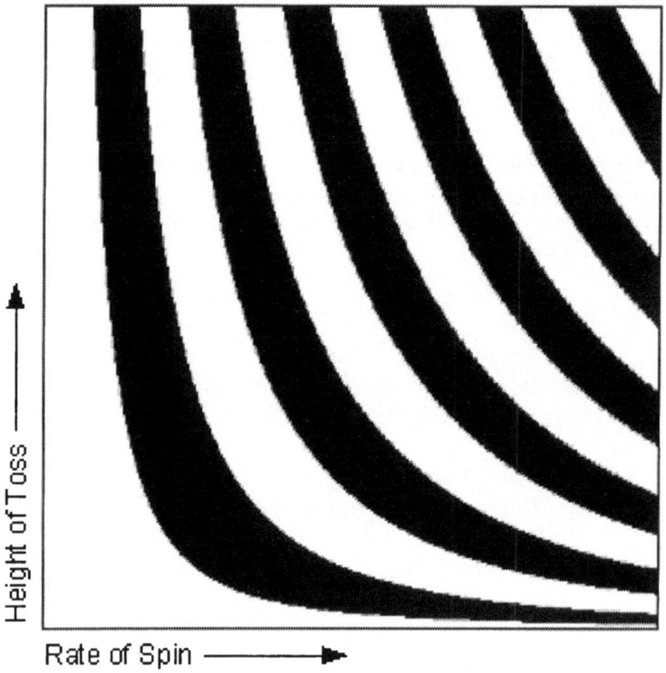

Figure 1-1. The outcome of a flipped coin. The dark regions indicate where the coin will come up "heads"; the light where it will be "tails."

May's equation (Crutchfield et al., 1986) is one of many that will generate apparent randomness:

$$x_{i+1} = \lambda x_i(1 - x_i)$$

for x_i between 0 and 1. Even though this equation is very simple, it exhibits the two requirements of randomness. The parameter λ controls the extent of the nonlinearity. If we generate various x_{i+1} starting with $x_0 = 0.5$, we find that the behavior of the x_i for i greater than about 20 depends strongly on the value of λ. For small λ, the x_i settle down to one of two values. But, as Fig. 1-2 shows, for large λ, the terminal x_i can take on several values.

In fact, for λ between 3.800 and 3.995, there are so many values that the x_i begin to take on the character of an experimental data set. Furthermore, an entirely different set of numbers would result for values of x_0 other than 0.50; therefore, this equation is also very sensitive to initial conditions.

Remember that this behavior emanates from the completely deterministic May's equation, from which (and from several like it) we learn the following:

1. The sensitivity to initial conditions means we cannot infer the initial conditions (that is, tell the difference between $x_0 = 0.50$ and another value) from the data set alone. Thus,

while geologists can classify a given medium with the help of modern analogues and facies models (e.g., Walker, 1984), they cannot specify the initial and boundary conditions.

2. The sensitivity also means that it is difficult (if not impossible) to infer the nature of the physical law (in this case, May's equation) from the data. The latter is one of the main frustrations of statistics—the inability to associate physical laws with observations. It also means that the techniques and procedures of statistics are radically different from traditional science and engineering approaches.

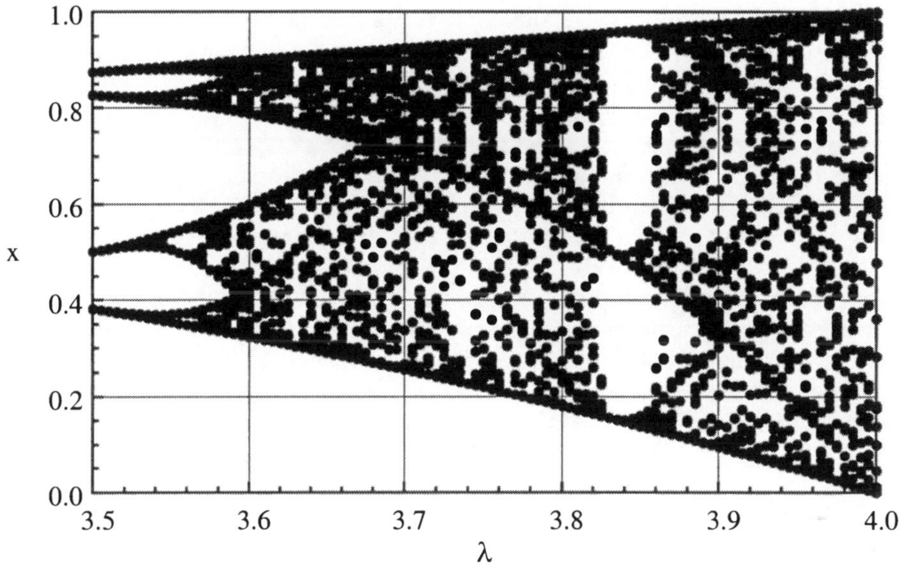

Figure 1-2. Bifurcation diagram for May's equation.

The latter point is especially significant for earth scientists. It seems that if we cannot infer a physical law for sedimentology, how then can we make sedimentologically based predictions? The answer lies at the heart of statistics. We must take the best and most complete physical measurements possible, ensure that sampling is adequate, derive the best statistical estimates from the data, and then use these to make predictions (in some fashion) that are consistent with the inferred statistics. Each of these points is touched upon at various places in this book.

1-2 THE MEASUREMENTS

Statistical analysis commonly begins after data have been collected. Unfortunately, for two reasons this is often a very late stage to assess data. First, measurements will already have been made that could be inappropriate and, therefore, represent wasted time and money. Second, if the analysis indicates further measurements are needed, the opportunity to take more data may be lost because of changed circumstances (e.g., the wellbore being cased off or the core having deteriorated). Therefore, it is more effective to perform statistical analysis before or while measurements are made.

How can we analyze and calculate statistics before we have numbers to manipulate? The answer lies largely in the fact that geology is a study of similarities in rock features. If a formation property is being measured for a particular rock type, similar rock types can be assessed beforehand to guide the present collection scheme. In that way, we will have some idea of the magnitude of the property, how variable it may be, and the scale of its variation. This information can then guide the sampling scheme to make sufficient measurements at an appropriate scale.

While it may appear obvious, it is important to make more measurements where the property varies much than where it varies little. This procedure conflicts with common practice, which is to collect samples at fixed times or spacings. The result of the latter approach is that estimates of reservoir properties and their variation are poorest where they vary most.

It is less obvious that the scale of the measurement is important, too, because variations in rock properties often occur at fairly well-defined scales (see below for further discussion of this). Any measurement, representing the effective value on the measurement volume, may not correspond to any of the geological scales. This mismatch between the scales of measurement and geology often leads to poor assessment and predictions. The assessments of properties are poor because the measurements are representing the values and variabilities of multiples of the geological units, not of the unit properties. Predictions are degraded because knowledge of the geometries of variations can no longer be incorporated into the process. When information is ignored, the quality of predictions deteriorates.

There is also often an uncertainty in the measurement values arising from interpretation and sampling bias. Each measurement represents an experiment for a process having a model and a given set of boundary conditions. The instrumentation then interprets the rock response using the model to determine the constitutive property for the volume of rock sampled. Two examples of volumes and models are pressure-transient tests and core-plug permeability measurements. The transient test has poorly defined boundary conditions and the predicted permeability depends on solutions to radial flow equations. A core plug fits in a rubber sleeve, ensuring a no-flow boundary, and the end faces have constant pressures applied. This simplifies the geometry so that Darcy's law can be invoked to obtain a permeability. The well test has a large, poorly defined volume of investigation while the plug has a very specific volume. Any errors in the assumed boundary conditions or interpretative model will lead to imprecise estimates of the rock properties.

Sampling bias does not necessarily produce imprecise data. It may instead produce data that do not represent the actual rock property values. For example, the mechanical considerations of taking core plugs means that the shalier, more friable regions may not be sampled. Similarly, samples with permeabilities below the limit of the measuring device may not be measured at all. During coring, some regions may disintegrate, leading to uncertainties in the locations of recovered rock and their orientations.

Thus, measurements have many failings that may mislead statistical analysis. The approach advocated in this book emphasizes the exploratory power of statistics in the knowledge of the geology and measurements. Through examples and discussion, we illustrate how dependence solely on statistical manipulations may mislead. At first, this lessens the confidence of the neophyte, but with practice and understanding, we have found that the models and analyses produced are more robust and consistent with the geological character of the reservoir.

1-3 THE MEDIUM

The majority of reservoirs that are encountered by petroleum geoscientists and engineers are sedimentary in origin. Sedimentary rocks comprise clastics (i.e., those composed of detrital particles, predominantly sand, silt, and clay) and carbonates (i.e., those whose composition is primarily carbonate, predominantly limestones and dolomites). Both clastic and carbonate reservoirs have different characteristics that determine the variability of petrophysical properties.

1-3.1 Variability in Geological Media

Clastic rocks are deposited by a variety of depositional processes. The accumulation of detrital sediments depends on sediment transport (Allen, 1985). When the transport rate changes, erosion or deposition occur. Sediment transport is a periodic phenomenon; however, for all preserved sequences, deposition prevailed in the long term. Petrophysical properties are controlled at the pore scale by textural properties (grain size, grain sorting), regardless of depositional environment (Pettijohn et al., 1987). Texture is controlled by many primary parameters: the provenance (source of sediments) characteristics, the energy of deposition, climate, etc., and their rates of change. Secondary phenomena such as compaction and diagenesis can also modify the petrophysical properties. Nevertheless, the influence of primary depositional texture usually remains a strong determining factor in the petrophysical variability in clastic reservoirs.

In carbonate rocks, primary structure is also important. Carbonate sediments are biogenic or evaporitic in origin. Primary structure can be relatively uniform (e.g., pelagic oozes, oolitic grainstones) or extremely variable (coralline framestones). Carbonates, particularly aragonite and calcite, are relatively unstable in the subsurface. Pervasive diagenetic phenomena can lead to large-scale change of the pore structure. The change from calcite to dolomite (dolomitization) leads to a major reduction in matrix volume and development of (micro) porosity. Selective dolomitization of different elements of the carbonate can preserve the primary depositional control. More often the effective reservoir in carbonates derives entirely from diagenetic modification, through dissolution (stylolites), leaching (karst), or fracturing. As a result of these phenomena (often occuring together), carbonates have very complex petrophysical properties and pose different challenges to the geoscientist, petrophysicist, or engineer. See Lucia (1999) for more discussion.

As we have outlined, the fundamental controls on petrophysical properties in clastic reservoirs (textural) are generally quite different from those in a carbonate (diagenetic or tectonic). It is to be expected that different statistical techniques and measures are needed to address the petrophysical description of the reservoir medium. Other rock types occuring less commonly as reservoirs, such as fractured igneous rocks and volcanics, might require different combinations of the statistical techniques. However, the approach to statistical analysis is similar to that presented in this book: prior to statistical analysis, *attempt to identify the controlling phenomena* for the significant petrophysical parameters.

1-3.2 Structure in Geological Media

We noted in the previous section that the controls on petrophysical properties are different in clastics and carbonates. These differences manifest themselves in different large-scale performance of the medium to an engineered project.

In clastic media, the textural variability is arranged into characteristic structures. These (sedimentary) structures are hierarchical, and the elements are variously described as stratal elements, genetic units, and architectural elements for various depo-systems (Miall, 1988; van Wagoner et al., 1990). These elements are essentially repetitive within a single reservoir, characteristic of the depositional process, and their association forms the basis of sedimentological and environmental interpretations. Distinctive elements are lamina, beds, and parasequences in the marine envronment at the centimeter, meter, and decameter scales, respectively (Fig. 1-3). Groups of these elements are generally called *sets*.

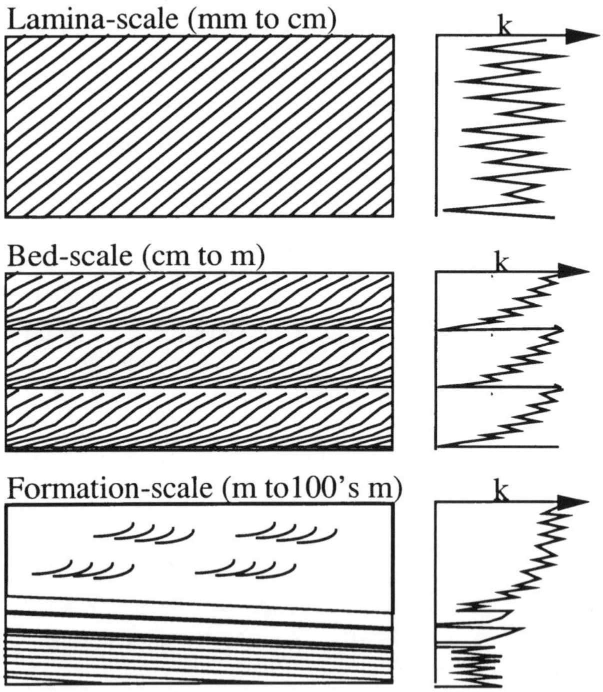

Figure 1-3. Lamination, bed, and parasequence/formation scales of sedimentary structure with schematic permeability variations. After Ringrose et al. (1993).

At the larger scale, the stacking of these elements (the architecture) controls the characteristics of reservoir *flow units* (Hearn et al., 1988; Weber and van Geuns, 1990). The shapes and sizes of elements can be measured in outcrop (e.g., the laminaset elements in various shallow marine outcrops in Corbett et al., 1994). The repetitive nature of the elements results from cyclical processes such as the passing of waves, seasons, or longer-term climatic cycles because of orbital variations (House, 1995).

In carbonate media, similar depositional changes can also introduce an element of repetitive structure (Grant et al., 1994). Often, however, the permeability appears to be highly variable, lacking structure, and appears random or having a very short correlation structure (Dehghani et al.,1999). Measurement of properties in very vuggy carbonates is also challenging (Lucia, 1999; Dehghani et al., 1999).

The degree of structure and the degree of randomness often require careful measurement and statistical analysis. Clastic reservoirs comprise a blend of structure (depositionally controlled) overprinted by a random component (because of natural "noise" or postdepositional modification). Clastics tend to be dominated by the former, and carbonates by the latter. This is by no means a rigorous classification, as a very diagenetically altered clastic might exhibit a random permeability structure, and, alternatively, a carbonate may exhibit clear structure. The degree of structure, however, is one that must be assessed to predict the performance of the medium (Jensen et al., 1996).

1-3.3 Assessing Geological Media

There are various methods for assessing the variability and structure of geological media. Variability can be assessed by measurements at various increasing scales (Fig. 1-4): thin-section, probe or plug permeameter, wireline tool, drill-stem test, and seismic. The difficulty in comparing measurements across the scales is compounded by the various geological elements that are being assessed, the different measurement conditions, and the interpretation by various experts and disciplines. In general, the cost of the measurements increases with scale, as does the volume of investigation. As a result, the variability decreases but representativeness becomes more of an issue.

Structure is usually assessed by statistical analysis of series of measurements. Geometrical and photogeological data are also used to assess structure at outcrop (Martino and Sanderson, 1993; Corbett et al., 1992). These structure assessments have become increasingly quantified by the application of semivariograms and other methods (e.g., Davis, 1973). This quantitative approach has raised awareness of the techniques and issues addressed in this book.

1-3.4 Support and Stationarity Issues in Geological Media

Dehghani et al. (1999) show that, for some geological media, it may be very difficult to obtain meaningful measurements. Their example is a carbonate, in which such problems are usually more manifest, but the example raises important considerations for all geological media.

Meaningful measurements in geological media might be judged from their usefulness to reservoir modelling. We would define meaningful measurements as

1. being associated with a particular rock type or facies in addition to a depth location and well identifier, and
2. having an assessment of support and stationarity with reference to the geology.

Prior to entering the main text of this book to obtain a statistical definition of these concepts, it is useful at this time to have a "soft" definition. By an assessment of support, we mean that a nearby measurement *at the same volume scale* should have a *similar value of a property* (e.g., horizontal permeability). By assessment of stationarity, we mean that *similar rock types* have *similar statistical properties* (e.g., permeability averages, variabilities, and distributions).

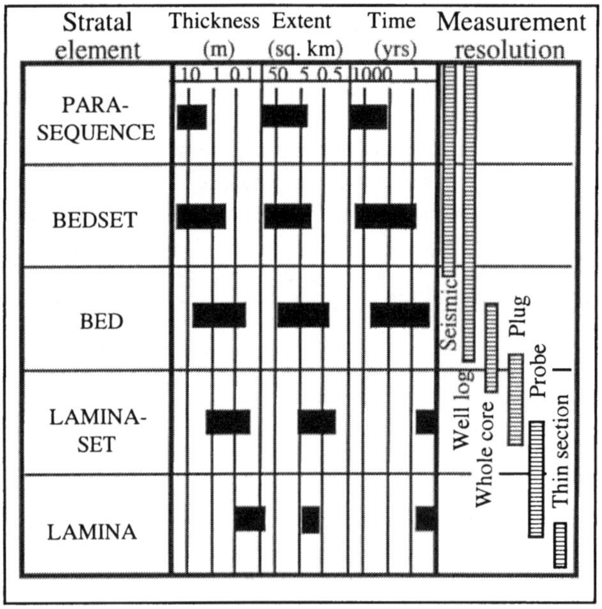

Figure 1-4. Shallow marine stratal elements and measurement scales. After van Wagoner et al. (1990).

These sampling considerations are important because satisfying (1) above allows us to link the property with an object or grid block within a geological model. Satisfying (2) helps statistical assessment of the data and assignment of properties for numerical flow simulation. This is particularly important in carbonates, but also in certain heterogeneous clastic reservoirs (e.g., variably cemented sandstones and conglomerates). It is also a major issue in fractured reservoir characterization. Statistical methods alone cannot provide a meaning to data.

1-3.5 Geological versus statistical independence

It is important to assess the statistical relationships (correlation) of the properties measured in geological media. Correlated data may carry less information than uncorrelated data. The degree of correlation is often measured by an autocorrelation function (e.g., the semivariogram). We have seen that multiple relationships may exist in geological media at several different scales (lamina, beds, parasequences) and that these can be measured by targeted sampling schemes. It is very unlikely that a geological sedimentary medium is characterized by a single correlation length. This suggests that each measurement (providing that the support volume is appropriate) carries information because it reflects a separate geological event separated by seconds, hours, years, or millions of years from measurements made at adjacent locations. This is particularly the case with vertically spaced measurements, and suggests that care should be taking in following a strict statistical definition of (in)dependence. Rocks are possibly more independent than the data sometimes suggest.

1-4 THE PHYSICS OF FLOW

Much of the discussion in this book is aimed at developing credible petrophysical models for numerical reservoir simulators. There are excellent texts on petrophysics and reservoir simulation (e.g., Aziz and Settari, 1979; Peaceman, 1978), so we give only a rudimentary discussion here of those aspects that will be needed in the ensuing chapters of this book.

Simulators solve conservation equations for the amount of a chemical species at each point in a reservoir. In some cases, an energy balance is additionally solved, in which case temperature is a variable to be determined. The conservation equations are relatively simple differential equations (Lake, 1989), but complexity arises because each component can exist in more than one phase; the degree of the distribution among the phases as well as fluid properties themselves are specified functions of pressure. Pressure is itself a variable because of a requirement on overall conservation of mass. In fact, each phase can manifest a separate pressure, which leads to additional input in the form of capillary pressure relations. The most common type of simulator is the black oil simulator that consists of three phases (oleic, gaseous, and aqueous) and three components (oil, gas, and water).

The power of numerical simulation lies in its generality. The previous paragraph spoke of this generality in the types of fluids and their properties. But there is immense generality otherwise. Most simulators are three-dimensional, have arbitrary well placements (both vertical and horizontal), can represent arbitrary reservoir shapes, and offer an exceedingly wide range of operating options. Many simulators, such as dual porosity and enhanced oil recovery simulators, have physical mechanisms for specialized applications. There are no limits to the size of the field being simulated except those imposed by the computer upon which the simulation is run.

The combination of all of these aspects into one computer program leads to so much complexity that numerous approximations are necessary. The most important of these is that the original conservation equations must be discretized in some fashion before they are solved. Discretization involves dividing the reservoir being modeled into regions—*cells* or *grid blocks*. The number of such cells determines the ultimate size of the simulation; modern computers are capable of handling upwards of one million cells.

The discretization, though necessary, introduces errors into a simulation. These are of three types.

1-4.1 Numerical Errors

These arise simply from the approximation of the originally continuous conservation equations with discrete analogues. In the simulator adjacent grid blocks should not have extreme variations in permeability (i.e., less than an order of magnitude) if significant numerical errors are to be avoided. This suggests that the grid block sizes should not be coincident with the scale of the geological facies if these are strongly variable. The grid blocks should exhibit local stationarity - be similar or change gradually - for optimum numerical solution. Numerical errors can be mitigated with specialized discretization techniques (e.g., finite-elements vs. finite-differences), grid refinement, and advanced approximation techniques.

1-4.2 Scale-Up Errors

Even with modern computers, the gridblocks in a large simulation are still quite large–as much as several meters vertically and several 10's of meters areally. (The term *grid block* really should be *grid pancake*.) This introduces an additional scale to those discussed in the previous section, except this scale is artificial, being introduced by the simulation user. Unfortunately, the typical measurement scales are usually smaller than the gridblock scale; thus some adjustment of the measurement is necessary. This process, called *scaling-up* or *upscaling*, is the subject of active research. Statistical procedures (e.g., averages) are often used in upscaling.

1-4.3 Input Errors

Each gridblock in a simulation requires several data to initialize and run a simulator. In a large simulation, nearly all of these are unmeasurable away from wells. The potential for erroneous values then is quite large. The point of view of this book is that input errors are the largest of the three; many of the techniques to be discussed are aimed at reducing these errors.

The conservation equations alone must be augmented by several constitutive relationships. These relationships deserve special mention because they contain petrophysical quantities whose discussion will occur repeatedly throughout this book.

Porosity is the ratio of the interconnected pore space of a permeable medium to the total volume. As such, porosity, usually given the symbol ϕ, represents the extent to which a medium can store fluid. Porosity is often approximately normally distributed, but it must be between 0 and 1. In the laboratory, porosity can be measured through gas-expansion methods on plugs taken from cores (ϕ_{plug}) or inferred from electrical-log measurements. Each has its own sources of error.

Like all petrophysical properties, porosity depends on the local textural properties of the medium. If the medium is well-sorted, it depends primarily on packing. As the sorting becomes poorer, porosity begins to depend on grain size as well as sorting.

The most frequently discussed petrophysical quantity in this book is *permeability*. This is defined from a form of Darcy's law:

$$u = -\frac{k\Delta P}{\mu L} \tag{1-1}$$

where u is the interstitial velocity of a fluid flowing through a one-dimensional medium of length L, ΔP is the pressure difference between the inlet and outlet, μ is the fluid viscosity, and k is the permeability. Permeability has units of (length)^2, usually in μm^2 or Darcys (D);

conventionally 10^{-12} m^2 = 1 μm^2 = 10^{-3} mD. mD means milliDarcies. The superficial velocity is $u = Q/A\phi$, where Q is the volumetric flow rate and A is the cross-sectional area perpendicular to the flow. Note that flow rates and pressure drops are linear functions of permeability, not the logarithm of permeability. It may be appropriate to transform permeability to its logarithm for some statistical analysis, but it is permeability that occurs in the flow equations.

Equation (1-1), in which k is a scalar, is the simplest form of Darcy's law. Simulators use a more general form:

$$\boldsymbol{u} = -\lambda\,(\nabla P + \rho \boldsymbol{g}) \tag{1-2}$$

where $\boldsymbol{u}$ is a vector, ρ the fluid density, and ∇P the vector gradient of pressure. λ in this equation is the fluid *mobility* defined as $\lambda = \boldsymbol{k}/\mu$. This illustrates that the type of fluid, represented by its viscosity, is also an important consideration. To simplify the statistical manipulations, it may be important to normalize the data for fluid effects.

Permeability in this representation is now a tensor that requires nine scalar components to represent it in three Cartesian dimensions.

$$\boldsymbol{k} = \begin{bmatrix} k_{xx} & k_{xy} & k_{xz} \\ k_{yx} & k_{yy} & k_{yz} \\ k_{zx} & k_{yz} & k_{zz} \end{bmatrix}$$

where each of the terms on the right side are scalars. The tensorial representation is present so that $\boldsymbol{u}$ and its driving force $(\nabla P + \rho \boldsymbol{g})$ need not be colinear. Most flow simulators use a diagonal version of the permeability tensor:

$$\boldsymbol{k} = \begin{bmatrix} k_x & 0 & 0 \\ 0 & k_y & 0 \\ 0 & 0 & k_z \end{bmatrix} \tag{1-3}$$

k_x and k_y are the x- and y- direction permeabilities; k_z is usually the vertical permeability and is frequently expressed as a fraction (k_v/k_h) of the x-direction permeability. k_v/k_h is commonly less than one and very variable. k_v/k_h is usually much less than one in a typical gridblock. Unless otherwise stated, when we say *permeability*, we mean k_x.

Permeability is among the most important petrophysical properties and one of the most difficult to measure. On a very small scale, it can be measured with the probe permeameter. The most common measurement is on plugs extracted from the subsurface; on certain types of media, larger pieces of the core may be used—the so-called whole-core measurements. All of these measurements are quite small-scale compared to their application volume. Far larger volumes—of the order of several cubic meters—of a reservoir may be sampled through pressure transient analysis. These tests measure the pressure drawdown and buildup, following periods of well flow and shut-in, respectively. The permeability is interpreted from models of the flowing regime around the well. These models are open to misapplication, so it becomes important to seek corroboration in intervals where core data are available. The comparison of core estimates (averaged and upscaled) to well test estimates of permeability require careful consideration (e.g., Warren et al., 1961; Zheng et al., 2000). In each case, the support volumes of the measurements are different - the largest scale may not be the most appropriate for reservoir evaluation and modelling.

All measurements are fraught with difficulties and permeability measurements are no exception. Consequently, we inevitably see correlations of permeability with other petrophysical quantities. It is easier to measure the correlant properties and estimate subsurface permeabilities (by models or empirical relationships), than it is to measure the permeability directly. The primary correlant here is porosity; the reader will see numerous references to log(k) vs. ϕ throughout this book. Permeability has also been correlated with electrical resistivity, gamma ray response, and sonic velocity. Newer techniques (e.g., nuclear magnetic resonance and Stonely wave amplitude) are showing promise for providing reasonable estimates of in-situ permeabilities.

Permeability varies much more dramatically within a reservoir than does porosity; it tends to have a skewed (often log-normal) distribution. Permeability depends locally on porosity, and most specifically on grain size (see Chap. 5). It also depends on sorting and strongly on the mineral content of the medium and the cement (Panda and Lake, 1995).

The version of Darcy's law appropriate for multiphase immiscible flow is

$$\boldsymbol{u}_j = -\lambda_j (\nabla P_j + \rho_j \boldsymbol{g}) \tag{1-4}$$

where the terms are analogous to those in Eq. (1-2) except that they refer to a specific phase j. The mobility term is now $\lambda_j = \boldsymbol{k}\, k_{rj}/\mu_j$. Again the terms $\boldsymbol{k}$ and μ are analogous to those in the previous definition of mobility. The new term k_{rj} is the *relative permeability* of phase j. Although there is some evidence for the tensorial nature of k_{rj}, this book treats it as a scalar function.

Relative permeability is a nonlinear function of the saturation of phase j, S_j. This relationship depends primarily on the wetting state of the medium and less strongly on textural properties. Under some circumstances, k_{rj} also depends on interfacial tension, fluid viscosity, and velocity. It is difficult to measure; all techniques involve some variant of conducting floods through core plugs, which raises issues of restoration, *in situ* wettability, experimental artifacts, and data handling.

Relative permeability is among the most important petrophysical quantitites in describing immiscible flow. Some measure of this importance is seen in the fractional flow of a phase j. The fractional flow of a phase is the flow rate of the phase divided by the total flow rate. With suitable manipulations (Lake, 1989), the fractional flow of water in two-phase oil-water flow is, in the absence of gravity, given by

$$f_w = \left(1 + \frac{k_{ro}\mu_o}{k_{rw}\mu_w}\right)^{-1}$$

f_w is, therefore, a nonlinear function of k_{ro} and k_{rw}, which are themselves nonlinear functions of saturation. This relationship between saturation and flow is perhaps the most basic representation of multiphase flow there is. The statistical treatment of permeability functions that depend on saturation is not simple; numerical simulation is often used to "average" these properties.

In addition to the ones mentioned above—porosity, permeability, and relative permeability—there are a host of other petrophysical properties that are important to specific applications. In miscible displacements, diffusion and dispersivity are important; in immiscible displacements, capillary pressure plays a strong role in determining ultimate recovery efficiency. Here again, though, the disparateness of scales manifests itself; capillary pressure is exceedingly important on the grain-to-grain scale, but this scale is so small that we represent its effects through the residual saturations in the k_{rj} functions. For this reason also, capillary pressure is an excellent characterization tool for the small scale. In each case, the statistical assessment of these properties has to be made with consideration of the physics

1-5 ESTIMATION AND UNCERTAINTY

The basic motivation for analyzing and exploring data must be to predict properties of the formation being sampled. This requirement thus necessitates the production of estimates (statistics) that amount to some guess about the value of properties at unsampled locations in the reservoir. These estimates are often the sole focus of attention for many users of statistics.

Because estimates are guesses, they also have an uncertainty associated with them This is because each estimate is the product of the analysis of a limited set of data and information. The limited nature of this information implies that our assessment is incomplete and that, with further information, we could supply a "better" estimate. Two aspects of statistical analysis help the user to be aware of and to mitigate this uncertainty.

The first aspect is that, for every estimate we produce, we can also produce an estimate of its associated uncertainty. This provides a powerful method for assessing the impact of the limited nature of our data sets. We will find that many of the statistics used in reservoir characterization have uncertainties related to two features of the data. The first is the number of data in the analysis. Clearly, data represent information and the more data we have, the less uncertain we expect the estimates to be. The second is the variability of the property under study. This is a parameter over which we have no control but, with geological information and experience, we may have some notion of its magnitude. For example, the arithmetic average has an accuracy that is related to the ratio (data variability) $/\sqrt{\text{number of data}}$. Such expressions are very helpful in understanding what is causing estimates to vary from the "right value" and what we gain by increasing the number of samples. Since information (i.e., number of data) is expensive, we can find how much an improved estimate will cost. In the case of the arithmetic average, doubling the accuracy requires quadrupling the number of samples. When samples are correlated, there is a loss of information contained in a given number of samples. Thus, the accuracy of estimates will decrease as correlation strength increases.

The second aspect is that we can guide the analysis according to the measurement physics and geological subdivisions of the reservoir. This affords the ability to incorporate into the analysis at every possible opportunity the geological and physical knowledge of the reservoir. The inclusion of such information is less quantifiable, unfortunately, than the effects of data numbers and variability, but is nonetheless important in producing statistically relevant estimates. As an extreme example, we can take the average of the temperature and porosity of a rock sample, 20°C and 0.20, to give a result of 10.1. The mathematics does not know any different but the result is uninterpretable. We have no confidence in being able to apply the result elsewhere. Ridiculous? Yes, but it is extremely easy to ignore information and suppose that the statistical procedures will iron out the inconsistencies. The result will be poor estimates of quantities that are only weakly related to the reservoir.

Interpretability and accuracy of estimates also helps in another common engineering function, comparison of estimates obtained from different kinds of measurements. Comparisons and the comparability of estimates will be discussed in this book, but the subject can be very complex. Estimates of average permeability from, for example, pressure transient analysis and core-plug data are often only weakly comparable. For transient analysis, the statistical uncertainties of these measurements may be only a small part of the total uncertainties, which include the domain of investigation, well-bore effects, and model uncertainty. Yet, these latter aspects are not well-quantified and, therefore, cannot be easily translated into an uncertainty of the estimated permeability. Nonetheless, error bands for estimates can at least convey the statistical causes of uncertainty. These can then assist careful, geo-engineering judgments to compare estimates.

1-6 SUMMARY

Statistics is a powerful tool that can supply estimates and their uncertainties for a large number of reservoir properties. The discipline also provides important exploratory methods to investigate data and allow them to "tell their story." This tool is best used in concert with the geological and physical information we have about a reservoir or aquifer.

Measurements are a vital part of statistical analysis. While providing a "window" to examine system response, they have several shortcomings. To gain the most from the data, these shortcomings should be understood. Statistical analysis should parallel data collection to guide the choice of sample numbers and measurement type. Geological information concerning the scale and magnitude of variability can improve the measurement process.

Sedimentary systems are the primary subject for reservoir characterization. Both clastic and carbonate systems often exhibit regularities that can be exploited during measurement and prediction. These features appear at several scales and, therefore, measurements with differing volumes of investigation will respond differently to the rock. These scales and the structure thus should be understood because the understanding will help the statistical analysis during both the data-collection and analysis phases.

CHAPTER 2

BASIC CONCEPTS

Up to this point, we have relied on your intuitive grasp of what probability means to make the case for the roles for probability in reservoir description. Before we give details of methods and applications that use probability, however, we must agree on what probability is. Therefore, we will devote some space to defining probability and introducing terms used in its definition.

Central to the concept of probability is the notion of an experiment and its result (outcome).

Experiment $\mathfrak{I}$ - The operation of establishing certain conditions that may produce one of several possible outcomes or results.

Sample Space Ω - The collection of *all* possible outcomes of $\mathfrak{I}$.

Event E - A collection of *some* of the possible outcomes of $\mathfrak{I}$. E is a subset of Ω.

These definitions are very general. In many statistical texts, they are illustrated with cards or balls and urns. We attempt to make our illustrations consistent with the basic theme of this book, reservoir description. An experiment $\mathfrak{I}$ could be the act of taking a rock sample from a reservoir or aquifer and measuring its porosity. The sample space Ω is the collection of all porosities of such specimens in the reservoir. Figure 2-1 illustrates that the event E may be the value of a specific porosity (ϕ) that is actually observed in a core plug.

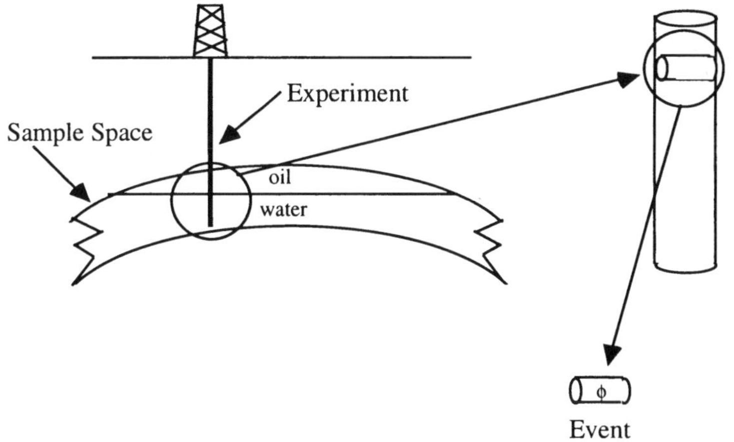

Figure 2-1. Sketch of an experiment measuring the porosity (ϕ) of a core plug rock sample.

Some books call the sample space the *population*. An individual outcome (a single specific porosity value) is an *elementary or simple event*, sometimes just called an *event*. The *null set*, $\varnothing$ (no outcomes), is also an event.

An experiment is always conducted to determine if a particular result is obtained (realized). If the desired result is obtained, then the event defined by the result has occurred.

Occurrence of E - An event E has occurred if the outcome of an experiment belongs to E.

Example 1a - Defining Events. Suppose there are two wells with the measured core plug porosities in Table 2-1.

This reservoir has had 14 experiments performed on it; there are therefore 14 elementary events. If we take the event E to be all porosities between 0.20 and 0.25, then the elementary event has occurred four times in Well 1 and four times in Well 2. We denote the event as $E = [0.20, 0.25]$. Square brackets [ ] represent intervals that include the extremes.

The frequency of these occurrences is important because below we will associate probability with frequency. The event must be well-defined before the frequencies can be calculated.

Table 2-1. Core plug porosities for Example 1.

Well 1	Well 2
0.19	0.25
0.15	0.20
0.21	0.26
0.17	0.23
0.24	0.19
0.22	0.21
0.23	
0.17	

Slight changes in the event definition can greatly alter its frequency. To see this, recalculate the frequencies from the data in Table 2-1 when a new event definition of $E = [0.19, 0.25]$.

2-1 COMBINATIONS OF EVENTS

Consider the case of two events, E_1 and E_2. Both represent outcomes of the experiment $\mathfrak{I}$. We can combine these events in two important ways to produce new events. The first combination includes *all* the elements of E_1 and E_2 and is called the *union* of E_1 and E_2.

Union of Two Events - $E_1 \cup E_2$ is the set of outcomes that belong to *either* E_1 or E_2.

The second combination involves only the elements common to E_1 and E_2. This is called the *intersection* of E_1 and E_2.

Intersection of Two Events - $E_1 \cap E_2$ is the set of outcomes that belong to *both* E_1 and E_2.

Both union and intersection are themselves events. Thus, the union of two events E_1 and E_2 occurs whenever E_1 or E_2 occurs. The intersection of two events E_1 and E_2 occurs when both E_1 and E_2 occur.

Example 1b - Illustrating Union and Intersection. Referring to Table 2-1 of core plug porosities, if $E_1 = [0.15, 0.19]$ and $E_2 = [0.18, 0.24]$, then $E_1 \cup E_2 = [0.15, 0.24]$ contains eight elements from Well 1 and four elements from Well 2. $E_1 \cap E_2 = [0.18, 0.19]$ contains one element from each well.

Of course, both definitions can be extended to as many events as you like.

Example 2 - Facies Detection by Combining Events. Suppose that core description and wireline log analysis produced the diagram in Fig. 2-2 for a reservoir. Each shaded circle represents one rock type and indicates where specific gamma ray and porosity readings occur. We can label events associated with the three rock types through their gamma ray and porosity values: $E_1 = [GR_3, GR_4]$, $E_2 = [GR_1, GR_2]$, $E_3 = [\phi_1, \phi_2]$, and $E_4 = [\phi_3, \phi_4]$. The three facies can be distinctly identified using *both* *GR* and ϕ data: $E_1 \cap E_3$ = Facies 3, $E_2 \cap E_3$ = Facies 1, and $E_2 \cap E_4$ = Facies 2. If only Facies 2 and 3 are present in portions of the reservoir, then *either* measurement would be sufficient to discriminate these facies since $E_1 \cup E_3$ = Facies 3 and $E_2 \cup E_4$ = Facies 2.

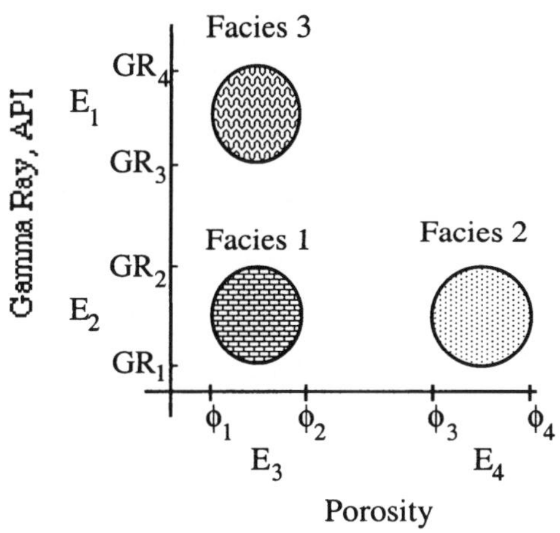

Figure 2-2. Hypothetical wireline measurement patterns for three reservoir facies. Gamma ray response in API units.

We will return to these ideas in Example 3 below.

Intersection and union lead to one highly useful concept, mutual exclusivity, and a rather trivial one, exhaustive events.

Mutually exclusive events - Events E_1 and E_2 are mutually exclusive if both E_1 and E_2 cannot occur simultaneously; that is, if $E_1 \cap E_2 = \varnothing$.

Exhaustive sequence of events - The sequence $E_1, E_2, \ldots, E_n$ accounts for all possible outcomes such that $E_1 \cup E_2 \cup \ldots \cup E_n = \Omega$.

For mutually exclusive events, the intersection contains no outcomes. For example, E_1 and E_4 in Fig. 2-2 are mutually exclusive, since the intersection of these two events has no outcomes (i.e., no lithofacies identified) and can only be the null event $\varnothing$. If all the lithofacies in the reservoir have been identified on the figure, $E_1 \cap E_4 = \varnothing$. If, during the analysis of well data, there were portions of the reservoir satisfying both E_1 and E_4, then there may be a previously unrecognized facies present or the measurements may be in error (e.g., poor borehole conditions).

In an exhaustive sequence, at least one of the constituting elementary events must occur. The term exhaustive sequence refers to the sample space and not the experiment. This is the reason that it is trivial; if we knew the exhaustive sequence of events of all the porosities in the reservoir (the sample space), there would be no need of estimation or statistics, and the reservoir would be completely described.

Complementary event E^c - The subset of Ω that contains all the outcomes that do not belong to E.

The complementary event for E_1 in Example 2 is all of the gamma ray values below GR_3 or above GR_4, i.e., $E_1^c = (0, GR_3) \cup (GR_4, \infty)$. This definition also applies to the sample space. The parentheses ( ) indicate intervals that do not contain the extremes.

We note in conclusion that both the union and intersection are commutative. That is, $E_1 \cup E_2 = E_2 \cup E_1$ and $E_1 \cap E_2 = E_2 \cap E_1$.

2-2 PROBABILITY

We are now prepared to consider a working definition of probability. It also involves the connection between statistics and probability.

Probability - Let E be an event associated with an experiment $\mathfrak{I}$. Perform $\mathfrak{I}$ n times and let m denote the number of times E occurs in these n trials. The ratio m/n is called the *relative frequency* of occurrence of E. If, as n becomes large, m/n approaches a limiting value, then we set $p = \lim_{n \to \infty} \left(\frac{m}{n} \right)$. The quantity p is called the *probability of event E*, or Prob(E) = p.

The relative frequency of occurrence of the event E is a *statistic* of the experiment $\mathfrak{I}$. The ratio m/n is obtained by conducting n experiments. Also associated with the event E is the notion that, as n approaches infinity, the statistic m/n will approach a certain value, p. The statistic m/n only tells us about things that have happened, whereas the probability allows us to make predictions about future outcomes (i.e., other times and/or other places). The link between the two occurs when we say "we know what m/n is, but we will never know what p is, so let us assume that p is about equal to m/n." In effect, we assume that future performance replicates past experience while recognizing that, as n increases, the statistic m/n may still vary somewhat.

Example 3a - Frequencies and Probabilities. Table 2-2 lists the grain densities (ρ_g) measured from 17 core plugs.

Table 2-2. Core-plug grain densities for Example 3.

Plug no.	ρ_g, g/cm^3	Plug no.	ρ_g, g/cm^3
1	2.68	10	2.69
2	2.68	11	2.70
3	2.68	12	2.74
4	2.69	13	2.69
5	2.69	14	2.68
6	2.70	15	2.68
7	2.69	16	2.68
8	2.70	17	2.70
9	2.68		

Estimate the relative frequencies of these events:

1. E_1: $2.65 \leq \rho_g \leq 2.68$;
2. E_2: $2.68 < \rho_g \leq 2.71$;
3. E_3: $2.65 \leq \rho_g \leq 2.71$;
4. E_4: $0 \leq \rho_g < \infty$; and
5. E_5: $-\infty < \rho_g < 0$.

First, the experiment $\mathfrak{I}$ consists of taking a core plug and measuring its grain density. So, in this case, 17 experiments ($n = 17$) have been performed and their outcomes listed. The sample space, Ω, consists of all possible values of grain density. We now consider the relative frequencies of these events.

1. $E_1 = [2.65, 2.68]$ occurred for plugs 1, 2, 3, 9, 14, 15, and 16. $m = 7$ so $\text{Prob}(E_1) = 7/17 = 0.412$;
2. $E_2 = (2.68, 2.71]$ occurred for plugs 4, 5, 6, 7, 8, 10, 11, 13, and 17. $m = 9$ so $\text{Prob}(E_2) = 9/17 = 0.529$;
3. $E_3 = [2.65, 2.71]$ occurred for plugs 1, 2, 3, 4, 5, 6, 7, 8, 9, 10, 11, 13, 14, 15, 16, and 17. $m = 16$ so $\text{Prob}(E_3) = 16/17 = 0.941$;
4. $E_4 = [0, \infty)$ occurred for all the plugs. $m = 17$ and $\text{Prob}(E_4) = 17/17 = 1.0$; and
5. $E_5 = (-\infty, 0]$ occurred for none of the plugs. $m = 0$ and $\text{Prob}(E_5) = 0/17 = 0$.

While we have labeled these frequencies as probabilities, this may not be accurate; they are in fact only estimates of the probabilities. All five of these frequencies are computed from the data obtained by conducting only 17 experiments. How accurate are they? That is, assuming that the experimental procedure for grain-density determinations is accurate, how representative are these frequencies of the whole reservoir? If we obtained more data, would the ratio m/n change for any of the listed events? For example, observing that other names for E_4 and E_5 are Ω and $\varnothing$, respectively, we would claim that $\text{Prob}(E_4) = 1$ and $\text{Prob}(E_5) = 0$ are exactly the same values as their relative frequencies. This observation, however, requires knowledge that is not based solely upon the data set.

The probability definition encompasses what we believe to be true about any kind of modelling. If we can quantify and understand the performance of any set of observations (here these are the relative frequencies), then we assume that we can estimate, to some extent, the future performance.

Assessing probabilities is relatively simple by most modelling standards because the observations need not be interpreted, just recorded. This is both the bane and the strength of statistical analysis. It is a weakness because all subsequent analyses will be largely unrelated to a physical cause. For example, we will not be able to say what caused the grain densities in Example 3; statistics does not distinguish between SiO_2 ($\rho_g = 2.65$ g/cm^3) and $CaCO_3$ ($\rho_g = 2.71$ g/cm^3). It is a strength, however, when we realize that a very large number of observations, especially those of geological origin, simply defy quantitative analysis because of their complexity (Chap. 1).

The association of probabilities with frequencies is by no means agreed upon even among statisticians. This is because it is entirely possible for future events to take a direction different from that in the past, particularly if there is an outside stimulus. To proceed, of course, we must accept the hypothesis given by the definition. However, with further data, it can be tested to determine if the hypothesis still holds. Chapters 5 and 7 treat this topic further.

2-3 PROBABILITY LAWS

The above definitions lead directly to probability *axioms* or *laws*. In most cases, they are rather obvious, but they provide a bridge between occurrences (or event space) and mathematics.

Axiom 1: For every event E, $0 \le \text{Prob}(E) \le 1$

Axiom 2: For the special case $E = \Omega$, $\text{Prob}(E) = 1$

Axiom 1 says that the probability of any event must be between 0 (no chance of happening) and 1 (certainty of happening). Axiom 2 says that some event must occur from the experiment.

The addition laws involve mutual exclusivity.

Axiom 3 (Addition Law): Let E_1 and E_2 be two events, then

$$\text{Prob}(E_1 \cup E_2) = \text{Prob}(E_1) + \text{Prob}(E_2) - \text{Prob}(E_1 \cap E_2)$$

Axiom 3 is sometimes known as the fundamental law of addition. It simply says that the probability of E_1 *or* E_2 is the sum of the individual probabilities less the probability of E_1 *and* E_2. We also use the law most in the following special cases.

Axiom 3': If E_1 and E_2 are two mutually exclusive events (i.e., $E_1 \cap E_2 = \varnothing$), then

$$\text{Prob}(E_1 \cup E_2) = \text{Prob}(E_1) + \text{Prob}(E_2)$$

Axiom 3": If $E_1, E_2, \ldots, E_i, \ldots$ is a mutually exclusive sequence of events, then

$$\text{Prob}(E_1 \cup E_2 \cup \ldots E_\infty) = \text{Prob}(E_1) + \text{Prob}(E_2) \ldots \text{Prob}(E_\infty) = \sum_{i=1}^{\infty} \text{Prob}(E_i)$$

Axiom 3" is simply a generalization of Axiom 3'.

> *Example 3b - Illustrating the Basic Axioms.* Let's reconsider the grain density events E_1, E_2, and E_3 defined in Example 3a. Recall that $E_1 = [2.65, 2.68]$, $E_2 = (2.68, 2.71]$, and $E_3 = [2.65, 2.71]$. E_1 and E_2 are mutually exclusive events since there is no value of ρ_g that will satisfy both E_1 and E_2 (i.e., $E_1 \cap E_2 = \varnothing$). E_3 is the combination of both the events E_1 and E_2. From the data, we obtained $\text{Prob}(E_1) = 7/17$ and $\text{Prob}(E_2) = 9/17$. If we apply Axiom 3', we must have $\text{Prob}(E_1 \cup E_2) = 16/17$, which is precisely the same value we obtained for $\text{Prob}(E_3)$.

Combining Axioms 1 and 3 gives *Boole's inequality*:

$$\text{Prob}(E_1 \cup E_2) \le \text{Prob}(E_1) + \text{Prob}(E_2)$$

With a little algebra, this expression becomes a *Bonferroni inequality*

$$1 - \text{Prob}(E_1 \cup E_2) \ge 1 - \text{Prob}(E_1) - \text{Prob}(E_2)$$

The left-hand side is the probability of neither event occurring and is useful for calculating the confidence level of joint hypotheses (see Sec. 9-6). For example, E_1 might represent a shale exists at a given depth, while E_2 is the event that the gamma-ray exceeds 100API. If $\text{Prob}(E_1) = 0.05$ and $\text{Prob}(E_2) = 0.05$, then the probability of having neither event is at least 0.9, irrespective of whether the events are related or not.

We note also that the null set may be written as the complement of the sample space, since

$$\text{Prob}(\varnothing) = 1 - \text{Prob}(\Omega) = 0$$

The null set is often called the *impossible event.*

2-4 CONDITIONAL PROBABILITIES

When Ω contains a finite number of outcomes, calculation of event probabilities is straightforward, if occasionally tedious. It is easiest for cases involving a finite number of equally likely outcomes, such as throwing dice, flipping coins, or drawing balls from the proverbial urn or cards from a deck where the laws of combinations and permutations can be used. Even such cases require care in conducting the experiment, however, to avoid influencing the outcomes of later experiments. We first consider a simple card example.

Example 4a - Conditional Probabilities. Calculate the probability of selecting two spades in two draws from a normal 52-card deck. Each card is returned to the deck and the deck thoroughly shuffled before the next draw.

The sample space consists of all pairs of cards: $\Omega = \{(A\heartsuit, A\heartsuit), (A\heartsuit, 2\heartsuit), \ldots, (A\heartsuit, K\heartsuit), (A\heartsuit, A\clubsuit), \ldots\}$. Since each draw could produce any one of 52 cards, the total number of outcomes is $N = (52)(52) = 2704$. Remember that each card could be any one of 52 and that 2704 is just the number of ways to take 52 things, two at a time (with replacement).

The event E is the drawing of two spades and the number of outcomes favorable to the event is (13)(13) = 169, since there are 13 spades in the deck and the second draw is with replacement. Therefore the probability of E is 169/2704 = 0.0625, a surprisingly small number.

The experiment is more complicated when the card is not replaced after each draw. This is because withholding the first card alters the probability in the second draw. Such changes lead to the notion of *conditional probability.* For two events, E_1 and E_2, the conditional probability of E_2 given that E_1 has occurred is defined to be

$$\text{Prob}(E_2 \mid E_1) = \frac{\text{Prob}(E_1 \cap E_2)}{\text{Prob}(E_1)}$$

provided that the $\text{Prob}(E_1) > 0$. (The vertical bar between E_1 and E_2 on the left of this definition means "given that.") Actually, we would not be interested in the probability of E_2 given that E_1 never occurs because $\text{Prob}(E_2 \mid E_1)$ would not make sense. If E_1 and E_2 are mutually exclusive, then if E_1 has occurred, E_2 cannot occur and we have $\text{Prob}(E_1 \mid E_2) = 0$.

Usually, we use the conditional probability definition to estimate $\text{Prob}(E_1 \cap E_2)$ so that the definition is sometimes referred to as the fundamental law of multiplication. This is illustrated in the next example.

Example 4b - Conditional Probabilities Revisited. We repeat the task considered in Example 4a of calculating the probability of selecting two spades in two draws from

a normal 52-card deck. This time, however, the first card is held out before the second draw. The sample space now consists of fewer pairs of cards: Ω = {(A♥, 2♥), . . ., (A♥, K♥), (A♥, A♣), . . .}. Pairs with identical cards, such as (A♥, A♥) and (3♠, 3♠), are no longer possible.We also define the events differently from Example 4a. Let E_1 = {drawing a spade on the first card} and $E_2 \mid E_1$ = {drawing a spade on the second card given that the first card is a spade}. What we seek is Prob($E_2 \cap E_1$).
From the definition of probability, Prob(E_1) = 13/52 = 1/4 because there are 13 spades in 52 cards, and Prob($E_2 \mid E_1$) = 12/51 = 4/17 because there are now 12 spades in 51 cards. The second probability differs from the probability of drawing a spade in Example 4a (E_2 = 13/52) because the lack of replacement changes both the number of cards in the deck and the spades available to be chosen. Rearranging the definition of conditional probability, we have Prob($E_1 \cap E_2$) = Prob($E_2 \mid E_1$) Prob(E_1) = 1/17 = 0.059 or a slight decrease in probability caused by the lack of replacement.

The definition of conditional probability can be easily generalized to provide a multiplication rule for the intersection of several dependent events. Let E_1, E_2 and E_3 be three events. In terms of conditional probabilities, the probability of all three events happening is

$$\text{Prob}(E_1 \cap E_2 \cap E_3) = \text{Prob}(E_1 \mid E_2 \cap E_3)\text{Prob}(E_2 \cap E_3) =$$
$$\text{Prob}(E_1 \mid E_2 \cap E_3)\text{Prob}(E_2 \mid E_3)\text{Prob}(E_3)$$

You should compare this to the addition law for the union of events above, where sums, rather than products, are involved.

Example 5 - Shale Detection in a Core. A probe permeameter is set to take a measurement every 3 cm along a 10 cm section of core. The formation consists of 1-cm shale and sand layers interspersed at random, with 80% sand in total (i.e., a net-to-gross ratio of 0.8). What is the probability that only sand will be measured? Assume the probe falls either on sand or shale, not a boundary.

Because of the 3-cm spacing, only three measurements are made over the interval. Define the following events and their respective probabilities:

E_3 = probe measures a layer of sand, Prob(E_3) = 8/10;

E_2= probe measures sand given first layer is sand, Prob($E_2 \mid E_3$) = 7/9; and

E_1 = probe measures sand given first two layers are sand, Prob($E_1 \mid (E_2 \cap E_3)$) = 6/8.

We seek Prob($E_1 \cap E_2 \cap E_3$). From the definition of conditional probability,

$$\text{Prob}(E_1 \cap E_2 \cap E_3) = (8/10)(7/9)(6/8) = 0.47$$

So, with three measurements, there is a one-in-two chance of shale layers going undetected over a 10-cm sample.

Example 6 - Assessing Net Pay. Although the preceding concepts are basic, they can be used to provide practical insights. Figure 2-3 shows a permeability-porosity (k-ϕ) crossplot, representative of a West Texas carbonate reservoir. All samples represent identical volumes.

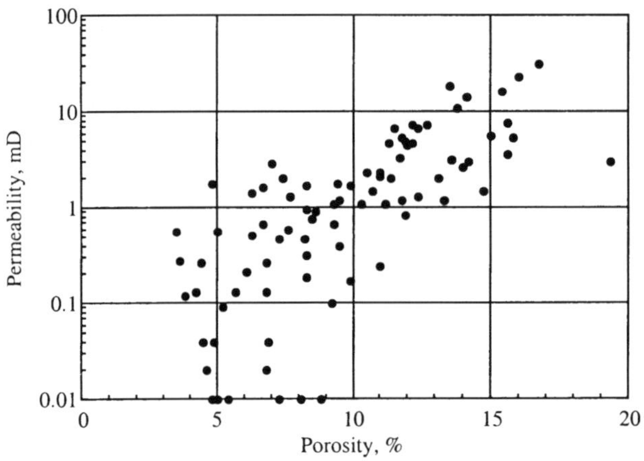

Figure 2-3. Permeability-porosity correlation. $I = 86$ points.

Treating these data as the sample space (that is, taking relative frequencies to be probabilities), estimate the following from the figure:

1. $\text{Prob}(k > 1) = \dfrac{47}{86} = 0.547$. This value represents, for this data set, the portion of the formation which is potentially most productive. This result indicates nearly 45 % of the reservoir has a low permeability. These low-permeability regions are called "non-pay" because they contribute little or nothing to production. The decision to set 1 mD as the cutoff is reservoir specific and determined by several engineering factors, including the type of fluid being produced. For example, a gas reservoir might have a lower cutoff than an oil reservoir.

2. $\text{Prob}(\phi > 10) = \dfrac{38}{86} = 0.442$ This value reflects how much of the reservoir contains significant levels of porosity and, perhaps, producible hydrocarbons. Notice that the probabilities of the events ($\phi > 10$) and ($k > 1$) are similar. This, along with the fact that porosity is easier to measure than permeability, has led to the common practice of predicting the proportion of net pay using porosity data instead of permeability data. For this reservoir and the 10% cutoff value, a net-to-gross ratio of 0.442 is indicated.

3. $\text{Prob}((k > 1) \cap (\phi < 10)) = \dfrac{11}{86} = 0.128$ i.e., about 13% of the defined non-pay portion of the reservoir has fluids that are readily producible.

This last probability result, however, will show some problems with using porosity to predict net pay. Recalling the definition of conditional probability, we combine this result with that of the previous part to give

$$\text{Prob}((k > 1) \mid (\phi < 10)) = \frac{\text{Prob}((k > 1) \cap (\phi < 10))}{\text{Prob}(\phi < 10)} = \frac{11/86}{46/86} = 0.239$$

This tells us that porosity is not a perfect predictor of low permeability in this reservoir. Using either ϕ or k, we can estimate the net-to-gross ratio for the reservoir. However, if we use ϕ to determine whether a specific location is non-net (i.e., $k < 1$), there is only a 24 % chance of being correct.

2-5 INDEPENDENCE

We can now define what is meant by independent events.

Independent Events - Two events, E_1 and E_2, are independent if

$$\text{Prob}(E_1 \cap E_2) = \text{Prob}(E_1)\,\text{Prob}(E_2)$$

This result is known as the multiplication law for independent events. When this equation is true, the conditional probability of E_1 is exactly the same as the unconditional probability of E_1. That is, E_2 provides no information about E_1, and we have from the above

$$\text{Prob}(E_2 \mid E_1) = \frac{\text{Prob}(E_1 \cap E_2)}{\text{Prob}(E_1)} = \text{Prob}(E_2)$$

The two-card draws in Example 4a were independent events because of the replacement and thorough reshuffle. We now apply the independence law to a geological situation.

Example 7 - Independence in Facies Sequences. Suppose there are three different facies, A, B, and C, in a reservoir. In five wells, the sequence in Table 2-3 is observed.

Table 2-3. Facies sequences for five wells.

Position	Well 1	Well 2	Well 3	Well 4	Well 5
Top	*C*	*B*	*A*	*B*	*A*
Middle	*A*	*A*	*B*	*A*	*B*
Bottom	*B*	*C*	*C*	*C*	*C*

Let event E_1 = {A is the middle facies} and E_2 = {C is the bottom facies}. Based on all five wells, are E_1 and E_2 independent? If Well 1 had not been drilled, are E_1 and E_2 independent?

For all five wells, Prob(E_1) = 3/5, Prob(E_2) = 4/5, and Prob($E_1 \cap E_2$) = 2/5. Since Prob(E_1) Prob(E_2) = 3/5•4/5 = 0.48 ≠ Prob($E_1 \cap E_2$), E_1 and E_2 are dependent.

Excluding Well 1, Prob(E_1) = 2/4, Prob(E_2) = 1, and Prob($E_1 \cap E_2$) = 2/4. Prob(E_1) Prob(E_2) = 0.5 = Prob($E_1 \cap E_2$), so E_1 and E_2 are independent.

Clearly, the data in Well 1 are important in assessing the independence of these two events.

Statistical independence may or may not coincide with geological independence. There may be a good geological explanation why facies *C* should be lower in the sequence than either *A* or *B*. In that case, the results of Well 1 are aberrant and indicate, perhaps, an overturned formation, an incorrect facies assessment, or a different formation.

2-6 BAYES' THEOREM

We now take the idea of conditional probabilities a step farther. From the definition of conditional probability,

$$\text{Prob}(E_2 \mid E_1)\text{Prob}(E_1) = \text{Prob}(E_1 \cap E_2)$$

and

$$\text{Prob}(E_1 \mid E_2)\text{Prob}(E_2) = \text{Prob}(E_2 \cap E_1)$$

The right sides of these expressions are equal because the events are commutative, so

$$\text{Prob}(E_2 \mid E_1)\text{Prob}(E_1) = \text{Prob}(E_1 \mid E_2)\text{Prob}(E_2)$$

We can solve for one conditional probability in terms the other:

$$\text{Prob}(E_1 \mid E_2) = \frac{\text{Prob}(E_2 \mid E_1)\text{Prob}(E_1)}{\text{Prob}(E_2)}$$

This relationship is called *Bayes' Theorem* after the 18th century English mathematician and clergyman, Thomas Bayes. It can be extended to *I* events to give

$$\text{Prob}(E_i \mid E) = \frac{\text{Prob}(E \mid E_i)\text{Prob}(E_i)}{\sum \text{Prob}(E \mid E_i)\text{Prob}(E_i)}$$

where *E* is any event associated with the (mutually exclusive and exhaustive) events $E_1, E_2, \ldots, E_I$. The denominator of this expression follows because Prob(E_2) is the probability that E_2 can occur regardless of any prior events. It provides a way of incorporating previous (prior) experience into probability assessments for a current situation.

Example 8 - Exploration and Probabilities. Suppose there is a 60% probability of a structure at a given drill site. We are considering taking a seismic survey to pin this down further. What is the benefit of using seismic data?

Experience suggests that there is a 90% chance that seismic will indicate a structure when one exists and a 70% chance it will indicate no structure when one is absent. Using E_1 = a structure exists and E_2 = seismic indicates a structure, we calculate some probabilities.

$\text{Prob}(E_1) = 0.6$ and $\text{Prob}(E_1^c) = 0.4$.

Further, $\text{Prob}(E_2 \mid E_1) = 0.9$ and $\text{Prob}(E_2^c \mid E_1^c) = 0.7$

The combined probabilities lead to the following results.

Probability that a structure exists and the seismic will detect it:

$$\text{Prob}(E_1)\text{Prob}(E_2 \mid E_1) = (0.6)(0.9) = 0.54$$

Probability that a structure does not exist and the seismic will not detect it:

$$\text{Prob}(E_1^c)\text{Prob}(E_2^c \mid E_1^c) = (0.4)(0.7) = 0.28$$

Probability that a structure exists and the seismic will not detect it:

$$\text{Prob}(E_1)\text{Prob}(E_2^c \mid E_1) = (0.6)(0.1) = 0.06$$

Probability that a structure does not exist and the seismic will detect it:

$$\text{Prob}(E_1^c)\text{Prob}(E_2 \mid E_1^c) = (0.4)(0.3) = 0.12$$

The last two probabilities, which represent errors with the seismic, are not equal.

Now we apply Bayes' theorem.

The probability that a structure exists if the seismic survey indicates there is one.

$$\text{Prob}(E_1 \mid E_2) = \frac{\text{Prob}(E_2 \mid E_1)\text{Prob}(E_1)}{\text{Prob}(E_2)} = \frac{\text{Prob}(E_2 \mid E_1)\text{Prob}(E_1)}{\text{Prob}(E_2 \mid E_1)\text{Prob}(E_1) + \text{Prob}(E_2 \mid E_1^c)\text{Prob}(E_1^c)}$$

$$\text{Prob}(E_1 \mid E_2) = \frac{(0.9)(0.6)}{(0.9)(0.6) + (0.4)(0.3)} = 0.82$$

The probability that a structure exists if the seismic survey says it is absent:

$$\text{Prob}(E_1 \mid E_2^c) = \frac{\text{Prob}(E_2^c \mid E_1)\text{Prob}(E_1)}{\text{Prob}(E_2^c)} = \frac{\text{Prob}(E_2^c \mid E_1)\text{Prob}(E_1)}{\text{Prob}(E_2^c \mid E_1)\text{Prob}(E_1) + \text{Prob}(E_2^c \mid E_1^c)\text{Prob}(E_1^c)}$$

$$\text{Prob}(E_1 \mid E_2) = \frac{(0.1)(0.6)}{(0.1)(0.6) + (0.7)(0.4)} = 0.176$$

The use of the seismic survey has significantly enhanced the probability of success compared to the original estimate (0.6 vs. 0.82).

Example 9 - Geologic Information and Bayes' Theorem. Consider a problem where we have some geological information from an outcrop study and we want to apply these data to a current prospect. Suppose that we have drilled a well into a previously unexplored sandstone horizon. From regional studies and the wireline logs, we think we have hit a channel in a distal fluvial system, but we do not have enough information to suggest whether the well is in a channel-fill (low-sinuosity) or meander-loop (high-sinuosity) sand body. Outcrop studies of a similar setting, however, have produced Table 2-4 of thickness probabilities that could help us make an assessment.

Table 2-4. Outcrop data for Example 9 (based on Cuevas Gozalo et al., 1993).

Thickness less than, m	Low-Sinuosity Probability	High-Sinuosity Probability
1	0.26	0.03
2	0.38	0.22
3	0.56	0.60
4	0.68	0.70

If we drill a well and observe x meters of sand, what are the probabilities that the reservoir is a low- or high-sinuosity channel? The difference could have implications for reserves and well placement. In outcrop, high-sinuosity channels were observed to be generally wider and have higher connectivity than the low-sinuosity variety (Cuevas Gozalo et al., 1993).

We define E_1 = {low sinuosity} and E_2 = {high sinuosity}. Before any well is drilled, we have no way of preferring one over the other, so $\text{Prob}(E_1) = \text{Prob}(E_2) = 0.50$. After one well is drilled, the probabilities will change according to whether the observed sand thickness is less than 1 m, 2 m, etc.

Let's suppose $x = 2.5$ m. This is event E. From Table 2-4, we observe that x falls between two rows of thickness-less-than entries, $x \geq 2$ and $x < 3$. Since the event E falls within an interval ($2 \leq x < 3$), we can more precisely estimate the occurrence of E by calculating the interval probability of E as the difference in interval bound probabilities. In this example, we calculate conditional-interval probabilities as

$\text{Prob}(E \mid E_1) = \text{Prob}(E_1 \text{ at upper bound}) - \text{Prob}(E_1 \text{ at lower bound})$. From Table 2-4, $\text{Prob}(E \mid E_1) = 0.56 - 0.38 = 0.18$, while $\text{Prob}(E \mid E_2) = 0.60 - 0.22 = 0.38$.

Applying Bayes' Theorem with E associated to two mutually exclusive events E_1 and E_2, we obtain for the revised probabilities,

$$\text{Prob}(E_1 \mid E) = \frac{\text{Prob}(E \mid E_1)\text{Prob}(E_1)}{\text{Prob}(E)} = \frac{\text{Prob}(E \mid E_1)\text{Prob}(E_1)}{\text{Prob}(E \mid E_1)\text{Prob}(E_1) + \text{Prob}(E \mid E_2)\text{Prob}(E_2)}$$

$$\text{Prob}(E_1 \mid E) = \frac{(0.18)(0.50)}{(0.18)(0.50) + (0.38)(0.50)} = 0.32$$

and $\text{Prob}(E_2 \mid E) = 1 - \text{Prob}(E_1 \mid E) = 0.68$. So the outcrop information has tipped the balance considerably in favor of the high-sinuosity type channel.

We used the maximum information from the thickness measured in the previous solution. That is, we knew the thickness was less than 3 m *and* more than 2 m. If we only knew the thickness was less than 3 m, the probabilities would be

$$\text{Prob}(E_1 \mid E) = \frac{\text{Prob}(E \mid E_1)\text{Prob}(E_1)}{\text{Prob}(E)} = \frac{\text{Prob}(E \mid E_1)\text{Prob}(E_1)}{\text{Prob}(E \mid E_1)\text{Prob}(E_1) + \text{Prob}(E \mid E_2)\text{Prob}(E_2)}$$

$$\text{Prob}(E_1 \mid E) = \frac{(0.56)(0.50)}{(0.56)(0.50) + (0.60)(0.50)} = 0.48$$

and $\text{Prob}(E_2 \mid E) = 0.52$. Thus, the change in probabilities is not nearly so large because less prior information has been added to the assessment.

In Example 9, the probability values for E_1 and E_2, before measuring the formation thickness, are called *a priori* probabilities. $\text{Prob}(E_1 \mid E)$ and $\text{Prob}(E_2 \mid E)$ are called *a posteriori* probabilities. If there were further information with associated probabilities available (e.g., transient test data showing minimum sand body width), the *a posteriori* probabilities could be amended still further.

Example 9 has a feature that makes it appropriate for Bayes' Theorem: the result of the experiment (measuring the reservoir thickness) did not uniquely determine which of the two possible scenarios existed. Now consider the following example, where Bayes' Theorem may not be suitable.

Example 10 - Bayes' Theorem and Precise Quantities. We want to know the average reservoir porosity, ϕ_0, and some measure of possible variations of ϕ_0, for the Dead Snake reservoir. From well data, we measure porosity to be $\phi_m = 0.22$ for this reservoir. We also have available Table 2-5 of average porosities for eight reservoirs having a depositional environment and a diagenetic history similar to those of the Dead Snake. Can we use the information in Table 2-5 to provide a more appropriate estimate for ϕ_0?

Table 2-5. Porosity ranges for Example 10.

Average Reservoir Porosity	Number of Reservoirs
$0.00 \leq \phi < 0.20$	0
$0.20 \leq \phi < 0.24$	2
$0.24 \leq \phi < 0.28$	4
$0.28 \leq \phi < 0.32$	2
$0.32 \leq \phi < 1.00$	0

The answer depends upon how we view ϕ_m. If we view ϕ_m as being representative and error-free, then we have to set $\phi_0 = \phi_m = 0.22$ and any other information is irrelevant. On the other hand, we have Table 2-5, which suggests that ϕ_0 could be higher. That is, $\phi_0 = 0.22$ might not be representative for Dead Snake. To use the information in Table 2-5, however, we have to admit to the possibility that our estimate (0.22) may not be correct and quantify that uncertainty. If we can give some probabilities of error for ϕ_m, then Table 2-5 can help amend the *a priori* probabilities by use of Bayes' Theorem.

For example, suppose we determine from seismic data that the well is in a representative location but the measurement ϕ_m is prone to error. From previous experience, that error is $\text{Prob}(0.20 \leq \phi_m < 0.24 \mid 0.20 \leq \phi_0 < 0.24) = 0.7$ and $\text{Prob}(0.20 \leq \phi_m < 0.24 \mid 0.24 \leq \phi_0 < 0.28) = 0.3$. That is, there is a 30% chance we have $\phi_m = 0.22$ while the reservoir actually has porosity between 0.24 and 0.28. (For simplicity, we assume that the probabilities of other outcomes are zero.) Using Bayes' Theorem, we can calculate the probabilities for ϕ_0 as follows:

$$\text{Prob}((0.20 \leq \phi_0 < 0.24) \mid \phi_m = 0.22) = \frac{(0.7)(0.25)}{(0.7)(0.25) + (0.3)(0.5)} = 0.54$$

where

$$\text{Prob}((0.20 \leq \phi_m < 0.24) \mid (0.20 \leq \phi_0 < 0.24))\,\text{Prob}(0.20 \leq \phi_0 < 0.24) = (0.7)(0.25) = 0.175$$

and

$$\text{Prob}((0.20 \leq \phi_m < 0.24) \mid (0.24 \leq \phi_0 < 0.28))\,\text{Prob}(0.24 \leq \phi_0 < 0.28) = (0.3)(0.5) = 0.15$$

and $\text{Prob}(0.24 \leq \phi_0 < 0.28 \mid \phi_m = 0.22) = 1 - 0.54 = 0.46$. There are no other possible porosities for the Dead Snake that are both compatible with the measured $\phi_m = 0.22$ and the data in Table 2-5. This table provided the values for Prob(0.20 ≤

$\phi_0 < 0.24$) and Prob($0.24 \leq \phi_0 < 0.28$). Thus, while there is a 54% chance the Dead Snake does have a porosity as low as 0.22, there is a 46% chance that $\phi_0 \geq 0.24$.

This problem could be made more realistic with more intervals, but the principle remains the same. Without an error assessment for ϕ_m, Bayes' Theorem cannot be applied because we are strictly dealing with mutually exclusive events: the average porosity for a reservoir can only be one value and we claim to know that value to be 0.22 without error. We can derive error estimates for ϕ_m using methods in Chap. 5 (see, in particular, Examples 6 and 7).

The number of applications for Bayes' theorem in engineering and the geosciences is large (e.g., Newendorp, 1972). See Kapur et al. (2000) for an example application for assigning facies from interpretations of wireline logs.

2-7 SUMMARY

We have now covered several important foundation concepts regarding probability. Probability is defined in terms of a large number of experiments performed under identical conditions; it is the ratio of successful outcomes to the number of trials. Conditional probability allows the experimental conditions to be varied in stipulated ways. Bayes' Theorem permits additional information to be incorporated into probabilistic assessments. All these concepts will be exercised in the chapters to come.

CHAPTER 3

UNIVARIATE DISTRIBUTIONS

A random variable is the link or rule that allows us to assign numbers to events using the concepts in Chap. 2 by assigning a number–any real number–to each outcome of the sample space. We call the rule X, each outcome is called ω, and the result of applying the rule to an outcome is denoted $X(\omega)$. More formally, a random variable is defined as follows (Kendall and Stuart, 1977):

> A *random variable* $X(\bullet)$ is a mapping from the sample space Ω onto the real line $\Re^1$ such that to each element $\omega \in \Omega$ there corresponds a unique real number $X(\omega)$ with all of the associated probabilistic features of ω in Ω.

Sometimes $X(\omega)$ is called a *stochastic variable*. We can illustrate what the definition means using Fig. 3-1.

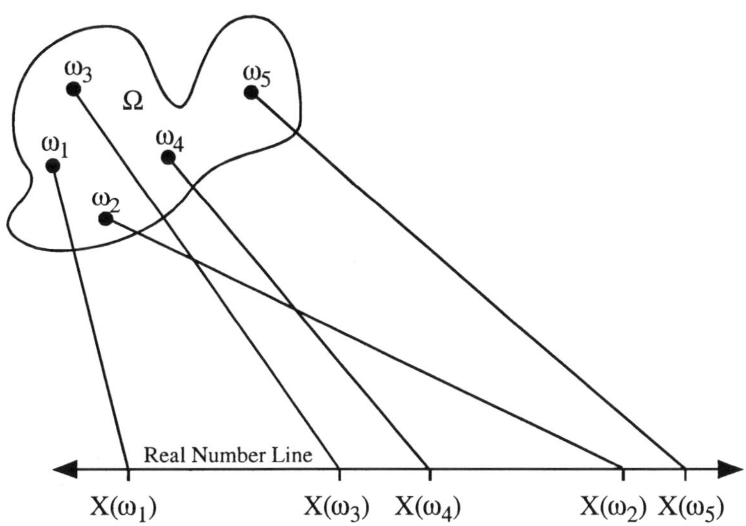

Figure 3-1. The random variable X assigns a real number to each possible outcome ω.

The idea this should convey is one of uncertainty, not about the rule X but about the outcome of the experiment. The random variable incorporates the notion that (1) certain values will occur more frequently than others, (2) the values may be ordered from smallest to largest, and (3) although it may take any value in a given range, each value is associated with its frequency of occurrence through a distribution function.

The value $X(\omega)$ associated with each element (ω) may not necessarily have any relationship to ω's intrinsic value. Examples 1 and 2 concern two properties of three reservoir rock samples.

Example 1 - Rock Sample Porosities. We measure (with no error) the porosities of the rock samples and call the measured values ϕ_1, ϕ_2, and ϕ_3. The sample space (Ω) is all values in the range 0 to 1 (i.e., $\Omega = [0, 1]$). Each outcome, ϕ, has a numerical value corresponding to the porosity of the rock sample. In this case, it makes sense to have X be the identity function. So the event "the porosity of sample one is ω" is assigned the value ω. We can order the values (e.g., $\phi_3 \leq \phi_1 \leq \phi_2$) and establish that some samples have a larger fraction of pore volume than others.

Example 1 deals with continuous data. Such variables have a distinct ordering because they relate to a property that can be "sized." That is, one sample can have a larger or smaller amount of that property. If the ordering is on a scale that has a well-defined zero point, where there is none of the property, *and* the variable can always be meaningfully added or multiplied, then the variable is on a *ratio* scale, and it is called a *ratiometric* variable. Length falls nicely into this category, along with other extensive properties such as volume, mass, and energy. For example, the combined volume of two incompressible systems is the sum of the system volumes. Porosity is a ratio of volumes so it has a well-defined zero point, but only under conditions where the associated rock volumes are known can porosities be added meaningfully.

A second type of continuous variable (on an *interval* scale) is the one that has an arbitrary zero point *or* cannot always be meaningfully added. Temperature, time, position, density, pressure, porosity, permeability, resistivity, gamma ray API, and spontaneous potential deflection are some examples. Intensive variables are interval types. Sums, differences, and multiplication/division of such variables may or may not be meaningful. For example, consider the temperatures of two samples of similar material. Let $T_1 = 30$°C and $T_2 = 10$°C, so $\Delta T = T_1 - T_2 = 20$°C. Sample 1 is hotter than sample 2 by 20°C, but sample 2 may have more energy than sample 1 because sample 2 may have greater thermal mass than sample 1. Combining the two samples would not give $T = 40$°C. Similarly, let $\phi_1 = 0.10$ and $\phi_2 = 0.20$ for two rock samples. If these porosities refer to identical volumes of rock, we can deduce that sample 1 has less void volume than sample 2. In general, if we merge the two samples, the total porosity ϕ_T is not equal to $\phi_1 + \phi_2$ but to some weighted average of ϕ_1 and ϕ_2: $\phi_T = \lambda_1\phi_1 + \lambda_2\phi_2$, with $\lambda_1 + \lambda_2 = 1$ and $\lambda_i \geq 0$.

Example 2 - Rock Sample Depositional Environments. We identify the depositional environments of three samples: sabkha, dune, and interdune. The sample space (Ω) consists of all depositional environments, including lacustrine, aeolian, fluvial, and shoreface deposits among others. Here, we choose X to be X(sabkha) = 0, X(dune) = 1, and X(interdune) = 2. While we can order the numbers 0, 1, and 2, as we did in Example 1, the ordering has no significance. The differences (1 - 0) or (2 - 1) have no significance either. Furthermore, we may even have trouble differentiating some cases where sands might have features of two or more environments.

Example 2 deals with *categorical* or discrete data on a *nominal* (or "name") scale. There is no quantitative meaning to the numbers attached to the events (categories); dune is not twice as much of anything as the interdune and there is no natural zero. Similarly, *ordinal* scale data, such as bit wear, hardness, and color lack the proportion information of continuous data. Ordinal data, however, do have an order to the data because the variable has a size associated with it; sample A is harder than sample B, for example, but we cannot say that it is twice as hard.

All types of data occur in geoscientific problems, and we must be aware of the ways that we can use the information they contain. Ratiometric variables contain the most information (highest level), and nominal variables contain the least information (lowest level). A wide variety of statistical methods is available for continuous variables, while a more restricted range exists for categorical variables. Categorical variables can still give vital information about a system, however, so that conditional probabilities may be useful (e.g., Kapur et al., 2000). Recall, for example, the implications for reserves potential if a fluvial reservoir is from a low- or high-sinuousity part of the system (Chap. 2). There are many examples in this text where categorical geological information is incorporated, either explicitly or implicitly, as a conditional probability to help improve assessments.

Higher-level variables can also be changed into lower-level variables if information is ignored. (The reverse is not possible.) For example, indicator variables can be generated from ratiometric or interval variables by setting cutoff values. An ordinal variable, P, might be defined as

$$P = \begin{cases} 0 & k < 1\text{mD} \\ 1 & k > 1\text{mD} \end{cases}$$

based on the permeability k, an interval variable. P contains less information than k, but it is still useful: $\text{Prob}(P = 1)$ is the formation net-to-gross ratio and P is the net-pay indicator. Recall that net pay is that portion of the formation thought to have sufficient permeability that it will contribute to economic production. See Chap. 2 for more discussion of net pay and Chap. 12 for information on indicator Kriging.

For the methods in this book to apply, variables should be one of the four types just discussed, else they are deterministic. There are other types of variable, however, and you should take care to ensure that methods discussed in this book are appropriate for the variables being considered. Fuzzy variables, for example, are one type that is not suitable for the methods described here. Fuzzy variables have values that may partly belong to a set, whereas the variables we use definitely either do or do not belong to a set. For probabilities of events to be defined, outcomes of experiments must be distinct and recognizable as either having occurred or not. See Klir and Fogler (1988) for further details.

It would be nice to distinguish by notation between random variables, which have uncertain values, and deterministic variables. For example, random variables might be denoted by capital letters while deterministic variables could be denoted by lower case letters. We shall use this convention for generic variables (e.g., x and X). However, the common usage of certain symbols for particular reservoir properties, e.g., k for permeability, R for resistivity, ϕ for porosity, and T for transmissivity, does not obey any particular rule of this kind. Therefore, we can not strictly adhere to this; we will expect the reader to understand from the context of the problem which variable is which.

The definitions we have introduced do not rule out dependencies between variables in different sample spaces or between variables within the same sample space. We now consider how the concept of a random variable provides the essential link between probabilities and distributions.

3-1 THE CUMULATIVE DISTRIBUTION FUNCTION

The most common of generic distribution types is the *cumulative distribution function* (CDF). Given a random variable X, the cumulative distribution function $F(x)$ is defined as

Cumulative Distribution Function (CDF): $F(x) = \text{Prob}(X \le x)$.

In words, $F(x)$ is the probability of finding a value of a random variable X that is less than or equal to x. The argument of F is x, the deterministic bounding value, not X the random variable. Thus, F says something only about the probability of X being less than a certain value, but says nothing precisely about what X is. Sometimes we find the CDF defined as the probability of a random variable greater than or equal to x; this is just one minus the definition given above because of Axiom 3 (Chap. 2): $F^c(x) = 1 - F(x)$ is the *complement* of F.

The CDF uniquely defines all of the probabilistic properties of a random variable. This might seem to rob X of some of its randomness because X must now conform to a deterministic function, namely $F(x)$. Remember that the adjective *random* refers to X, not $F(x)$. The form of the CDF can range from cases where there are an infinite set of X's, to a finite set of discrete X's, to where there is only one X. The latter is the degenerate case where the random variable becomes deterministic.

CDF's have the following properties:

1. $0 \le F(x) \le 1$ for all x since $F(x)$ is a probability.

2. $\lim_{x \to -\infty} F(x) = 0$ and $\lim_{x \to +\infty} F(x) = 1$.

3. $F(x)$ is a nondecreasing function: $F(x+h) \ge F(x)$ for any $h \ge 0$ and all x.

4. $F(x)$ is a continuous function from the right for all x: $\lim_{x \to 0^+} F(x+h) = F(x)$, where $h \to 0^+$ means h approaches 0 through positive values.

An important use of the CDF is that it can be used to find out how often events within a given range of values will occur (i.e., interval probabilities). Suppose we wish to investigate the random variable X and its frequency of occurrence between a lower bound a and an upper bound b ($a \le b$). If we let the events $E_1 = (-\infty, a]$ and $E_2 = (a, b]$, then clearly E_1 and E_2 are mutually exclusive. Axiom 3' (Chap. 2) applies and gives

$$\text{Prob}(E_1 \cup E_2) = \text{Prob}(E_1) + \text{Prob}(E_2)$$

or

$$\text{Prob}(X \le b) = \text{Prob}(X \le a) + \text{Prob}(a < X \le b)$$

which becomes, upon rearrangement,

$$\text{Prob}(a < X \le b) = \text{Prob}(X \le b) - \text{Prob}(X \le a) = F(b) - F(a)$$

In words, the probability of a random variable having a value between a and b is given by the difference between the values of the CDF evaluated at the bounds.

All of statistics exhibits a dichotomy between discrete and continuous mathematics. The former is far more practical than the latter, but continuous statistics are usually more theoretically tractable.

3-1.1 Discrete CDF's

These are defined by a function $F(x)$ with a set of jumps of magnitude p_i at points x_i for $i = 1, 2, 3, \ldots, I$ such that

$$\text{Prob}(X = x_i) = F(x_i) - \lim_{x \to x_i^-} F(x)$$

and

$$\sum_{i=1}^{I} p_i = 1$$

The former statement implies that, if we take any two values a and b ($a \le b$) such that the interval (a, b) does not contain any of the jump points x_i, then

$$\text{Prob}(a < X \le b) = F(b) - F(b) = 0$$

Thus, a random variable X cannot fall within the flat segments of the CDF (Fig. 3-2). This means that X can take on values only at the jump points. In this case, X is a purely discrete or categorical random variable. Nominal or ordinal data have this sort of CDF.

If the difference between adjacent x_i's (i.e., $x_{i+1} - x_i$) is the same for all i, this difference is called the *class size*. If we connect the upper corners of this plot with straight-line segments, the resulting curve is called a frequency polygon.

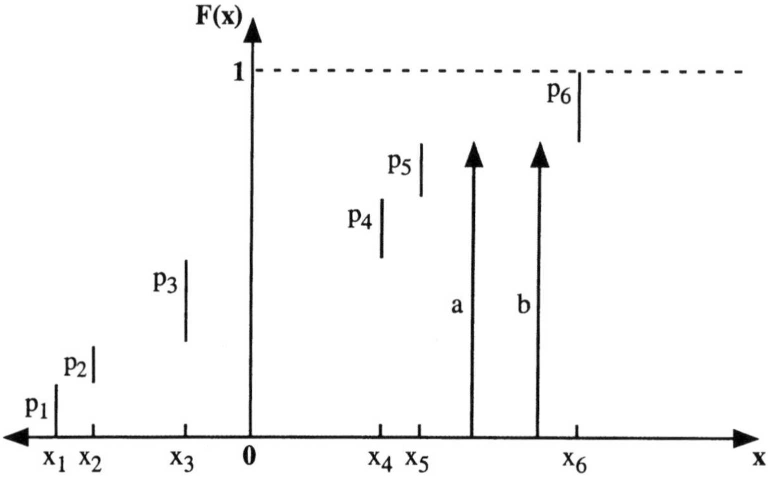

Figure 3-2. An example CDF for a discrete random variable.

3-1.2 Continuous CDF's

If we consider a CDF for a random variable that is absolutely continuous and differentiable (this rules out the frequency polygon) for all x, a continuous CDF $F(x)$ will be defined mathematically as

$$F(x) = \int_{-\infty}^{x} f(u)du$$

with

$$F(+\infty) = \int_{-\infty}^{+\infty} f(u)du = 1$$

A typical shape is sketched in Fig. 3-3.

The random variable X pertaining to this CDF is a purely continuous random variable. Continuous CDF's lend themselves readily to analysis, but they cannot be developed directly from experimental observations. What we usually end up doing is fitting a theoretical smooth curve to the frequency polygon or to the data themselves.

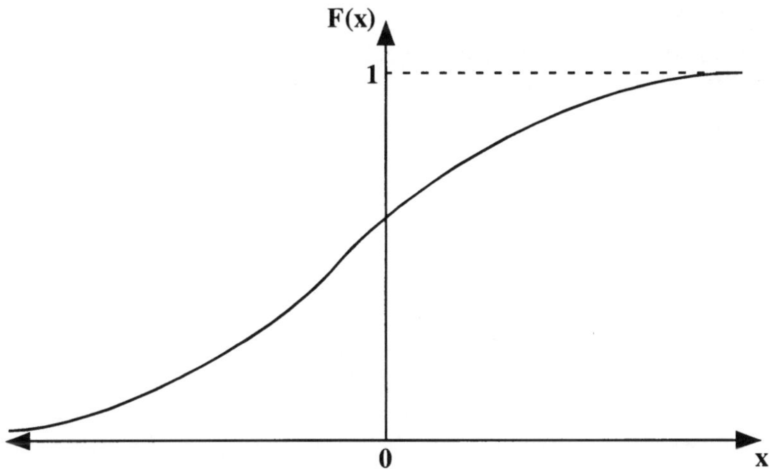

Figure 3-3. An example CDF for a continuous random variable.

The function $f(u)$ in the definition of $F(x)$ is the *probability distribution function* (PDF) whose properties we consider below. But let us first consider an example of how CDF's can be used to quantitatively convey geologic information.

Example 3 - Stochastic Shales. Shales can influence several aspects of reservoir performance, including coning retardation, gravity drainage, and waterflood sweep. In these cases, it is important to know the locations and areal extent of shales in a reservoir.

The simplest case is when a shale extends over an entire reservoir interval. It is observed in every well, and its effect on the reservoir flow can be simply modeled in a reservoir simulator. Such shales are often called *deterministic shales.*

The *stochastic shale*, on the other hand, might appear only in one or two wells, and its areal extent between wells is unknown. It may not even be the same shale that appears in several wells. Its effects on reservoir performance might be quite small or significant. If a number of such shales are dispersed throughout the reservoir, their impact can be quite large. To simulate the effects of such shales, some idea must be obtained about their locations and sizes. This is where the CDF can assist.

It has been recognized (e.g., Weber, 1982) that shale dimensions vary with the environment of deposition. In general, the areal extent of shales is greater in low-energy environments. The exact size of any particular shale will be determined by a large number of factors governed by the circumstances under which deposition occurred. With the aid of CDF's such as Fig. 3-4, we can begin to quantify the observed size-environment relationship. Each curve in Fig. 3-4 is a CDF and, because the probabilities vary with depositional environment, these curves are *conditional CDF's*. If D is a discrete variable identifying the environment and X is the shale length, each curve can be represented as $F(x \mid d_i) = Prob(X \leq x \mid D = d_i)$ where $d_1, d_2, \ldots, d_I$ are the depositional environments.

Such CDF's can be used in a qualitative sense, giving the chances that it is the same shale that appears at two wells. For example, suppose the d_i = deltaic or barrier and the distance between two vertical wells is 1500 ft. There is observed a single shale in each. In the absence of other information, there is approximately a 50% chance that shales observed in these two wells are the same. CDF's can be used in a more quantitative manner, using the *Monte Carlo simulation method* (Haldorsen and Lake, 1982).

The data required to produce CDF's of geological-feature geometries may be very costly to obtain. Field studies often involve considerable effort and have to overcome numerous problems (e.g., poor access and limited exposure) to obtain measurements. Consequently, one may only have a few points (e.g., the distributary channel CDF of Fig. 3-4) and have to make wide extrapolations. Statistical procedures may help in some respects (e.g., correcting for limited exposure in White and Willis, 2000), but considerable engineering judgment may still be needed.

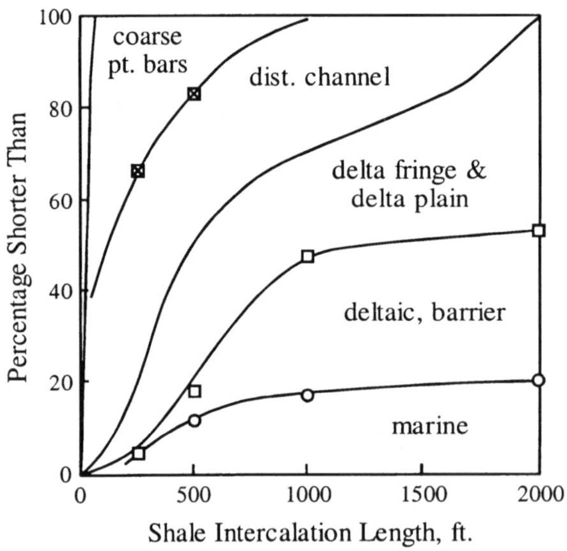

Figure 3-4. Shale size CDF's based on outcrop data. After Weber (1982). The lines are curve fits to data. Points shown are from Zeito (1965).

Stochastic simulation based on Monte Carlo methods involves using a random number generator to "scatter" a number of shales within a reservoir simulation model and calculate the performance. The locations of the shales are picked randomly subject to certain constraints. The sizes of the shales are chosen from the shale-size CDF. The calculated performance will, of course, depend on the shale distribution in the model. Consequently, a number of runs, all with different shale distributions, are usually needed to determine how variable the performance might be. Each new shale distribution is an *equiprobable realization*, one possible outcome of the infinite possible number of realizations (stochastic experiments). Given computer resource limitations, only a finite number of realizations are possible. The variability in reservoir performance from these realizations is claimed to be representative of outcomes from the underlying population. Monte Carlo simulation is discussed in more detail in Sec. 3-4.

3-1.3 Producing CDF's From Data

For some random variables, we may know from theory what the CDF is. Some common distributions are considered in Chap. 4. There are times, however, when we will not know the CDF of a variable. If we can obtain sample values of the variable, we can produce an approximate CDF from those data. Such a CDF is called an *empirical* or a *sample* CDF.

Empirical CDF's are usually produced for non-nominal variables where there is some natural ordering to the data. If there is no natural size to the variable, the empirical CDF might change shape, depending upon the number assigned (i.e., the random variable) to each category. For variables with ordering, we produce an empirical CDF as follows.

1. Order the data so that $X_1 \leq X_2 \leq X_3 \leq \cdots \leq X_I$ where I is the number of data.

2. Assign a probability, p_i, to the event Prob($X \leq X_i$). Here we have to make some assumptions about the assignment. If X is a continuous variable or is discrete and the uppermost categories may not have been sampled, $p_i = (i - {}^1/_2)/I$ is usually adequate. We make the assumption here, without any evidence to the contrary, that each sample value has an equiprobable chance of occurring. If X is discrete and all categories have been sampled, $p_i = i/I$.

3. Plot X_i versus p_i and, depending on whether the random variable is continuous or discrete, either connect the lines with stair-steps (e.g., Fig. 3-2) or a continuous curve (e.g., Fig. 3-4).

Clearly, this estimated CDF is better–more like the actual CDF–the more data we have.

The first probability assignment formula given above in Step 2 differs from the naive assumption that Prob($X \leq X_1$) = $1/I$, Prob($X \leq X_2$) = $2/I$, · · · , Prob($X \leq X_I$) = 1. This is because the second probability assignment formula, $p_i = i/I$, forces Prob($X > X_I$)= 0. It does not allow for the possibility that, if we are dealing with a continuous variable, there may be unsampled regions of the reservoir where $X > X_I$. A similar argument holds if X is discrete with possible unsampled upper categories. Each step in p, $p_{i+1} - p_i$, is still $1/I$ but all the p_i's have been reduced by $I/2$ so that Prob($X > X_n$) = $I/2$. Clearly, this is still an approximation without data to justify it. There are other formulas for assigning probabilities, such as $p_i = i/(I+1)$, which give different values to Prob($X > X_I$). If the population CDF is known, the optimal probability assignment can be computed.

Example 4a - Producing Empirical CDF's. Draw empirical CDF's for the following data (X) and their logarithm ($X^* = \ln(X)$): 900, 591, 381, 450, 430, 1212, 730, 565, 407, 440, 283, 650, 315, 500, 420, 714, and 324.

We first rank the data. One ranking suffices for both variables, X and X^*, since $\ln(X)$ is a monotonic function. We then define a probability for each value ($x_i = X_i$ and $x_i^* = \ln(X_i)$), using the formula $p_i = \text{Prob}(X \le x_i) = \text{Prob}(X_i^* \le x_i^*) = (i - {}^1/_2)/I$ for the $I = 17$ points. Table 3-1 shows the calculations.

Figure 3-5 shows the empirical CDF's. The untransformed CDF (left) is more curved than the logarithmic CDF (right), which is closer to a straight line. This is because the data set has mostly moderate but a few large values, whereas taking the logarithm has evened out the spread of the variable with probability.

Table 3-1. Data for Example 4a.

No.	x_i	x_i^*	Prob p_i
1	283	5.65	0.029
2	315	5.75	0.088
3	324	5.78	0.147
4	381	5.94	0.206
5	407	6.01	0.265
6	420	6.04	0.324
7	430	6.06	0.382
8	440	6.09	0.441
9	450	6.11	0.500

No.	x_i	x_i^*	Prob p_i
10	500	6.21	0.559
11	565	6.34	0.618
12	591	6.38	0.676
13	650	6.48	0.735
14	714	6.57	0.794
15	730	6.59	0.853
16	900	6.80	0.912
17	1212	7.10	0.971

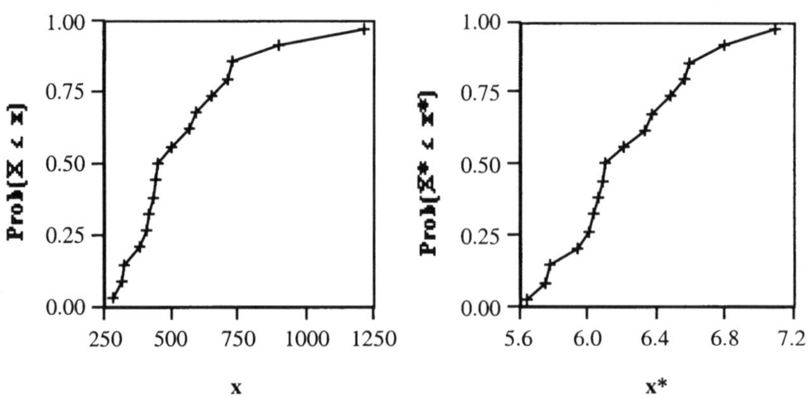

Figure 3-5. Empirical CDF's for the data of Example 4a.

3-1.4 Order Statistics

Some values of x for the CDF $F(x)$ have special names. $x_{0.50}$ is the value where $x = F^{-1}(0.50) \equiv x_{0.50}$ and is called the *median*. $x_{0.25}$ and $x_{0.75}$ are the first and third *quartiles*, respectively, of the CDF $F(x)$. The difference $x_{0.75} - x_{0.25}$ is sometimes used as a measure of the variability of X and is called the *interquartile range.*

A *quantile* of the CDF $F(x)$ is any value x_p such that $F^{-1}(p) = x_p$. If we do not know the population CDF and have only the empirical CDF, we can provide only estimates of the actual x_p's. These estimated values are termed *order statistics*. For the data in Example 4a, $x_{0.50}$ is approximately 450.

3-2 PROBABLITY DISTRIBUTION FUNCTIONS

Probability distribution functions (PDF's) are a very common statistical tool. They represent exactly the same information as contained in the CDF, but it is displayed differently. The CDF takes a "global" view of X, conveying the probability of X being less than some stipulated value x. The PDF takes a "local" view and describes how the probability of occurrence of X changes with x. As with CDF's, there is the distinction between discrete and continuous properties.

3-2.1 Discrete PDF's

Consider the discrete CDF $F(x)$ discussed previously. For each "jump" i and any small $h > 0$,

$$\text{Prob}(x_i - h < X \leq x_i + h) = F(x_i + h) - F(x_i - h) = p_i$$

If we let $h \to 0$, we get $\text{Prob}(X = x_i) = p_i$

Thus the jump at the end of each interval represents the probability that X has the value x_i. The set of numbers $(p_1, p_2, p_3, ...)$ plotted against $(x_1, x_2, x_3,...)$ is called the discrete PDF of the random variable X (Fig. 3-6).

If a horizontal line is drawn through the top of each vertical line to a mid-point between the neighboring vertical lines, or their extension, on the right (left), the plot is a bar chart or a histogram.

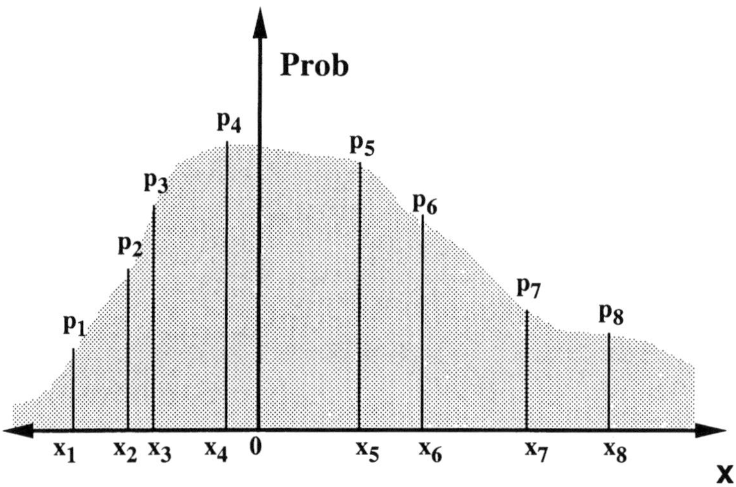

Figure 3-6. Example of a PDF for a discrete random variable.

3-2.2 Continuous PDF's

Consider a small interval $(x, x + \delta x)$ from the x axis of a purely continuous random number X with CDF $F(x)$. Then we have

$$\text{Prob}(x < X \leq x + \delta x) = F(x + \delta x) - F(x) = \int_{x}^{x+\delta x} f(u)du$$

This forms the basis for interpreting $f(x)$ as a probability: $f(x)$ represents the frequency of occurrence of a value of x in the neighborhood of x. The best physical interpretation of $f(x)$ is as a derivative of $F(x)$, because we see from above that $f(x) = dF(x)/dx$. A typical continuous PDF is shown in Fig. 3-7.

The basic properties of continuous PDF's are

1. $f(x) \geq 0$ for all x since $F(x)$ is nondecreasing

2. $\int_{-\infty}^{+\infty} f(u)du = 1$

3. for any a, b $(a \leq b)$, $\text{Prob}(a < X \leq b) = \int_{a}^{b} f(u)du$

A rather curious consequence of the last property is that the probability that X takes a particular value x is zero. If we let b approach a, then we have

$$\text{Prob}(X = x) = \int_{a}^{a} f(u)du = 0$$

This is a bit disquieting but entirely consistent with the notion of a random variable. Recall that we should be able to say nothing about a specific value of a random variable.

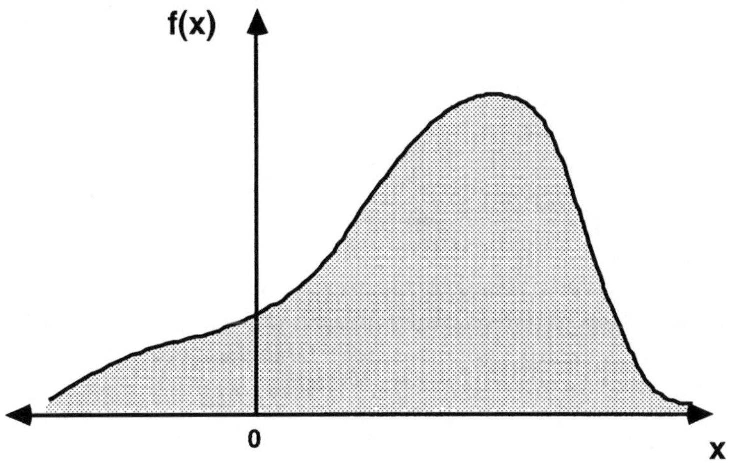

Figure 3-7. Example of a PDF for a continuous random variable.

Both continuous and discrete PDF's will have peaks and troughs. Those values of x where a peak occurs are called *modes*. A PDF with one mode is *unimodal*; a PDF with two modes is *bimodal*, etc.

3-2.3 Empirical PDF's

Otherwise known as histograms, these PDF's are based on samples. They are a common method for assessing data behavior but, as we will see in Example 5, they can mislead.

Empirical PDF's can be produced for any type of variable. The way we produce an empirical PDF is as follows.

1. If the variable is not nominal, order the data so that $X_1 \le X_2 \le X_3 \le \cdots \le X_I$ where I is the number of data.

2. For continuous variables, divide the interval $X_I - X_1$ into convenient intervals. We call these intervals *bins* or *classes*. Usually the bins are all of equal size, Δx, so the height of each interval is proportional to the probability. If we choose too few bins, the histogram has little character; if we choose too many bins, the histogram is too bumpy. A rule of thumb is $\Delta x \approx 5(X_I - X_1)/I$. For categorical variables, each bin is one category.

3. Count the number of data in the ith bin, I_i, and set $\text{Prob}(x_i \le X < x_i + \Delta x) = p_i = I_i/I$ for $i = 1$ to I-1.

4. Plot p_i versus $x_i \leq X < x_i + \Delta x$ for $I = 1$ to I-1.

As with the empirical CDF, the empirical PDF more closely approaches the population PDF as I increases. If unequal-sized classes are used, the height of each interval must be determined so that its area (not just the height) is proportional to the probability. This is because the area of the histogram must be unity.

Example 4b - Drawing PDF's. Draw empirical PDF's for the data (X) and their logarithm ($X^* = \ln(X)$) given in Example 4a: 900, 591, 381, 450, 430, 1212, 730, 565, 407, 440, 283, 650, 315, 500, 420, 714, and 324.

By ordering the data, we find that $X_1 = 283$, $\ln(X_1) = 5.65$, $X_{17} = 1212$, $\ln(X_{17}) = 7.10$. The bin size, using the rule of thumb, is about 250 for the untransformed data while it is 0.40 for the logarithmic data. Table 3-2 shows the probabilities.

Table 3-2. Data for Example 4b.

No.	Value	Freq.	p_i
1	250-499	9	0.53
2	500-749	6	0.35
3	750-999	1	0.06
4	1000-1249	1	0.06

No.	Value	Freq.	p_i
1	5.60-5.99	4	0.23
2	6.00-6.39	8	0.47
3	6.40-6.79	3	0.18
4	6.80-7.19	2	0.12

The histograms are shown in Fig. 3-8.

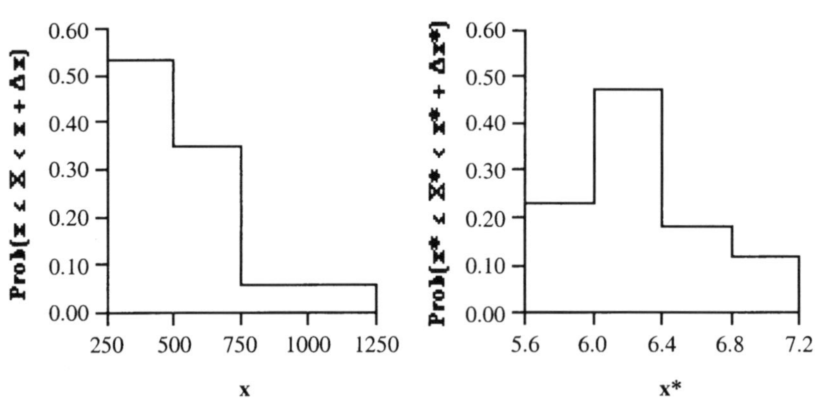

Figure 3-8. PDF's for the data and log-transformed data of Example 4b.

The untransformed data have a few large values (2 out of 17) whereas the logarithmic histogram has a more even spread across the range. Both data sets have only one peak (unimodal PDF's), and the untransformed data have the more strongly negatively (right) skewed PDF.

For some applications we would like to draw a smooth curve through the empirical PDF's. In these cases, nonlinear transforms, as the ln in Example 4b, could cause difficulties since the area under the smooth curve deviates from one.

Example 5 - Detecting Modes with PDF's. We now experiment with two empirical PDF's using the same data set. The data are 6.5, 7.5, 10.5, 12.5, 13.5, 20.5, 25.5, 26.5, 27.5, 29.5, and 37.0. We violate the rule of thumb for bin size and set $\Delta x = 10$. One histogram will begin at $x = 0$ while the second will begin at $x = 5$. Figure 3-9 shows the results.

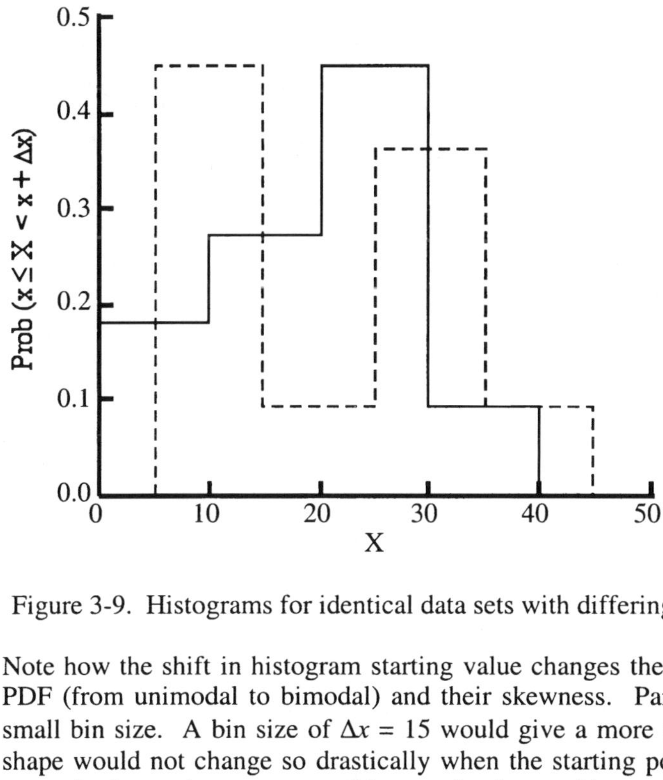

Figure 3-9. Histograms for identical data sets with differing starting values.

Note how the shift in histogram starting value changes the apparent shape of each PDF (from unimodal to bimodal) and their skewness. Part of the problem is the small bin size. A bin size of $\Delta x = 15$ would give a more stable histogram, so the shape would not change so drastically when the starting point was changed. This example shows that even something as simple as a histogram can be deceptive and care should be exercised when interpreting them. The problems with empirical PDF's, as illustrated in this example, mean that they are rather difficult to deal with in practice. For this reason, the CDF is the more practical tool, even though the PDF is conceptually more familiar and more easily understood.

3-2.4 Why Are CDF's and PDF's Important?

There are several reasons why we want to know the distribution of a reservoir property. They are:

(1) *Modeling.* Knowing the CDF/PDF, we can produce models of how the property varies within the reservoir. For example, using stochastic shale models as input to reservoir simulation permits the presence of shales to be evaluated for their impact upon reservoir performance. Reserves distributions are often modeled using the CDF's of several reservoir properties. See Examples 7 and 8 in this chapter and Chap. 12 for more details.

(2) *Estimation.* Chapter 5 is largely devoted to estimating certain parameters based on their CDF/PDF. We can make better use of the available data when we know the population PDF. An example is when we want to estimate the average formation permeability. An estimate using the PDF may easily have one-half or less of the variability of an estimate that is calculated ignoring the PDF.

(3) *Diagnosis.* We can test data to determine whether they are "good" or "bad" by comparing the sample PDF with other sample PDF's or some theoretical reference PDF. PDF's have been used to assess the accuracy of core measurements (Thomas and Pugh, 1989). Significant changes of facies throughout a reservoir may also be detected by comparing permeability PDF's from within a well or between different wells. This next example illustrates this use, along with pointing out some limitations to PDF's.

Example 6 - PDF's and Geological Elements - The Lower Brent Group (Middle Jurassic) is a major North Sea oil-producing interval and commonly comprises a thick sand-dominated reservoir, without significant shale breaks (Corbett and Jensen, 1992b). As a single reservoir group, the shape of its porosity and permeability histograms (sample PDF's) can be used to help diagnose the presence of units (e.g., facies) and confirm their geological identification.

The core porosities for this interval (Fig. 3-10) show a predominantly unimodal, negatively skewed distribution (i.e., most of the data values lie to the high end of the data range) clustering at 27%. A second cluster at 3% suggests the presence of a second, minor grouping. Core-plug horizontal permeabilities, in contrast, are strongly positively skewed. A logarithmic transformation of the permeabilities results in a more symmetrical distribution that is clearly multimodal, suggesting the presence of several different groups. That is, each peak represents a substantial number of permeability data near that value, and the peaks could represent different geological units.

Geologically, there are good reasons for splitting the Lower Brent into two parts, known as the Etive and Rannoch Formations. These geological elements (formations) are defined on the basis of wireline log characteristics (e.g., gamma ray) and descriptions of cores. They are correlateable at interwell (km) distances. The Etive is a characteristically medium- to coarse-grained, cross-bedded, stacked channel sandstone. The Rannoch is a fine-grained, highly micaceous, cross-laminated, shoreface sandstone. The grain-size difference produces the contrasting permeabilities, even though these units have similar porosities.

A plot of permeability versus depth (Fig. 3-11) shows the distinction between formations. The spatial association of the permeability is an important geological factor. Most of the high permeability values are in the Etive and low values are from the Rannoch. This information is not readily apparent from the PDF's because they ignore the sample locations.

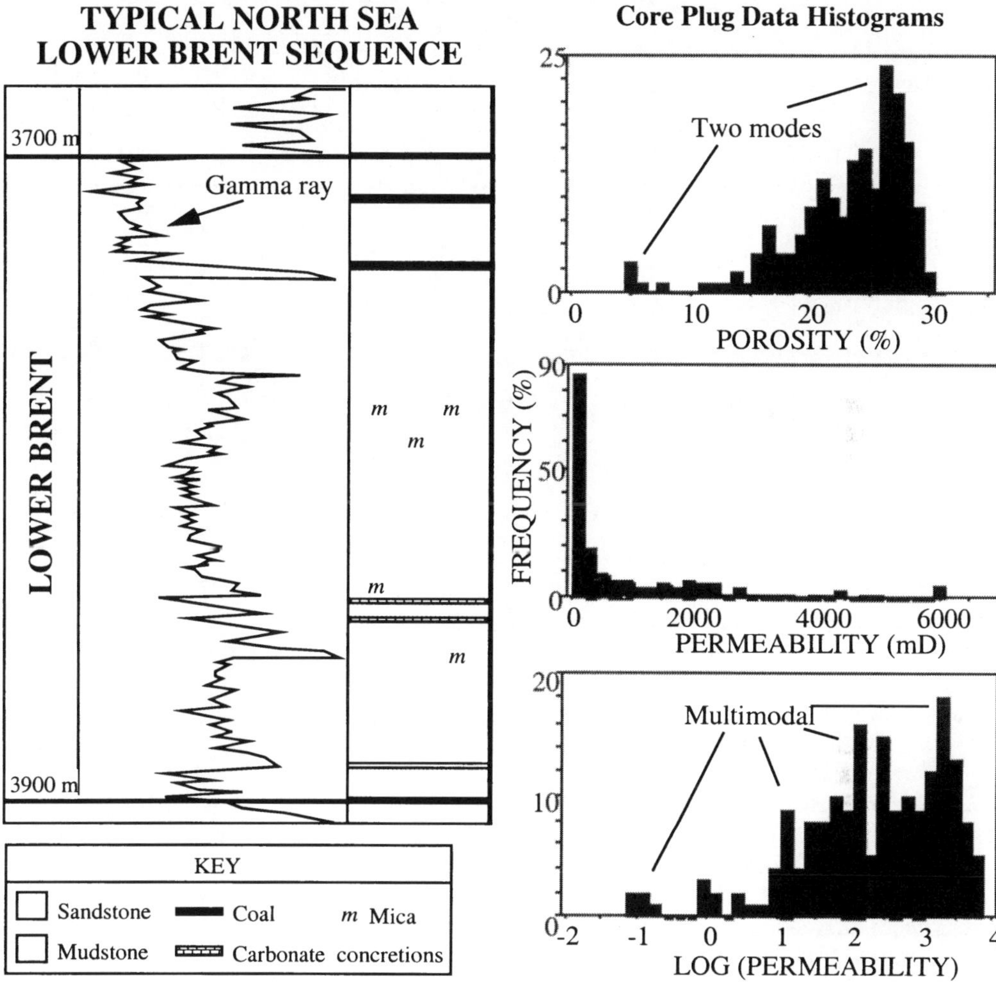

Figure 3-10. Geological section and histograms of core porosity and permeability for the Lower Brent Group.

The high-permeability Etive is clearly a separate population from the low-permeability Rannoch. The Etive is characterized by unimodal, slightly negatively skewed porosity and permeability distributions. The Rannoch, which still has bimodal PDF's, could be further subdivided into carbonate-cemented and carbonate-free populations. The low-porosity and low-permeability intervals are discrete carbonate-cemented concretions (often known as "doggers"). A stratigraphic breakdown on this basis results in two flow units with approximately symmetrical porosity and log-permeability distributions.

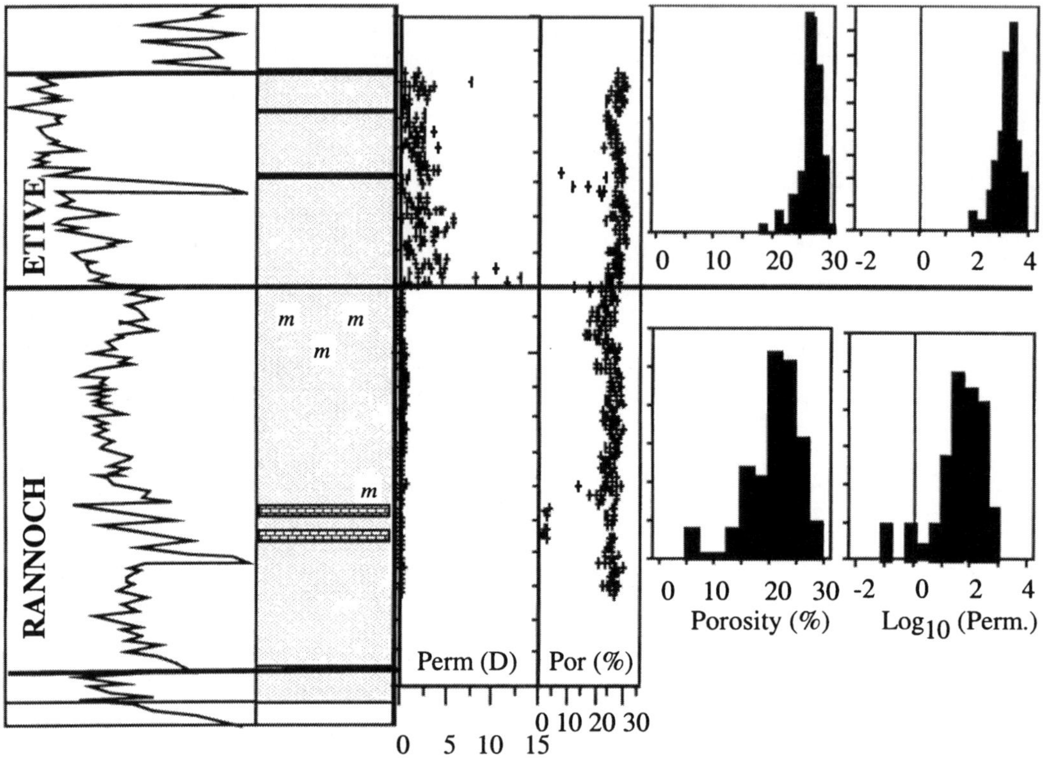

Figure 3-11. Porosity and permeability histograms for the Etive and Rannoch formations in the Lower Brent.

The Etive, uncemented Rannoch, and cemented Rannoch are thus "sub-populations" of the Lower Brent population. The histograms and plots of the petrophysical data versus depth can be used to distinguish and separate the geological units. In this example, the permeability data are a more powerful discriminator than the porosity. This means that, while hydrocarbon reserves might be similar, the elements will have different flow and recovery characteristics.

Within the Rannoch sequence, with the help of a series of very fine-scale probe-permeameter measurements (Corbett and Jensen, 1992b), a further breakdown of the permeability distribution can be made. Some details of this analysis will be discussed in Chap. 13.

It is possible to determine the basic elements of geological architecture (the laminae), which tend to have unimodal PDF characteristics (Goggin et al., 1988; Corbett and Jensen, 1993b). Much of the reservoir characterization effort in recent years is driven by the need to identify and provide petrophysical PDF's for the basic geological elements (Dreyer et al., 1990; Lewis and Lowden., 1990; Willis et al., 1998).

The PDF alone may be limited in the determination of effective flow properties, however. For example, well tests in a braided fluvial reservoir having a symmetrical log-permeability

PDF with an arithmetic average of approximately 500 mD showed the effective permeability to be approximately 1000 mD in one well and approximately 50 mD in a second (Toro-Rivera et al., 1994). These differences have been explained by the geological organization of the system into large, flow-dominating channels and small-channel, matrix- dominated flow, respectively. The differences in effective permeability are a function of the spatial distribution of the permeability and cannot be ascertained from the PDF alone. The petrophysical PDF's are, however, good descriptors and help to confirm geological assessments.

Consequently, the PDF and CDF can be powerful devices for making better models and estimates. For example, if PDF and CDF of porosity and/or permeability are multimodal, more geologic analysis may be required to further subdivide the medium into flow units with unimodal petrophysical properties. Carefully defined flow units at early development stages help earth scientists and engineers optimize and manage production throughout the life of a reservoir. Empirical PDF's and CDF's are the first step towards exploring and assessing the geological/flow units that make up a reservoir zone. This process can also highlight aspects of the reservoir that might otherwise be overlooked.

3-2.5 Transforming CDF's

During the statistical modeling of reservoir properties, random variables with several different CDF's may need to be simulated. Random-number generators in calculators and computers often have only one fixed CDF (e.g., uniform random over [0, 1]), so that a method is needed to transform the computer random variable to one with the desired CDF.

Let X be the continuous random variable, with CDF $F_X(x)$, produced by the computer and let Y be the reservoir random variable with desired CDF $F_Y(y)$. For the moment, we will assume that $F_X(x)$ and $F_Y(y)$ are known functions; they are invertable by virtue of the general properties of the CDF. The equation $y = F_Y^{-1}[F_X(x)]$ will convert the computer variable to one with the desired CDF. The reasoning behind this relationship can best be explained using Fig. 3-12.

The key concept in changing one CDF to another is that the probabilities should be the same ($F_Y = F_X$). The value of y should correspond to the x with the same probability. Hence, given an x, $F_X(x) = F_Y(y)$ for the suitable y. We then apply the inverse transformation F_Y^{-1} to both sides of the equality to obtain $y = F_Y^{-1}[F_X(x)]$. This relationship applies also to random variables, since $X \leq x$ is satisfied when x has the same value as X.

Computer random-number generators often have the CDF of Fig. 3-12 (left). It is the Rand() function in some spreadsheets. There are also tables of four- or five-digit uniformly distributed random numbers in many statistics books and mathematical handbooks (e.g., Abramowitz and Stegun, 1965) with this CDF. If the CDF on the right of Fig. 3-12 is the standard normal distribution (Chap. 4), the transformation is called the normal score transformation.

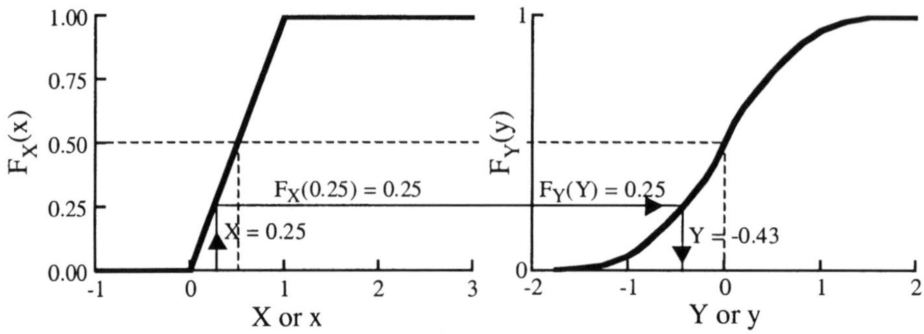

Figure 3-12. Transformation from random variable X to Y with change in CDF.

Since the PDF and CDF contain the same information, the PDF's of the supplied and desired variables could be specified instead of their CDF's. The conversion procedure, however, first requires converting the PDF's to CDF's.

Example 7 - Reservoir Property Distributions. A reservoir property has the following PDF:

$$f_Y(y) = \begin{cases} 0 & y < 0 \\ 2y & 0 \le y \le 1 \\ 0 & y > 1 \end{cases}$$

Three numbers, x = 0.33, 0.54, and 0.89, were generated on a computer with the PDF

$$f_X(x) = \begin{cases} 0 & x < 0 \\ x & 0 \le x \le 1 \\ 0 & x > 1 \end{cases}$$

Transform the x's to y's.

We first convert the PDF's to CDF's.

$$F_Y(y) = \int_{-\infty}^{y} f(u)du = \begin{cases} 0 & y < 0 \\ 2\int_0^y y du = y^2 & 0 \le y \le 1 \\ 1 & y > 1 \end{cases} \quad \text{and}$$

$$F_X(x) = \int_{-\infty}^{x} f(u)du = \begin{cases} 0 & x < 0 \\ \int_0^x 1du = x & 0 \le x \le 1 \\ 1 & x > 1 \end{cases}$$

These CDF's are shown in Fig. 3-13.

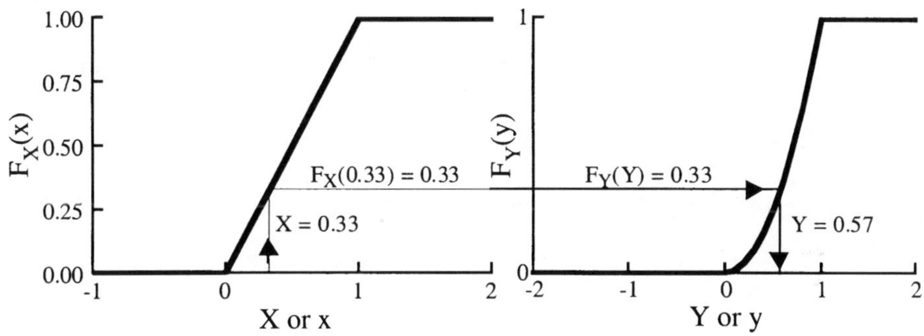

Figure 3-13. Transformation for first random number in Example 7.

X is called a *uniformly distributed* random variable because its PDF is constant over the interval [0, 1]. Since $F_X(x) = x$, $F_X(0.33) = 0.33$, $F_X(0.54) = 0.54$, and $F_X(0.89) = 0.89$. $F_Y(y) = y^2$ so $F_Y^{-1}(y) = \sqrt{y}$. Thus, $Y = 0.57$, 0.73, and 0.94, respectively.

3-3 FUNCTIONS OF RANDOM VARIABLES

So far, we have dealt only with "simple" random variables. There are instances, however, when a random variable is presented as a transformed quantity. For example, permeability and resistivity are often plotted on a logarithmic scale, so that we are actually seeing a plot of log(k) or log(R). It is evident that, since k and R are random variables and logarithm is a monotonic function, log(k) and log(R) are random variables. What is the PDF of k or R, given that we know the PDF of log(k) or log(R)?

If $Y = h(X)$ is a monotonic function of X and has a known PDF, $g(y)$, then the PDF of X is given by (Papoulis, 1965)

$$f(x) = g(y)\left|\frac{dy}{dx}\right|$$

This expression can be rearranged to determine $g(y)$ if $f(x)$ is known. If $Y = h(X)$ is not monotonic, then X has to be broken up into intervals in which $h(X)$ is monotonic. Papoulis (1965, Chap. 5) has the details with a number of examples.

Determining the PDF when a variable is a function of two or more random variables, e.g., $Z(X, Y)$, is quite complicated. For example, if $Z = X + Y$, where $f(x)$ and $g(y)$ are the PDF's of two independent random variables, X and Y, then $h(z)$, the PDF of Z, is given by

$$h(z) = \int_{-\infty}^{+\infty} f(z-y)g(y)dy = \int_{-\infty}^{+\infty} f(x)g(z-x)dx$$

These integrals represent the convolution of the PDF's of X and Y. Papoulis (1965, Chap. 7) proves this result and considers a number of examples. It is clear that determining the PDF of a function of two or more random variables is not, in general, a simple task. There are some results that are useful when $f(x)$ and $g(y)$ have certain forms, e.g., have Gaussian PDF's, and these will be discussed in Chap. 4.

3-4 THE MONTE CARLO METHOD

There exists a powerful numerical technique, the *Monte Carlo method*, for using and characterizing random variables in computer programs. If we know the CDF's of the variables, the method enables us to examine the effects of randomness upon the predicted outcome of numerical models. Monte Carlo requires that we have a model defined that relates the input variables (e.g., reservoir properties) to the feature of interest (e.g., oil recovery, breakthrough time, water cut, or dispersion). The model is not random; only the input variables are random. The distribution (CDF) of the output quantity and, in particular, its variability are used to make decisions about economic viability, data acquisition, and exploitation strategy.

Monte Carlo methods can be quite numerically demanding. If many input variables are random and they all have large variabilities, a large number of runs or iterations of the model may be needed to appreciate the range of model responses. For example, an input numerical model for a waterflood flow simulation could require massive amounts of computer time if all the input parameters, including porosity, permeability, capillary pressure, and other properties, are considered to be random variables.

Example 8 - Reserves Estimates Using Monte Carlo. In petroleum economics, by far the most frequent use of Monte Carlo is in reserves calculations. The stock-tank oil initially in place (STOIIP) is given by

$$STOIIP = \frac{\phi(1 - S_w)Ah}{B_{oi}}$$

where Ah is the net reservoir volume, B_{oi} is the initial oil formation volume factor, S_w is the initial water saturation, and ϕ is the porosity. In this case, this equation is the model and we are interested in the STOIIP CDF (or PDF) as ϕ, S_w, and Ah may all be considered as independent random variables. In particular, ϕ and S_w can have meaning as independent random variables only if they represent average values over a given net reservoir volume, Ah.

A random-number generator in the computer generates random numbers for all the variables in the STOIIP equation for which the user specifies a CDF or PDF (Fig. 3-14). For each value of ϕ, Ah, and S_w, the STOIIP is computed. This process is repeated several hundred times. The output is a series of STOIIP values that, using the empirical CDF procedure described earlier, can give the STOIIP CDF. From that CDF, summary statistics such as the average or median can be calculated.

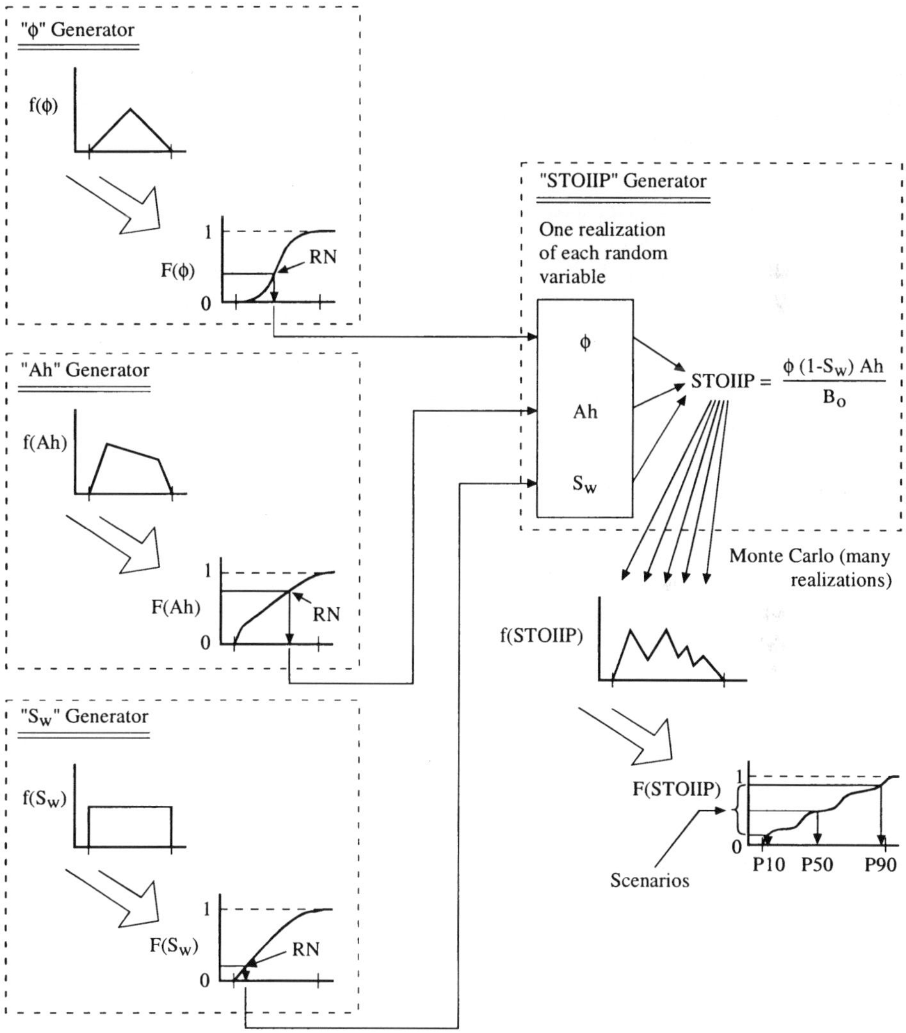

Figure 3-14. Schematic procedure for producing STOIIP estimates using Monte Carlo.

To use the Monte Carlo method, the distributions for all the input variables have to be determined. Experience from other fields, data from the field under study, and geological knowledge all contribute to the selection of the CDF for each variable.

Interdependence between the variables (e.g., small ϕ and large S_w) can be accommodated if it is known how the variables are interrelated.

Some governments require Monte Carlo simulation results to be reported when a company wants to develop a prospect.

While Monte Carlo may seem like an easy option to theoretical approaches, you should recognize that the results are sensitive to the input CDF's and consequently variability in the input data will affect the results. For example, in reserves estimation, if ϕ and S_w do not vary much while the rock volume *Ah* varies considerably, the STOIIP PDF will be virtually identical to the *Ah* PDF. In Chap. 6, we will discuss further the effects of variabilities of the arguments upon the resultant variation.

3-5 SAMPLING AND STATISTICS

In closing this chapter, there remain two philosophical issues: measurement volumes and the role of geology.

The statistical view of the reservoir is entirely from probability distributions of the random variable *X* (Fig. 3-15), representing some property. We imagine that this property, e.g., permeability, porosity, or grain size, exists as a random variable in a physical region. Each point in the region has associated with it several PDF's: one for each random variable of interest. When we take samples and measure them, the properties are no longer random at the sampled locations. We have specific values of the reservoir properties obtained over a certain volume of the region. For example, the volume of a core plug is about 10 cm^3, whereas the volume investigated by a well-logging device may be 0.05 m^3. In any case, our measurements are averages of the random variables over a given volume and they cannot reflect the point-to-point variation of the properties.

By measuring samples of nonzero volume, our characterization of the reservoir is based on averages of the random variable, not the variable itself. When we estimate a parameter at a point, we are estimating the most likely value of the parameter's average. In the range of scales of the geological variability present in the reservoir, some scales may be contained within the measurement volume and, thus, not be recognized in the result. For example, core plugs might contain several laminations and, therefore, may not represent the lamination-scale variation present.

This point of view has two implications. The first is that all physical measurements are now actually averages, and that when we in turn average these, we are taking averages of averages. Thus, it is possible to have measurements over different volumes of material (e.g., core-plug permeability and transient-test permeability). Trying to reconcile measurements made on two different volumes can be very difficult, especially since the character of the PDF's and CDF's will change for different volume scales. This is especially true when comparing the statistics of porosity averages derived from core/log data with porosity estimates derived from seismic attributes, even when they are samples over common depth intervals.

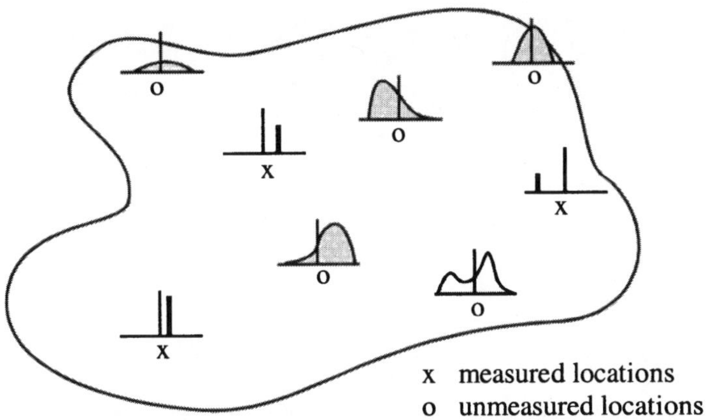

Figure 3-15. Reservoir property distributions at sampled and unsampled locations.

The second implication lies in regarding the original variable as being random. Many geological processes—sedimentation or diagenesis—are very well known and quite deterministic; thus, it is counter-intuitive to think that the result of these processes—the variable under consideration—is random. We don't regard the process as being random, however, just that the detailed knowledge of the result of the process is random. This is almost always the case in physics: the fundamental law is quite deterministic but, as discussed in Chap. 1, the detailed knowledge of the result of the process (or processes) is imprecise.

CHAPTER 4

FEATURES OF SINGLE-VARIABLE DISTRIBUTIONS

The probability distribution function (PDF) of a random variable tells quite a bit about the values the variable takes. For one thing, it says how frequently some values occur relative to other values. We can also extract "summary" information about the behavior of the random variable to determine, for instance, an "average" value, or just how variable the values are (e.g., the variance).

This chapter begins with definitions of the basic statistical operators, principally moments and their functions, and then moves on to a brief exposition of some of the more common PDF's. The operators are important because they allow description of any random variable, regardless of its PDF. But this chapter's main importance is because much of the subsequent development rests on the properties of these operators.

4-1 NONCENTERED MOMENTS

With the probability distribution function $f(x)$ known for a continuous random variable X, we can define the rth moment as

$$\mu_r = \int_{-\infty}^{+\infty} x^r f(x)dx$$

where r is a nonnegative integer. In making this definition, we have presumed that $f(x)$ is integrable.

4-1.1 The Expectation

By far the most important is the first (i.e., $r = 1$) moment,

$$E(X) = \int_{-\infty}^{+\infty} xf(x)dx \tag{4-1}$$

referred to as the *expected value* or *expectation* of X. $E(X)$ is the mean for a continuous PDF.

A discrete random variable X can take on one of a series of values: $X_1, X_2, \ldots, X_I$ with corresponding probabilities $p_1, p_2, \ldots p_I$. That is, $p_i = \text{Prob}(X = X_i)$. The expected value of X is defined to be

$$E(X) = \sum_{i=1}^{I} X_i p_i$$

So $E(X)$ is the sum of the *possible* values of X weighted by their respective probabilities of occurrence. It is related to, but not necessarily the same as, the arithmetic average. The arithmetic average, $\overline{X}$, is computed from *M samples* $X_1, X_2, \ldots .X_m \ldots, X_M$ taken from the parent population, which has possible values $X_1, X_2, \ldots, X_I$.

$$\overline{X} = \frac{1}{M} \sum_{m=1}^{M} X_m$$

Comparing the equations for $\overline{X}$ and $E(X)$, we see that the arithmetic average appears to equate p_i with $1/M$. This is only partly correct, however, since $X = X_i$ may appear several times among the samples. For example, several samples might all have the value X_6. When this happens, $p_i = k/M$ for the arithmetic average, where k is the number of times X_i appeared. Because of the sample size and sampling variation, the sample values X_m may not represent all the I possible values of the parent population. For example, if $M < I$, all possible values of X could not have been sampled. Hence, $\overline{X}$ and $E(X)$ may not be equal. However, for a finite sample there is no other alternative than to regard $\overline{X}$ as an estimate of $E(X)$.

For discrete variables, $E(X)$ need not be one of the values X_i. For example, we have all seen newspaper reports stating that the average number of children per household is 2.6, or some such number. While no household has a fraction of a child, the expected-value computation does not account for the fact that X is a discrete variable in this case and treats children like lengths of cloth.

Since there are two definitions for the expectation, corresponding to continuous and discrete PDF's, we should probably have two symbols. Conventional usage does not make this distinction, probably because the properties of the two are essentially identical.

4-1.2 Properties of the Expectation

Most of the properties follow from the properties of the PDF in Chap. 3 (which in turn followed from the Axioms in Chap. 2), or the properties of a sum or integral.

Expectation of a constant c: $E(c) = c$

Expectation of a constant c plus (minus) a random variable: $E(X \pm c) = E(X) \pm c$

An immediate consequence of this property is that the expectation of a random variable with zero mean is zero. Let $[X-E(X)]$ be such a variable, then

$$E[X - E(X)] = E(X) - E[E(X)] = E(X) - E(X) = 0$$

Such transformations are common in statistical manipulations. Other properties are

Expectation of a constant c times a random variable: $E(cX) = c\,E(X)$

Expectation of the sum (difference) of two random variables: $E(X \pm Y) = E(X) \pm E(Y)$

The last two properties can be generalized to multiple random variables and their sums or differences:

$$E\left(\sum_i a_i X_i + \sum_j b_j Y_j\right) = \sum_i a_i E(X_i) + \sum_j b_j E(Y_j)$$

As intuitive as these formulas seem, the last actually involve some subtlety. When the expectation deals with more than one random variable, the definition involves the *joint* probability distribution function for X and Y, $f(x, y)$:

$$\begin{aligned} E(X+Y) &= \int_{-\infty}^{+\infty}\int_{-\infty}^{+\infty}(x+y)f(x,y)dxdy \\ &= \int_{-\infty}^{+\infty}\int_{-\infty}^{+\infty}xf(x,y)dxdy + \int_{-\infty}^{+\infty}\int_{-\infty}^{+\infty}yf(x,y)dxdy \\ &= \int_{-\infty}^{+\infty}xf(x)dx + \int_{-\infty}^{+\infty}yf(y)dy = E(X)+E(Y) \end{aligned}$$

where $\int_{-\infty}^{+\infty} f(x,y)dy = f(x)$ and $\int_{-\infty}^{+\infty} f(x,y)dx = f(y)$ are the *marginal* PDF's for Y *and* X, respectively.

The properties described above carry over to more general functions of random variables, but the sum/difference property is all that is needed here. Discussion of the properties of $f(x, y)$ falls under the subject of bivariate distributions, covered in Chap. 8.

There is, however, a final property that requires a further restriction on the random variables.

Expectation of the product of two independent random variables: $E(XY) = E(X)E(Y)$

For this to be true we must be able to write the joint PDF as $f(x, y) = f(x)g(y)$. You will no doubt recognize this property of the PDF and the definition of independence of random variables from the Multiplication Axiom in Chap. 2.

$$E(XY) = \int_{-\infty}^{+\infty}\int_{-\infty}^{+\infty}xyf(x)g(y)dxdy = \left[\int_{-\infty}^{+\infty}xf(x)dx\right]\left[\int_{-\infty}^{+\infty}yg(y)dy\right] = E(X)E(Y)$$

We defer further discussion of dependent random variables until Chap. 8.

4-2 CENTERED MOMENTS

The rth centered moment of a random variable is

$$\mu_r' = E[(X - E(X))^r] = \int_{-\infty}^{+\infty} (x - E(X))^r f(x)dx$$

The expectation inside the integral raises the possibility of several ways to estimate μ_r'. For example, the integral may be approximated as a finite sum and the *E*(*X*) from a continuous PDF. In practice, we will almost always be estimating both expectations from a finite data set. Of course, μ_r' must satisfy all of the properties of *E*(*X*) with respect to the PDF.

4-2.1 Variance

One of the central concepts in all of statistics is the second ($r = 2$) centered moment, known as the *variance* and denoted specifically by *Var*(*X*).

$$Var(X) = \int_{-\infty}^{+\infty} (x - E(X))^2 f(x)dx$$

Using the properties of the expectation, *Var*(*X*) can be written as the difference between the expectation of X^2 and the square of the expectation, of *X*, $E(X)^2$:

$$Var(X) = E[(X - E(X))^2] = E(X^2) - E(X)^2 \tag{4-2}$$

without loss of generality. (In this formula and elsewhere, we adopt the convention that $E(X)^2$ means that the expectation is squared.) This form is actually more common than the original definition. The discrete version of *Var*(*X*) for *I* values is

$$Var(X) = \sum_{i=1}^{I} (X_i - E(X))^2 p_i$$

Compare this with the formula for estimating the variance from *I* samples $X_1, X_2, \ldots, X_I$:

$$s^2 = \frac{1}{I}\sum_{i=1}^{I} (X_i - E(X))^2$$

which requires that the mean of *X* be known. Observations similar to those about the discrete expectation also apply to this estimation formula.

The variance is a measure of the dispersion or spread of a random variable about its mean. Other measures exist, such as the mean deviation, $E[|X - E(X)|]$ and the *L*-scale (see Chapter 5). The variance, however, is a fundamental measure and, as we shall see, many other measures of variability are derived from it.

4-2.2 Properties of the Variance

Many of the variance's properties follow directly from those of the expectation operator. However, its most important property is that it is nonnegative.

Intrinsic property: $Var(X) \geq 0$

The reason this is so (the equality holds only in the degenerate case of no variability) follows directly from the basic definition wherein only squares of differences are used inside an integral and $f(x) \geq 0$. Of course, an absolute value or any even r will yield a nonnegative moment, but these are too complex mathematically. The reason that this seemingly small observation is so important is that when we minimize the variance in the following chapters, we can be assured that our search will have a lower bound. The following example illustrates this.

Example 1 - Minimizing Properties of the Variance. You might ask why the centering point for the variance is the mean instead of some other point, say the median. We now show that using the mean will make the variance a minimal measure of dispersion.

For this exercise only, let us define the variance as

$$Var(X) = \int_{-\infty}^{+\infty} (x-a)^2 f(x)dx$$

where a is an arbitrary constant. To find the minimum in $Var(X)$, take its derivative with respect to a and set it to zero.

$$\frac{dVar(X)}{da} = -2\int_{-\infty}^{+\infty}(x-a)f(x)dx = 0$$

noting that the differentiation is not with respect to the integration variable. But using the definition for $E(\ )$, we find that the minimum variance occurs when

$$E(X) = a$$

or that the variance, as originally defined, is itself minimized.

To expose the remainder of the variance's properties, we parallel those of the expectation.

Variance of a constant c: $Var(c) = 0$

Variance of a random variable plus (minus) a constant c: $Var(X \pm c) = Var(X)$

These properties are consistent with the idea of the variance being a measure of spread, independent of translations. There is, of course, no spread in a constant.

Variance of a random variable times a constant c:

$$Var(cX) = E[(cX)^2] - E(cX)^2$$

$$= c^2E(X^2) - c^2E(X)^2$$

$$= c^2Var(X)$$

Unlike the analogous property for the expectation, this is not intuitive. However, since the constant can be negative, it is entirely consistent with the idea of a nonnegative variance, since $Var(cX) \geq 0$.

Variance of the sum (difference) of two independent random variables:

$$Var(X \pm Y) = E[(X \pm Y)^2] - E(X \pm Y)^2$$

$$= E[X^2 \pm 2XY + Y^2] - E(X)^2 \pm 2E(X)E(Y) - E(Y)^2$$

$$= E(X^2) - E(X)^2 + E(Y^2) - E(Y)^2$$

$$= Var(X) + Var(Y)$$

since $E(XY) = E(X)E(Y)$.

The last two properties can be combined into

$$Var\left(\sum_i a_i X_i\right) = \sum_i a_i^2 Var(X_i) \qquad (4\text{-}3)$$

Remember that this property is valid for independent X_i only.

Example 2 - Variance of the Sample Mean. We can use Eq. (4-3) to derive a traditional result that has far-reaching consequences. Consider the drawing of I samples from a population containing independent X_i. The variance of the X_i is σ^2.

The variance of the sample mean $\overline{X}$ is

$$Var(\overline{X}) = Var\left(\frac{1}{I}\sum_{i=1}^{I} X_i\right)$$

Since the X_i are independent, we can use Eq. (4-3) directly as

$$Var(\overline{X}) = \frac{1}{I^2} Var\left(\sum_{i=1}^{I} X_i\right) = \frac{1}{I^2}\left(\sum_{i=1}^{I} Var(X_i)\right)$$

But, since $Var(X_I) = \sigma^2$,

$$Var(\overline{X}) = \frac{1}{I^2}\left(\sum_{i=1}^{I} \sigma^2\right) = \frac{I}{I^2}\sigma^2 = \frac{\sigma^2}{I} \qquad (4\text{-}4)$$

Equation (4-4) says that the variance of the sample means decreases as the number of samples increases. This makes sense in the limit of I approaching infinity, since the variance must then approach zero, each draw now containing all of the elements in the population. Of course, the variance of the mean is σ^2 when $I = 1$.

Equation (4-4) also suggests that a log-log plot of the square root of the variance (the standard deviation) of the sample mean versus I will yield a straight line with slope of -1. Such a plot is the precursor of the rescaled range plot used to detect spatial correlation in geologic data sets.

The procedure used in Example 2–manipulating a sum until the term inside the summation contains no subscripts–is a common statistical approach. This operation changes a summation into a multiplication to yield a result that is substantially simpler than the original equation.

Properties beyond this, for example the product of two random variables, are even more complicated but are not necessary for this text. We will revisit many of these properties in the discussion of correlated random variables in Chaps. 8 and 11.

4-3 COMBINATIONS AND OTHER MOMENTS

There are a few other related functions of these moments.

4-3.1 Standard Deviation

The *standard deviation* is the positive square root of the variance

$$SD = +\sqrt{Var(X)}$$

The standard deviation is a very familiar quantity. While $Var(X)$ has the units of X^2, the SD has units of X, making it more comparable to the mean.

4-3.2 Coefficient of Variation

The *coefficient of variation* is the standard deviation divided by the mean:

$$C_V = \frac{SD}{E(X)}$$

C_V is one of the most attractive measures of variability. It is dimensionless, being normalized by the mean, and varies between zero and infinity. See Chap. 6 for more discussion.

4-3.3 Coefficient of Skewness

The *coefficient of skewness* is the third centered moment divided by the second centered moment:

$$\gamma_1 = \frac{\mu_3'}{Var(X)^{3/2}}$$

γ_1 is a measure of the skewness of a PDF. $\gamma_1 < 0$ indicates a PDF skewed to the right (i.e., positively skewed) and $\gamma_1 > 0$ indicates a PDF skewed to the left (i.e., negatively skewed). $\gamma_1 = 0$ indicates a symmetrical PDF.

4-3.4 Coefficient of Kurtosis

The *coefficient of kurtosis* involves the ratio of the fourth to second moments:

$$\gamma_2 = \left(\frac{\mu_4'}{Var(X)^2}\right) - 3$$

γ_2 is a measure of the peakedness of a PDF. The normal distribution ($\gamma_2 = 0$) is *mesokurtic*. A PDF flatter than the normal ($\gamma_2 > 0$) is *platykurtic*; one more peaked than the normal distribution ($\gamma_2 < 0$) is called *leptokurtic*.

These last two definitions are more for terminology than usage. They are not used at all in the remainder of this text. But the expectation and variance properties will recur frequently.

4-4 COMMON PDF'S

Let us now look at a few common PDF's and some of their properties. We will begin with a discrete variable.

4-4.1 The Binomial Distribution

Consider an experiment with only two possible outcomes: E_1 = a "success" with Prob(E_1) = p and E_2 = a "failure" with Prob(E_2) = 1 - p. E_1 and E_2 are obviously mutually exclusive in a single experiment, but several repetitions of the experiment would produce combinations of successes and failures. We also take the experiments to be mutually exclusive.

Consider the probability of getting s successes in I trials for a given p. Put another way, we calculate Prob($R = s$) where $R = \Sigma X_i$, $i=1, \ldots, I$, and Prob($X_i = 1$) = p and Prob($X_i = 0$)

= 1 - p. This sort of problem might arise, for example, when estimating net-to-gross ratios for a sand-shale sequence (X_i = 1 for sand and X_i = 0 for shale).

To answer this, we consider all possible ways of getting s successes in I trials with each combination multiplied by its appropriate probability p.

$$\text{Prob}(R = s) = \frac{I!}{s!(I-s)!} p^s (1-p)^{I-s} \tag{4-5}$$

The symbol $I!$ is read "I factorial" and it means $I(I-1)(I-2)\cdot\cdot\cdot 1$. This function is available on scientific calculators. Another way of writing $\left[\frac{I!}{s!(I-s)!}\right]$ is ${}_IC_S$, the number of combinations of I things taken s at a time.

Equation (4-5) is the binomial distribution, an important result in statistics. Each term in the equation has physical significance: the first term is the number of ways in which there can be exactly s successful outcomes, the second is the aggregate probability of s successes, and the third is the aggregate probability of (I - s) failures.

Example 3 - Using the Binomial Distribution for Thin-Section Porosity Estimation. Consider a point count on a vugular carbonate thin-section that has 50% porosity. The microscope cross hairs will fall either on the matrix ($X_0 = 0$) or on an opening ($X_1 = 1$). What is the estimated porosity ($\bar{\phi}$) based on the number of samples we take?

If I is the total number of samples taken and R the number of times we draw X_1 (i.e., ϕ = 1), then the observed or *estimated* sample porosity is

$$\bar{\phi} = \frac{RX_1 + (I-R)X_0}{I}$$

Because the thin-section has 50% porosity, Prob(X_1) = Prob(ϕ = 1) = 1/2. Hence, the probability of r points out of I samples falling on vugs is

$$p = \text{Prob}(R = r) = \frac{I!}{r!(I-r)!}\left(\frac{1}{2}\right)^I$$

For I = 2 we can have R = 0, 1, 2. From the binomial distribution the various probabilities can be plotted as in Fig. 4-1 (left). We can replace R on the abscissa with the ratio R/I (= $\bar{\phi}$) to get Fig. 4-1 (right).

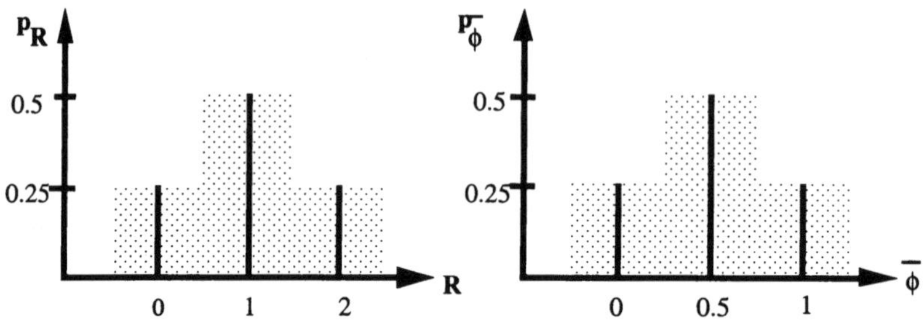

Figure 4-1. Probability distribution functions for R and $\bar{\phi}$ for I = 2 samples and Prob(ϕ = 1) = 0.5.

This is a very simple discrete PDF for $\bar{\phi}$. Clearly, we could repeat this for any I. For example, if I = 6 we have Fig. 4-2 (left) for R = 0, 1, 2, 3, 4, 5, 6 and the $\bar{\phi}$ PDF (Fig. 4-2 right).

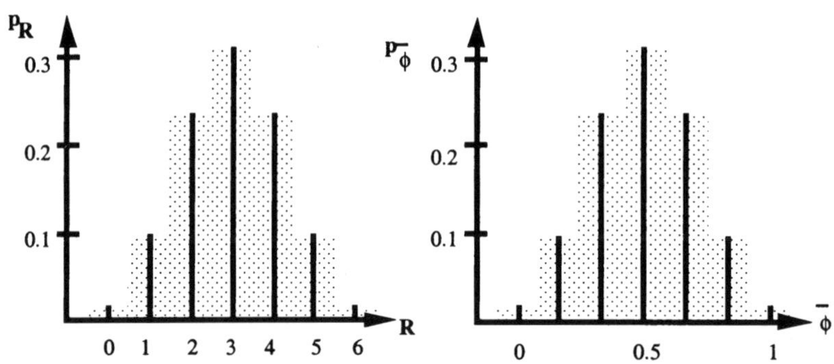

Fig‹‹re 4-2. Probability distribution functions for R and $\bar{\phi}$ for I = 6 samples and Prob(ϕ = 1) = 0.5.

Apart from showing an application of the binomial distribution, this example illustrates an important issue: the variability of estimated quantities ($\bar{\phi}$ in this case) that depend upon data. We always estimate quantities based on some number of samples and those estimates are themselves random variables.

Unless we counted an extremely large number of points, $\bar{\phi}$ would always be subject to some variability. We can get some idea of the variability by looking at the probabilities of the extreme events Prob ($\bar{\phi}$ = 0) and Prob ($\bar{\phi}$ = 1). For I = 2, Prob ($\bar{\phi}$ = 0) = Prob ($\bar{\phi}$ = 1) = 0.25, whereas when I = 6, Prob ($\bar{\phi}$ = 0) = Prob ($\bar{\phi}$ = 1) = 0.016. By tripling the sample size, the probabilities of the extremes have dropped considerably.

The mean value of R for I trials is

$$E(R) = \sum_{i=0}^{I} i({}_{I}C_{i})p^{i}(1-p)^{I-i} = Ip.$$

This result follows from substituting Eq. (4-5) and using the identity

$$\sum_{i=0}^{I} {}_{I}C_{i}X^{i} = (1+X)^{I}$$

By a similar process, the variance is given by $Var(R) = Ip(1 - p)$. These results are used in the following example.

Example 4 - Mixtures of Random Variables in Geological Sediments. Consider a formation unit consisting of two elements, 1 and 2, which interfinger in proportions p and $(1 - p)$, respectively. If each element has a random value of permeability, k_1 and k_2 with means $E(k_1) = \mu_1$ and $E(k_2) = \mu_2$ and variances $Var(k_1) = \sigma_1^2$ and $Var(k_2) = \sigma_2^2$, respectively, what are the mean and variance of samples taken from the combination? Assume that each measurement consists only of material from one of the two elements and that k_1 and k_2 are independent.

Let k be the sample permeability, so that

$$k = Xk_1 + (1 - X)k_2$$

where X is a binomially distributed variable. If $X = 0$, the sample is entirely from element 2; if $X = 1$, it is from element 1. Since the proportion p of the unit consists of element 1, $\text{Prob}(X = 1) = p$. We want $E(k)$ and $Var(k)$ in terms of p, μ_1, μ_2, and σ_2^2. The mean value is

$$E(k) = E[Xk_1 + (1 - X)k_2] = E(Xk_1) + E[(1 - X)k_2]$$

$$= E(X)E(k_1) + E(1 - X)E(k_2)$$

because the variables X, k_1, and k_2 are all independent. Thus,

$$E(k) = p\mu_1 + (1 - p)\mu_2$$

This result agrees with what we would expect. Similar reasoning gives

$$Var(k) = E[(Xk_1 + (1-X)k_2)^2] - E[Xk_1 + (1-X)k_2]^2$$

$$= E(X^2)E(k_1^2) + 2E(k_1)E(k_2)[E(X^2) - E(X)^2] + E(k_2^2)$$

$$- 2E(k_2^2)E(X) + E(k_2^2)E(X^2) - [p\mu_1 + (1-p)\mu_2]$$

$$= p\sigma_1^2 + (1-p)\sigma_2^2 + p(1-p)(\mu_1 - \mu_2)^2$$

The total variability of k arises from three factors: the variability of element 1 multiplied by the amount of 1 present; the variability of 2 also weighted by the amount present; and the additional variability arising because the mean permeabilities of the elements are different. The third factor is a maximum when $p = 1/2$.

Realistic sediments that conform to this example include eolian deposits. Dune slip-face sediments consist of grain-fall and grain-flow deposits. Each deposit has different grain size and sorting and, hence, will have a distinctly different permeability from the other deposit. Probe or plug measurements may be taken in a number of slip-face deposits, and the apparent variability could be quite different from the variability of either element.

Mixtures of random variables cause the resulting PDF to be a combination of the component PDF's. Such PDF's are called *heterogeneous* or *compound* distributions. Multimodal PDF's often arise because the measurements were taken in sediments with distinctly different properties. We cover some aspects of these types of distributions below.

4-4.2 The Central Limit Theorem

It takes considerable mathematics (Hald, 1952), but much less intuition, to see where the binomial PDF is going as I approaches infinity. The discrete PDF above becomes continuous and the PDF itself approaches a normal distribution, that bell-shaped curve we all have come to know and love. Actually, the approach to normality is quite close after only about $I = 20$ when $p = 0.5$. It takes considerably more points when p is closer to 0 or 1, but the normal distribution results as Fig. 4-3 illustrates. You can also see the progression of the mean and variance of the binomial distribution in Fig. 4-3.

According to a fundamental result of applied probability known as the *Central Limit Theorem* (CLT), a random variable X, generated by adding together a large number of independent random variables, will usually have an approximatly normal distribution irrespective of the PDF's of the component variables. This remarkable theorem says that a normal distribution will result from the binomial distribution as I approaches infinity regardless of the value of p. Conversely, when we observe a normal distribution in practice, we assume these attributes are satisfied.

An example of the CLT is core-plug porosity. Each plug consists of numerous pores that each contribute a small amount to the plug pore volume. Hence, the total porosity of plugs is likely to be nearly normally distributed. Indeed, in many clastic reservoirs, porosity is nearly normally distributed. The binomial distribution is not even necessary for this argument (we used it mainly as an illustration). As long as the underlying events are independent and additive and there are a large number of them, the PDF will be normal (Hald, 1952). The CLT is as close as we get to a physical basis for statistics and it accounts, in part, for the popularity of the normal distribution in analysis. Many of the simulation techniques and tools (to be discussed in Chap. 12) require normally distributed random variables.

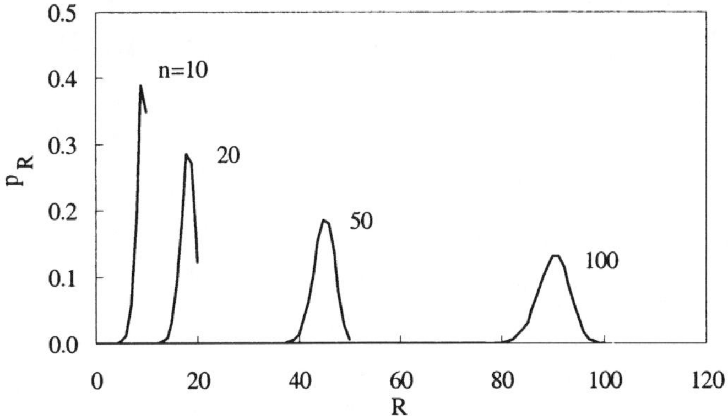

Figure 4-3. The progression of a binomial PDF to a normal distribution for $p = 0.9$.

In light of this result, we can also see how errors resulting from the measurement of a physical quantity could have a distribution close to normal. Typically, errors arise as a combined effect of a large number of independent sources of error. These errors, however, may not perturb the measurement very much, depending on the relative magnitude of the property and the associated noise. In addition, we would like the deviations (residuals) of a measurement away from a model to be normal, since this means that the model accounts for all of the features of the measurements, apart from unquantifiable influences.

4-4.3 The Normal (Gaussian) Distribution

In addition to the CLT, the *normal distribution* is hoary with statistical import. If a random variable X has a normal distribution, its PDF is uniquely determined by two parameters (μ and σ) according to

$$f(x;\mu,\sigma^2) = \frac{1}{\sqrt{2\pi\sigma^2}}\exp\left[-\frac{(x-\mu)^2}{2\sigma^2}\right], \quad -\infty < X < +\infty \tag{4-6a}$$

which graphs as shown in Fig. 4-4.

You can easily show that $E(X) = \mu$ and $Var(X) = \sigma^2$. We will represent the normal distribution with the short-hand notation N(mean, variance). So $X \sim N(3, 9)$ means X is normally distributed with a mean value of 3 and a variance of 9. The normal CDF is

$$F(x;\mu,\sigma^2) = \frac{1}{\sqrt{2\pi\sigma^2}}\int_{-\infty}^{x} exp\left[-\frac{(\eta-\mu)^2}{2\sigma^2}\right]d\eta \tag{4-6b}$$

The integral function above is called the *probability integral*.

Both the normal distribution and the probability integral can be made parameter-free by redefining the variable as $Z = (X - \mu) / \sigma$, known as the *standardized normal variate*. Thus $E(Z) = 0$ and $Var(Z) = 1$, and we say that $Z \sim N(0, 1)$ so that

$$F(z) = \frac{1}{\sqrt{2\pi}} \int_{-\infty}^{z} \exp\left[-\frac{\eta^2}{2}\right] d\eta \qquad (4\text{-}6c)$$

Extensive tables of F and f for $N(0,1)$ are printed in many statistics books; see Abramowitz and Stegun (1965). Functions for all the normal distribution functions are available on spreadsheet libraries. For the most part, our calculations using the normal distribution will require two results: $F(z = 1) - F(z = -1) = 0.68$ and $F(z = 2) - F(z = -2) = 0.95$. The first indicates that there is a 68% probability that a normal random variable will take a value within one standard deviation of the mean (i.e., $\mu - \sigma < X < \mu + \sigma$). The second result shows there is a 95% probability that the variable will lie within two standard deviations of the mean (i.e., $\mu - 2\sigma < X < \mu + 2\sigma$).

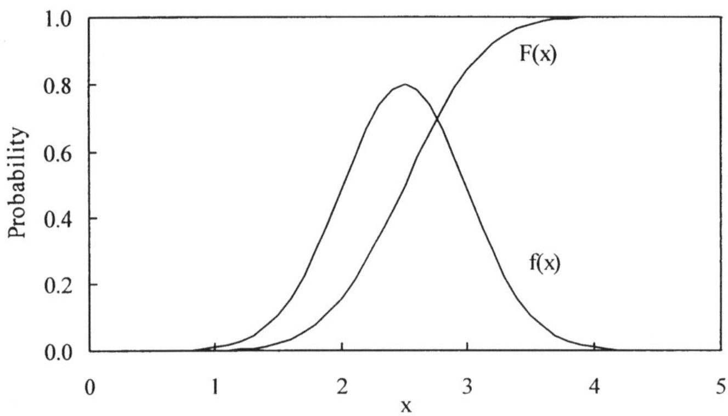

Figure 4-4. The continuous normal PDF and CDF.

The normal distribution occurs anywhere there are observations of processes that depend on additive random errors. A good example is in diffusion theory, where assuming a random walk of particles will lead to a concentration distribution having a normal shape (Lake, 1989. p. 157). In this case, the relevant function is the *error function*, defined as

$$erf(x) = \frac{2}{\sqrt{\pi}} \int_{0}^{x} \exp\left[-\eta^2\right] d\eta$$

which is related to Eq. (4-6c) as

$$F(z) = \frac{1}{2}\left[1 + erf\left(\frac{z}{\sqrt{2}}\right)\right] \qquad (4\text{-}6d$$

4-4.4 Probability Plots

How do we know when a variable is normally distributed? There are several fairly sophisticated means of testing data, but the most direct is to make a probability plot. We can invert the standard normal CDF given in Eq. (4-6d) to

$$z = -\sqrt{2}\, erf^{-1}[1 - 2F(z)] \tag{4-6e}$$

The quantity on the right of this equation is the *inverse error function*. A plot with the inverse error function on one axis is a *probability plot* (Fig. 4-5).

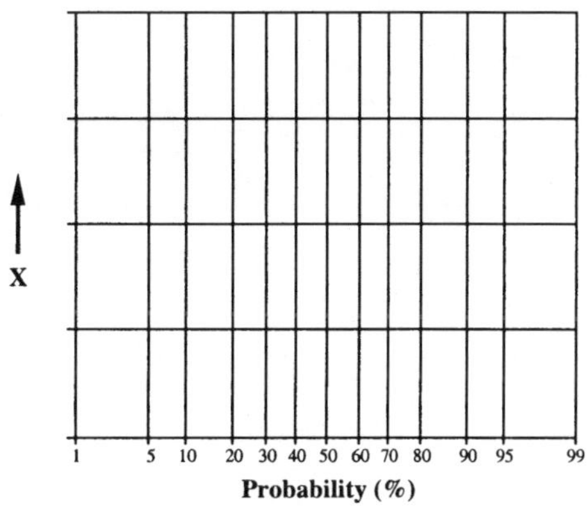

Figure 4-5. Probability coordinates.

There is a one-to-one relationship between the value of the normalized variable Z and the probability that some value less than or equal to Z will occur (i.e., $F(z)$). Consequently, we can talk in terms of variable values (z) or their associated probabilities $F(z)$. In Fig. 4-5, we mark off the horizontal axis in terms of F, expressed as a percent. If data from a normal CDF are plotted on the vertical axis, then a straight line will result with this type of paper. A log-normal distribution will similarly plot as a straight line if the vertical axis is logarithmic. Either Z or the original X may be plotted, since they are related by a simple arithmetic transformation. Plotting X is a commonly accepted way of testing a set of data for normality and for determining the variability of a set of data. A number of example plots will appear in later sections of this book.

How far points on a probability plot may deviate from a straight line and still be considered coming from a normal population is dealt with by Hald (1952, pp. 138-140). Clearly, however, if there are few (e.g., fewer than 15) points, the variation can be quite large while the underlying population PDF can still be considered normal. On the other hand, 100 or 200 points should fall quite closely on a straight line for the population to be considered normally distributed.

4-4.5 The Log-Normal Distribution

If errors multiply rather than add, the logarithms of the errors are additive. Since the logarithm of a random variable is a random variable, the CLT may apply to the sum of the logarithms to give a normal distribution. When this happens, the PDF of the sum is said to have a *log-normal distribution.*

Example 5 - An Implausible Reason for Permeability To Be Log-Normally Distributed. One of the most consistent assumptions regarding the distribution of permeability in a naturally occurring permeable medium is that it is log-normally distributed. One possible explanation for this is the theory of breakage.

Suppose we start with a grain of diameter D_{p0}, which fragments to a smaller grain in proportion f_0 to yield a grain of diameter D_{p1}. Repeating this process obviously leads to an immense tree of possibilities for the ultimate grain diameter. However, if we follow one branch of this tree, its grain diameter D_{pI} will be

$$D_{pI} = \prod_{i=1}^{I} f_i D_{pi}$$

which yields an additive process upon taking a logarithm. Thus, we should expect $\ln(D_{pI})$ to be normally distributed from the CLT. Since permeability is proportional to grain size squared, it is also log-normally distributed.

The log-normal PDF can be derived from the basic definition of the PDF and the standard log-normal variate:

$$z = \frac{\ln x - \mu_{\ln x}}{\sigma_{\ln x}}$$

Substituting this into Eq. (4-6c) gives the following for the CDF:

$$F(x;\mu_{\ln x},\sigma_{\ln x}) = \frac{1}{\sigma_{\ln x}\sqrt{2\pi}} \int_{\eta=0}^{\eta=x} \exp\left[-\frac{(\ln(\eta)-\mu_{\ln x})^2}{2\sigma_{\ln x}^2}\right] d\ln(\eta) \qquad \text{(4-7a)}$$

and using $y = \ln x$ in the transformation given in Chap. 3, we have for the PDF

$$f(x;\mu_{\ln x},\sigma_{\ln x}) = \frac{1}{x\sigma_{\ln x}\sqrt{2\pi}} \exp\left[-\frac{(\ln(x)-\mu_{\ln x})^2}{2\sigma_{\ln x}^2}\right] \qquad \text{(4-7b)}$$

The mean and variance of the log-normal distribution are given by

$$E(X) = \exp(\mu_{\ln x} + \sigma_{\ln x}^2/2)$$

and

$$Var(X) = exp(2\mu_{\ln x} + \sigma^2_{\ln x})[exp(\sigma^2_{\ln x}) - 1]$$

The coefficient of variation is independent of $\mu_{\ln x}$:

$$C_V^2 = \exp(\sigma^2_{\ln x}) - 1$$

You should note in the formulas that, for example, *E(X)* is the mean of *X*; $\mu_{\ln x}$ is the mean of the natural logarithm of X.

The PDF for the log-normal distribution is shown in Fig. 4-6. The significant features of the PDF are the skewness to the left and the long tail to the right. Thus, we see that most of the values in a log-normal PDF are small, but that there are a small number of large values. Of course, the random variable *X* in the log-normal distribution cannot be less than zero. Log-normal distributions appear to be at least as common in nature as normal distributions.

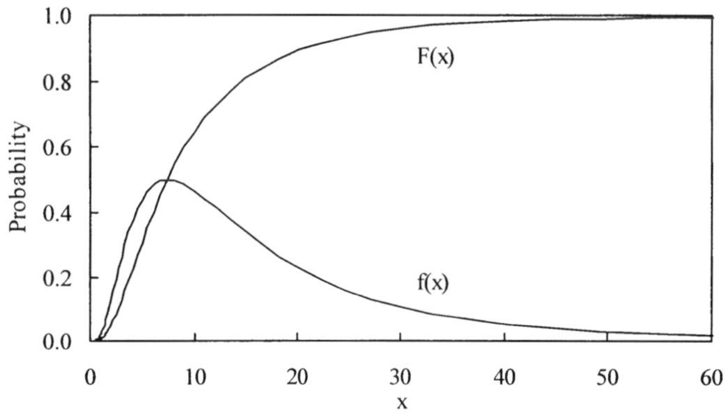

Figure 4-6. PDF and CDF for a log-normal distribution.

The case of the logarithmic transformation is an interesting example of how averages and nonlinear transforms can combine to produce non-intuitive results. If the random variable, *X*, is log-normally distributed, then $\ln X \sim N(\mu_{\ln x}, \sigma^2_{\ln x})$. Therefore, $E(\ln X) = \mu_{\ln x}$. We might be tempted to assume that $E(X) = \exp[E(\ln X)] = \exp(\mu_{\ln x})$. But, as we saw above, the variability of ln *X* also contributes to the mean: $E(X) = \exp(\mu_{\ln x})\exp(\sigma^2_{\ln x}/2)$. The $\exp(\sigma^2_{\ln x}/2)$ term in the product can be significantly larger than 1 for highly variable properties where a $\sigma^2_{\ln x}$ of around 4 is commonly observed (see Example 2 of Chap. 10).

The logarithmic transformation case is just one example of how we must be careful about using random variables in algebraic expressions. We will see later, especially in the chapters on estimation and regression, that the deterministic manipulation of variables that most of us learned in school is a special case. It is only in the special case that the effects of variability can be ignored.

4-4.6 p-Normal Distribution

We can go one step farther and define a more general power transformation of the form (Box and Cox, 1964)

$$y = \begin{cases} [(x+c)^{p}-1]/p & p \neq 0 \\ ln(x+c) & p = 0 \end{cases}$$

which will render a wide range of sample spaces normal with the appropriate selection of p and c. It is obvious that $p = 1$ is the normal distribution and $p = 0$ is log-normal. Commonly, c is taken to be zero. The skewness of the distributions, intermediate between normal and log-normal, corresponds to a value of p between 0 and 1. Figure 4-7 illustrates the shapes of the *p-normal* PDF.

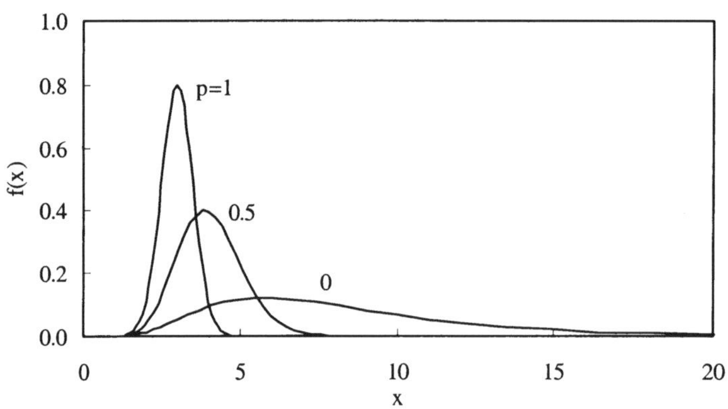

Figure 4-7. PDF shapes for various p.

Table 4-1 shows a few p-values estimated from permeability data from various media types. There are also various types or measurements, and each data set has varying number of data in it. We see from this that p can be quite variable, especially in sandstone sediments. For the carbonates in this table, the estimated p is always near zero. This reflects the greater variability and/or lower average permeability in these sediments. Note also that $p < 0$ is also possible, even though this does not fall within the range envisioned by the original transform.

Table 4-1. Estimates of p from various data sources. From Jensen (1986). plug = core plug and dst = drill stem test.

Data Set Label	Data Type	Number of Data	Media Type	p estimate, $\hat{p}$
Law	plug	48	sandstone	1.0
Sims Sand	plug	167	sandstone	0.5
Admire Sand	plug	330	sandstone	0.5
Lower San Andres	dst	112	dolomite	0.1
Pennsylvanian	dst	145	limestone	0.0
Nugget Sand	plug	163	sandstone	-0.3

Besides the binomial, normal, and log-normal distributions, other PDF's are sometimes used in reservoir characterization. For example, the exponential, Pareto, and gamma distributions have all been used to estimate the number of small fields in mature provinces (Davis and Chang, 1989). We will also meet the t-distribution in a later chapter. It is clear that there are a large number of possible PDF's.

4-4.7 Uniform Distribution

This distribution is the simplest of all the commonly used CDF's and PDF's. It consists of three straight lines (Fig. 4-8).

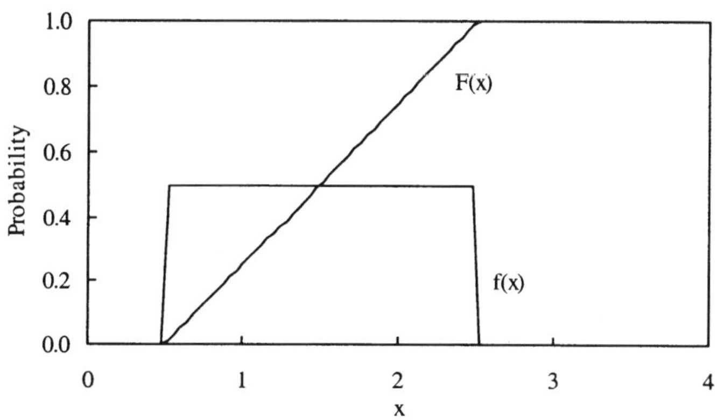

Figure 4-8. PDF and CDF for a uniform distribution.

The equation for the CDF is

$$F(x;a,b)=\begin{cases} 0 & X<a \\ \dfrac{x-a}{b-a} & a\le X<b \\ 1 & b\le X \end{cases} \tag{4-8a}$$

and, consequently, for the PDF is

$$f(x; a, b) = \begin{cases} 0 & X < a \\ \dfrac{1}{b-a} & a \le X < b \\ 0 & b \le X \end{cases} \tag{4-8b}$$

The uniform distribution is a two-parameter ($a \le b$) distribution. The mean of this distribution is $E(X) = (a + b)/2$ and the variance is $Var(X) = (b - a)^2/12$. The popularity of the uniform distribution arises from its simplicity and also from its finite bounds, $X = a$ and $X = b$. It is to be used when there is no *a priori* knowledge of the distribution type but there are physical reasons to believe that the variable cannot take values outside a certain range.

4-4.8 Triangular Distribution

The triangular distribution is the next step up in complexity from the uniform distribution. Its PDF and CDF are shown in Fig. 4-9. The triangular distribution is a three-parameter distribution ($a < b < c$). The equation for the PDF is

$$f(x;a,b,c) = \left(\frac{2}{c-a}\right) \begin{cases} 0 & X < a \\ \dfrac{x-a}{b-a} & a \le X < b \\ \dfrac{c-x}{c-b} & b \le X < c \\ 0 & c \le X \end{cases} \tag{4-9a}$$

The CDF is

$$F(x;a,b,c) = \begin{cases} 0 & X < a \\ \dfrac{(x-a)^2}{(c-a)(b-a)} & a \le X < b \\ 1 - \dfrac{(c-x)^2}{(c-a)(c-b)} & b \le X < c \\ 1 & c \le X \end{cases} \tag{4-9b}$$

with mean $E(X) = (a + b + c)/3$. The variance of this distribution is (McCray, 1975)

$$Var(X) = \frac{a^2 + b^2 + c^2 - ab - ac - bc}{18}$$

As for the uniform distribution, the triangular distribution is useful when there is reason to confine the variable to a finite range. However, it frequently occurs that there is a most likely value within this range, but still no *a priori* knowledge of the distribution form. In these cases, the triangular distribution is appropriate. As Fig. 4-9 suggests, the triangular distribution can be used as an approximation to both the normal and log-normal distributions.

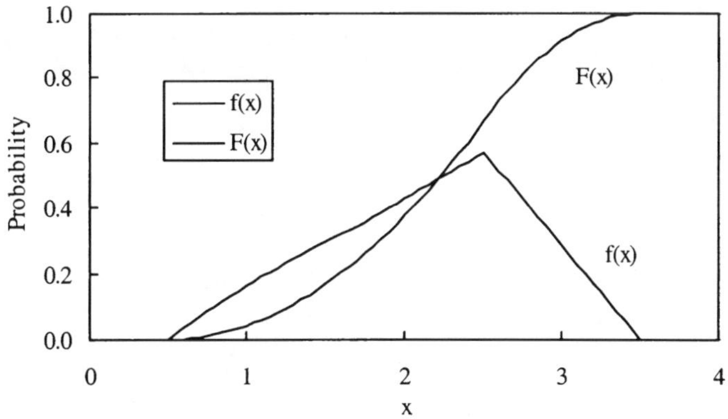

Figure 4-9. PDF and CDF for a triangular distribution.

4-4.9 Exponential (Boltzmann) Distribution

The exponential distribution is shown in Fig. 4-10. This distribution is a one-parameter function with the following form for the CDF:

$$F(x;\lambda) = 1 - \exp(-x/\lambda) \tag{4-10a}$$

and for the PDF

$$f(x;\lambda) = \lambda^{-1}\exp(-x/\lambda) \tag{4-10b}$$

both for $X > 0$ and $\lambda > 0$. The mean and variance are λ and λ^2, respectively. Both fracture aperture and spacing may have exponential distributions (see Dershowitz and Einstein, 1988, and references therein). Its slightly more complicated cousin, the Gibbs or Boltzmann distribution, is also used extensively in combinatorial optimization schemes (Chap. 12). The feature desired in these applications is that the mode of the exponential distributions occurs at the smallest value of the random variable.

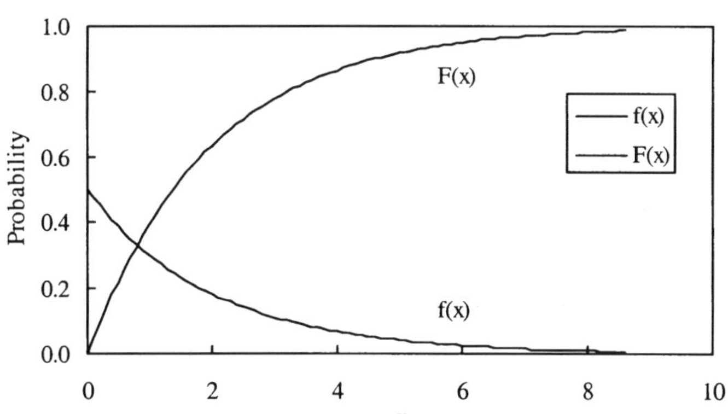

Figure 4-10. PDF and CDF for an exponential distribution.

The few distribution functions we cite here by no means cover all possibilities. There are literally hundreds of distribution functions. The ones we do list here are the ones to be used in the remainder of the text.

4-5 TRUNCATED DATA SETS

A truncated sample occurs when data values beyond some limit are unsampled or unmeasurable. Truncation can be the result of restrictive sampling, deficient sampling, or limitations in the measuring device. For example, some apparatus for measuring permeability cannot measure values less than 0.1 mD. A data set that is missing large-valued samples is *top-truncated*; if it is missing small-valued samples it is *bottom-truncated*. The upper plot in Fig. 4-11 illustrates these concepts with the random variable mapping outside of a limiting value, X_{lim}.

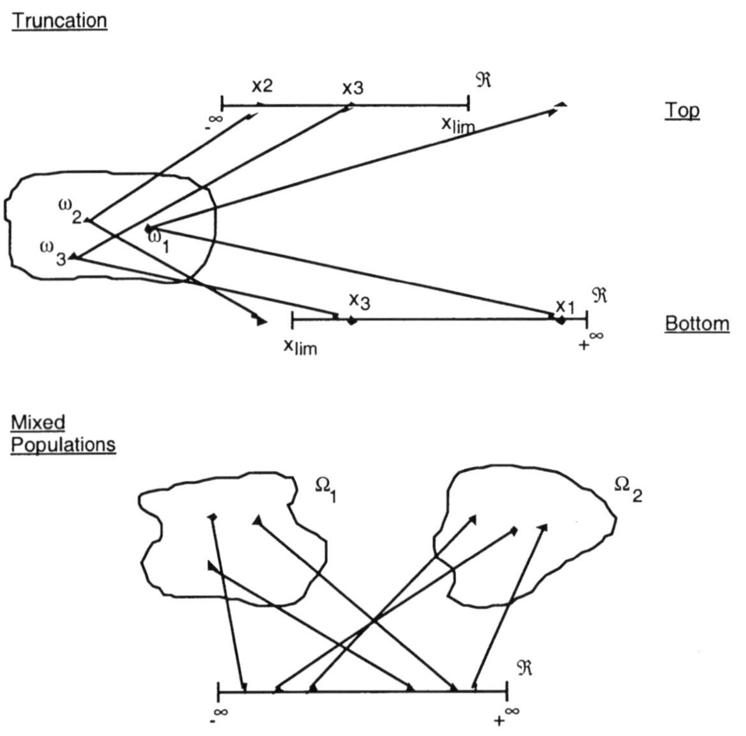

Figure 4-11. Schematic plots of truncation (upper) and mixed populations (lower).

Before proceeding, a word is in order about terminology. A data set is *truncated* if the number of unmeasured values is unknown; it is *censored* if they are known. Truncation in this section applies only to data; when applied to theoretical (population) CDF's, truncation can mean that a portion of the variable range is not attainable on physical grounds (porosities less than zero, for example).

As always in statistical manipulations, we cannot infer the underlying reason for the truncation (at least not from the statistics alone); we can only detect its occurrence and then attempt to correct for it. The correction itself is based on an application of the definition of conditional probabilities given in Chap. 3. See Sinclair (1976) for more details.

The basic tool is the CDF, $F(x)$, or the complementary CDF, $F^c(x)$. We assume that the untruncated sample CDF is normal or can be transformed to be normal; the data sets are log-normal here. Recall that the horizontal axes in both plots are probabilities; Prob($X \leq x$) in the CDF and Prob($X \geq x$) in the complementary CDF. We use F and F^c in percent in this section.

Figure 4-12 shows the effects of bottom and top truncation on a sample CDF. Omitting the data points, we show only lines in this figure. Truncation evidently flattens out the CDF in the region of the truncation. In fact, it is relatively easy to see the truncated value (x_{lim} in Fig. 4-11) by simply estimating on the value axis where the leveling out takes place. What is not so easy to see is what percentage of the data values are truncated, but two observations are pertinent.

1. The slope of the CDF (related to the variance) at the extreme away from the truncated region is essentially the same as for the untruncated sample.

2. The probability coordinate F^c where the leveling occurs is roughly the percentage of the data points that are omitted by the truncation.

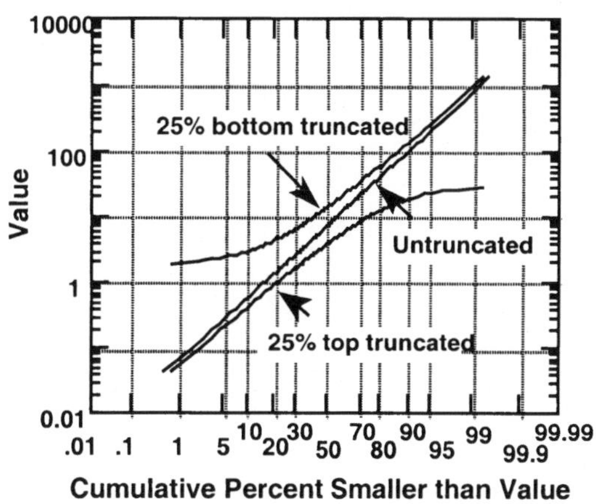

Figure 4-12. Schematic of the effects of truncation on a log-normal CDF.

In order to correct for truncation, we must somehow determine the untruncated CDF. For the CDF, this process is accomplished by the following manipulations:

Top truncation $\qquad F = F_a(1 - f_t) \qquad$ (4-11a)

Bottom truncation $\quad F = f_b + F_a(1 - f_b)$ (4-11b)

In these equations, F is the untruncated CDF, F_a the truncated (apparent) or original CDF, and f_b and f_t are the fraction of data points bottom- and top-truncated, respectively. From Eq. (4-11a), we have $F < F_a$ and from Eq. (4-11b) $F > F_a$ As mentioned above, f_b (or f_t) can be approximated by simple inspection of where the CDF flattens out. However, to be more certain several f_b (or f_t) values should be chosen and F vs. X plotted; the correct degree of truncation f_b (or f_t) is that which gives the best straight line. The similar relations for the complementary CDF apply and are left to the reader as an exercise.

Despite the trial-and-error nature of the process, determining the amount of truncation is fairly quick, especially on spreadsheet programs. Figure 4-13 illustrates a bottom-truncated data set of original oil in place from a region in the Western U.S. For this case, the reason for the truncation is apparent: reservoirs smaller than about 0.5 billion barrels were simply uneconomical at the time of the survey. However, the value of f_b from Eq. (4-11b) required to straighten out the data indicates that approximately 40% of the reservoirs are of this size or smaller. The corrections leave the untruncated portion of the curves unaffected. The following example indicates how a truncation correction can affect measures of central tendency.

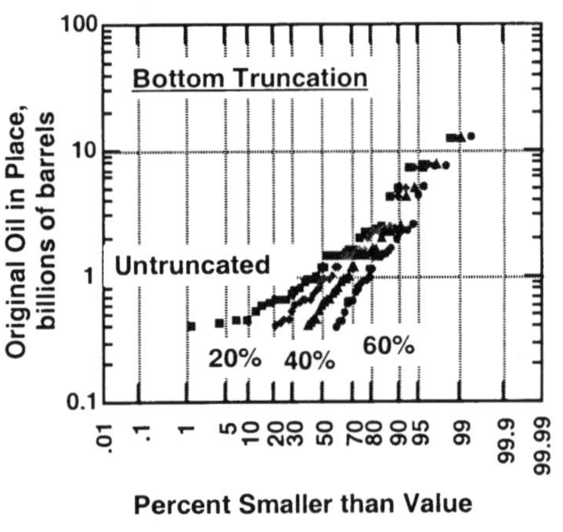

Figure 4-13. CDF for original oil in place showing f_b in percent.

Example 6 - Top Truncation of Core Permeability Data. A waterflood in a particular reservoir invariably exhibits greater injectivity than is indicated by the average core permeability data. Among the many possible causes for this disagreement is top truncation in the core data.

The first column of Table 4-2 shows an abridged (for ease of illustration in this exercise) set of permeability data from this reservoir. The second column shows the

$F_a = 100[(i - {}^1/_2)/I]$ values, where $i = 1, \ldots, I$ is an index of the data values and I is the total number of data points. The set is clearly top truncated (Fig. 4-14).

Table 4-2. Data for Example 6. Columns 3 and 4 show F values that have been corrected for $f_t = 0.2$ and 0.6 using Eq. (4-12a).

Permeability	F_a	F	F
mD	Original	$f_t = 0.2$	$f_t = 0.6$
0.056	3.570	2.857	1.429
0.268	10.714	8.571	4.286
0.637	17.857	14.286	7.143
1.217	25.000	20.000	10.000
2.083	32.143	25.714	12.857
3.343	39.286	31.429	15.714
5.149	46.429	37.143	18.571
7.722	53.571	42.857	21.429
11.389	60.714	48.571	24.286
16.641	67.857	54.286	27.143
24.243	75.000	60.000	30.000
35.423	82.143	65.714	32.857
52.244	89.286	71.429	35.714
78.358	96.429	77.143	38.571

The $f_t = 0.6$ curve (60%) has been overcorrected, as evidenced by the upward curvature. The best correction is slightly less than $f_t = 0.2$ (20%). But, taking $f_t = 0.2$ to be best, the geometric mean of this data set (the median) has been increased from around 6.5 mD to about 12 mD. This correction may be sufficient to explain the greater-than-expected injection rates. We should seek other causes to explain why more than 20% of the data are missing from the set.

A data set may be *both* top- and bottom-truncated. In this case we have

$$F = f_b + F_a(1 - f_b - f_t) \qquad (4\text{-}12)$$

Using Eq. (4-13) requires a two-step trial and error to determine both f_t and f_b. The reader may verify the analogous expression for F^C.

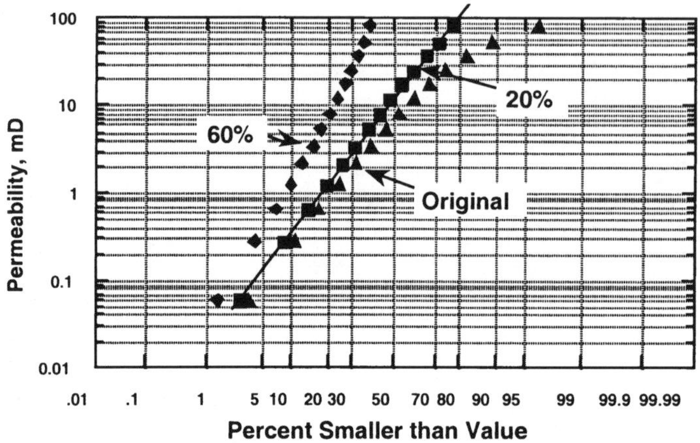

Figure 4-14. CDF illustrating correction for top truncation.

4-6 PARTITIONING DATA SETS

Nature is not always so kind as to give us data sets from only single PDF's. See the lower plot in Fig. 4-11 for a schematic illustration. In fact, many times each reservoir contains mixtures of materials with different parent distributions. The aggregate distribution is what we see when the reservoir is sampled (we called these heterogeneous distributions above). When this happens, the power of the PDF and the CDF gets distinctly blurred, but we can still gain some insight from them.

Partitioning is the separating of a mixed distribution function into its component (parent) parts. Partitioning is most efficient when the data values separate naturally into groups on the basis of variables other than the variable in the CDF. This is one of the primary roles geology serves in reservoir characterization.

When we partition, we obtain more homogeneous elements that we expect to be able to understand better. For example, in Example 4 we showed how combining two elements increases variability. If we know the variability of each element and its probability of occurrence, we can develop a model of the total variation that is more realistic than one that is based only on the mean and variance of the aggregate.

In the absence of geological information, partitioning is usually difficult, especially when the CDF consists of more than two parent populations. For these reasons, we limit the treatment here to partitioning into only two CDF's. Many of the basic assumptions here are the same as in the discussion on truncation: the parent populations are log-normally distributed and all operations will be on the cumulative distribution function (CDF) or its complement. We further assume that the parent populations are themselves untruncated.

Let the sample CDF of parent distributions A and B be $F_A(x)$ and $F_B(x)$, respectively. Then the probability coordinate of the mixed CDF is given by

$$F = f_A F_A + f_B F_B \tag{4-13}$$

where f_A and f_B are the fractions of the samples (data values) taken from populations A and B. This intuitive rule is actually a consequence of both the addition and multiplication rules for probabilities. It implies that data are selected for incorporation into F without regard to their value (independence) and a sample cannot be in both distribution A and distribution B (mutually exclusive). Of course, we must have $f_A + f_B = 1$.

Equation (4-13) is also a rule for combining the F's on the horizontal axis of the CDF. For example, if $F_A = 0.9$ and $F_B = 0.3$ for a particular value of x (recall $X < x$) and the mixture is 30% A and 70% B, then $F = 0.48$ of the samples in the mixed CDF are less than x. Of course, this mixing is not linear on the probability axis of the CDF. The essence of partitioning is actually the reverse of this: determine f_A and f_B given $F(x)$ and some means to infer F_A and F_B. Following Sinclair (1976), there are two basic ways to partition based on whether populations A and B are nonintersecting or intersecting.

4-6.1 Nonintersecting Parents

Nonintersecting means that the parent populations overlap very little. Figure 4-15 shows three mixed CDF's for parents A and B (A has the smaller variance) as a function of various proportions of A.

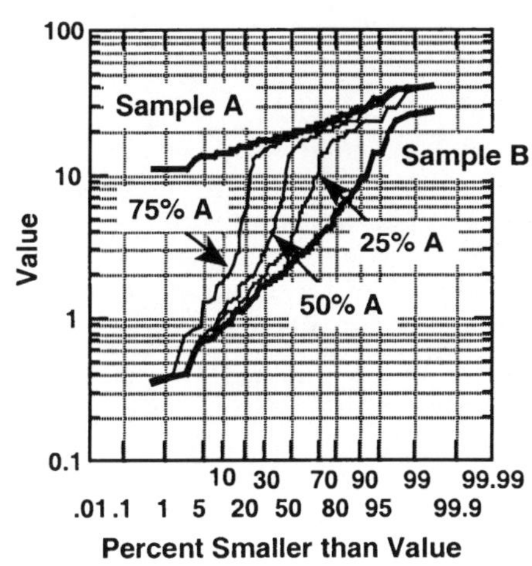

Figure 4-15. CDF's for nonintersecting parent populations.

The mixed CDF has the following attributes:

1. The central portion of the curve is the steepest. This observation is what distinguishes nonintersecting and intersecting CDF's. The steepness increases as the overlap in X values between A and B decreases.

2. The F value of the inflection points roughly corresponds to $f_B = 1 - f_A$.

3. The extremes (F approaching zero or one) become parallel to the CDF's for the parent samples.

These observations suggest the following graphical procedure for partitioning a sample drawn from two nonintersecting populations:

1. Plot the data points F vs. X and draw a smooth line through the points on probability coordinates. The smooth line represents the mixed population CDF.

2. Draw in straight lines corresponding to the parent populations. The parent population with the smallest variance (slope) will be nearly tangent to the mixed CDF at one of the extremes. This will also be true of the parent population with the largest variance, but less so.

3. Estimate the partitioning fraction from the inflection point on the smooth line.

4. The f_A and f_B are now known from step 3; F_A and F_B from step 2. Use Eq. (4-14) to reconstruct F.

If this $F(X)$ agrees with the original data, all inferred values are correct; if not, adjust the partitioning fraction and return to step 3. If it appears that a satisfactory match cannot be attained, it may be necessary to adjust F_A and/or F_B. In general, convergence is attained in three to five trials, as the following example shows.

Example 7 - Partitioning into Two Populations. The data set of 30 permeability values in Table 4-3 is believed to come from two parent populations.

The first column contains the permeability values, and the third and fourth columns the F_A and F_B values. These are the X coordinates of the two population lines in Fig. 4-16 corresponding to the permeability values in column one. Columns five and six contain F for two estimates of f_A.

The shape of the original experimental CDF in Fig. 4-16 clearly indicates a mixed population. Even though the X coordinate of the inflection point suggests $f_A = 0.7$, we use $f_A = 0.5$ and 0.8 for further illustration. However, neither f_A value yields a particularly good fit to the data (the solid line in Fig. 4-16), probably because the line for F_B is too low (mean is too small). For accurate work, F_B should be redrawn and the procedure repeated. However, the procedure is sufficiently illustrated so that we may proceed to the second type of mixed populations.

Table 4-3. Data for Example 7.

Permeability	F	F_A	F_B	F	F
mD	original	estimated	estimated	$f_A = 0.5$	f_A=0.8
1.70	1.67	3.00	0.80	1.90	2.56
1.75	5.00	6.00	0.90	3.45	4.98
1.80	8.33	9.00	1.00	5.00	7.40
1.90	11.67	12.00	1.00	6.50	9.80
1.95	15.00	15.00	1.00	8.00	12.20
1.95	18.33	45.00	1.00	23.00	36.20
1.95	21.67	46.00	1.00	23.50	37.00
2.00	25.00	51.00	1.00	26.00	41.00
2.10	28.33	54.00	1.50	27.75	43.50
2.10	31.67	57.00	1.50	29.25	45.90
2.10	35.00	60.00	1.50	30.75	48.30
2.15	38.33	63.00	2.00	32.50	50.80
2.15	41.67	66.00	2.00	34.00	53.20
2.15	45.00	69.00	2.00	35.50	55.60
2.20	48.33	72.00	3.00	37.50	58.20
2.25	51.67	78.00	3.40	40.70	63.08
2.30	55.00	85.00	3.50	44.25	68.70
2.50	58.33	91.00	3.90	47.45	73.58
2.60	61.67	94.00	4.00	49.00	76.00
2.70	65.00	96.00	4.00	50.00	77.60
2.70	68.33	99.00	4.50	51.75	80.10
3.50	71.67	99.99	8.00	54.00	81.59
4.95	75.00	99.99	20.00	60.00	83.99
8.10	78.33	99.99	50.00	74.99	89.99
11.50	81.67	99.99	70.00	84.99	93.99
14.50	85.00	99.99	80.00	89.99	95.99
15.50	88.33	99.99	85.00	92.49	96.99
16.50	91.67	99.99	86.00	92.99	97.19
22.50	95.00	99.99	95.00	97.49	98.99
28.00	98.33	99.99	98.00	98.99	99.59

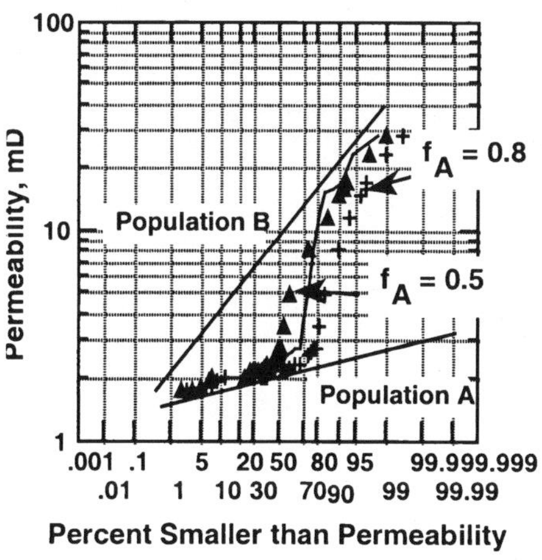

Figure 4-16. Experimental CDF for Example 7.

4-6.2 Intersecting Parents

Here the parent populations overlap to a significant degree. The signature for intersecting parents is not as clear as for nonintersecting parents; however, the differences in their respective CDF's is dramatic. Compare Figs. 4-15 and 4-17.

Just as for nonintersecting parents, partitioning these CDF's requires a trial-and-error procedure using Eq. (4-13). However, there is no longer an inflection point to guide the initial selection of f_A. Instead, the central portion of the curve is flatter than the extremes, a factor that makes partitioning of this type of mixture more involved. Nevertheless, following Sinclair (1976), there are some general observations possible from Fig. 4-17.

1. The central portion of the mixed CDF is flatter than the extremes; all three mixed curves intersect each other within this region. This is what makes the inflection point difficult to identify.

2. The value range (on the vertical axis) of the central segment is greater than the range of the parent population having the smallest range (sample A in Fig. 4-17).

3. The F range (on the horizontal axis) of the flat central segment is a coarse estimate of the proportion of the small-range population in the mixture (f_A in Fig. 4-17).

Observation 2 helps in drawing the CDF's for one of the parent populations; observation 3 gives an estimate of f_A. However, estimating the properties of the second parent CDF adds another level to the trial-and-error procedure.

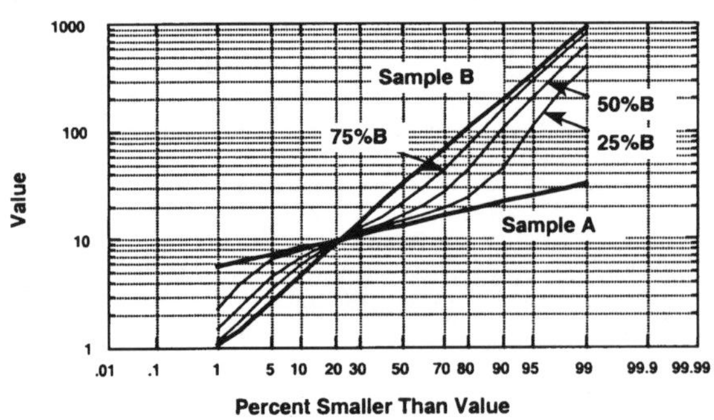

Figure 4-17. CDF's for intersecting populations.

4-6.3 Minimization Functions

Stable minimization algorithms, which have become standard features of spreadsheet applications, can add substantial generality to the procedures discussed above. To do this we re-write Eq. (4-13) as

$$\hat{F} = f_A \hat{F}_A + f_B \hat{F}_B = f_A \hat{F}_A + (1 - f_A)\hat{F}_B$$

where we have added a ^ over the *F*'s to identify them as the model (or estimate) of the CDF to which we fit the data *(F,X)*. Define now an objective function *L* as

$$L = \sum_{i=1}^{I} (F_i - \hat{F}_i)^2 = \sum_{i=1}^{I} (F_i - f_A \hat{F}(X_i; \mu_A, \sigma_A^2) + (1 - f_A)\hat{F}(X_i; \mu_B, \sigma_B^2))^2 \qquad (4\text{-}14)$$

A minimization of *L* with the parameters f_A, μ_A, σ_A^2, μ_B, and σ_B^2 will yield optimized estimates for these parameters. Results with using this procedure have been excellent. There are several obvious advantages over the graphical procedure.

1. The trial and error search for a solution is now computerized. This leads to less manual work and can be adapted to retain the graphical benefits of the fitting.

2. The minimization procedure is not restricted to log-normal distributions, nor is it restricted to two-parameter CDF models as shown in Eq. (4-14). In fact, it is not even necessary for *A* and *B* to have the same distribution types.

3. Partitioning into more than two constituting populations appears to be possible.

Of course, as with any minimization procedure, as the complexity of the objective function increases, so does the possibility of the detection of local minima. Thus, some care should be exercised in fitting to three or more populations or to CDF's that have multiple parameters. As with many procedures discussed in this text, the best answer to how much detail is needed lies in a parallel observation of the geology behind the data.

4-6.4 Summary

Although the corrections outlined above are based on probability laws, the lack of a physical base limits the insights to be derived from these procedures. In particular, it may not be clear if corrections are needed; deviations from a straight line might be caused by sampling error. Furthermore, deciding when a given degree of correction is enough is a matter of judgment. Sinclair (1976) recommends that, once a given CDF is partitioned, the practitioner should look elsewhere (more or other types of data, usually) for an explanation. For example, suppose that two sources of sediment feed a river that later deposits sand downstream. If the two distributions of, say grain size, are each normally distributed with the same variance but different means, then the deposited sand may have a sample PDF that has two peaks. Generally speaking, we can expect that mixtures of material will originate from parents having distributions with neither the same means nor variances.

Another uncertainty is the underlying assumption that the untruncated or parent populations are themselves transformed Gaussian distributions. As we have seen, there is little physical base for a given distribution type to prevail. As in most statistical issues, these types of decisions depend to a great extent on the end use.

Finally, neither the CDF nor the PDF say anything about the relationship of one sample space to another or about the spatial arrangement of the parameters. This is because both treat the observations as being ordered (sorted) without regard to where they came from. We will have something to say about such relations in Chaps. 8 to 11 that deal with correlation and autocorrelation.

Given these limitations, simply plotting the sample CDF's often gives diagnostic information without further analysis. For example, Fig. 4-18 gives schematic plots that represent various "type" curves for some common complications. Many data sets can be quickly classified according to this figure and such classification may prove to be sufficient for the problem under consideration.

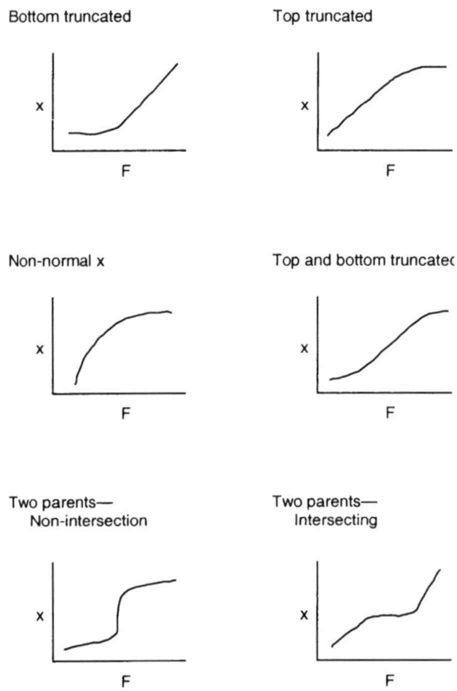

Figure 4-18. "Thumbprint" CDF's for truncated and mixed populations.

4-7 CLOSING COMMENT

Distribution functions provide the bridge between probability and statistics. The ideas of distribution functions and, in particular, the ways of summarizing them through the expectation and variance operations will recur in many places, as will the Central Limit

Theorem. The most important ideas deal with the manipulations of the expectation, as these are central to the notions of covariance and correlation. However, the reader should be reminded from the last few example PDF's that, even though statistics is a flexible and powerful tool, its benefit is greatly enhanced with an understanding of underlying mechanisms.

CHAPTER 5

ESTIMATORS AND THEIR ASSESSMENT

Since defining probability in Chap. 2, we have begun to appreciate the difference between the population and the sample. The population represents the entire body from which we could draw examples and make measurements. The sample is a limited series of measurements and represents the best we can do, given the physical and fiscal constraints of acquiring data.

We sample to assess the properties of the population. The assessment procedure is called *estimation*; there is a deliberately implied uncertainty in the name. The ultimate would be to obtain the population PDF; then we could calculate whichever parameters we sought. We have only a sample, however, and parameters must be estimated from this set of data based on the results of a limited number of experiments (Chap. 2). We also want to know, once we have an estimate in hand, how well it represents the true population value. This is where confidence limits are helpful.

A simple example may help to highlight the issues in estimation. Suppose that we decide to calculate the arithmetic average of five porosity values from similar rock volumes: 21.1, 22.7, 26.4, 24.5, and 20.9. Since the formula for the arithmetic average is

$$\bar{\phi} = \frac{1}{I} \sum_{i=1}^{I} \phi_i$$

where I is the number of data, we obtain $\bar{\phi} = 23.1$. We have a number. But, what does the value 23.1 represent? What have we calculated and what does it represent? Is 23.1 any "better" than 21.1, 24.5, or any of the other numbers? At least 21.1, 22.7, etc. are measured values, whereas there may not be any part of the reservoir with $\phi = 23.1$. As we shall soon see, this procedure has several assumptions that have gone unstated. Consequently, the value 23.1 may or may not be suitable for our purposes.

5-1 THE ELEMENTS OF ESTIMATION

Estimation involves four important elements:

1. The population quantity (the statistic) to be estimated, θ, (e.g., arithmetic mean, standard deviation, geometric mean) from a set of data. θ can represent one or more parameters. If it represents more than one parameter, we can consider it to be a vector instead of a scalar.

2. The estimator, W. This is the method by which the data values will be combined to produce the estimated or sample quantity.
3. The estimate, $\hat{\theta}$, probably the most familiar component to the reader.
4. The confidence interval or *standard error*, $[Var(\hat{\theta})]^{1/2}$. This allows us to judge how precise is the estimate $\hat{\theta}$.

In the porosity example, we chose the arithmetic average for the one-parameter estimator W and $23.1 = \hat{\theta}$. No confidence interval was calculated. Furthermore, the most important step–choosing θ—was completely ignored.

The number of parameters required depends on the form of the PDF and the subsequent application of the statistic. Recall from Chap. 4 that the binomial distribution has a two-parameter PDF: p and n, the probability of success in each trial and the total number of trials, respectively. In many experimental situations, it is advantageous to estimate p from a limited number of experiments with n trials each and then use this estimate to characterize all future experiments. In that sense, $\hat{p}$, the estimate of the population value p, is an average probability of success for a particular type of experiment against which other similar experiments may be judged. On the other hand, we may have an application that requires only the variance of the distribution, which is $np(1\text{-}p)$, without explicitly needing to know what p is. In this case, we may decide to estimate p–a parameter of the PDF–or we may decide to estimate the quantity $p(1\text{-}p)$ directly.

Similarly, the normal distribution has two parameters, the mean and variance, which may be estimated to characterize a random variable suspected of being normally distributed (e.g., porosity). If a measure of a physical process is normally distributed, the mean and variance precisely describe the PDF. Other distributions may require more (or fewer) parameters or statistics.

Recall that a statistic is any function of the measured values of a random variable (Chap. 2). It is an estimate of a parameter that could be computed exactly if the PDF were known. To distinguish between the estimate and the estimated quantity, the estimate is often prefixed with the terms *empirical* or *sample*. For example, some common statistics are *measures of central tendency* (e.g., averages or sample means) and *measures of dispersion* (e.g., sample variance or interquartile range), but they are by no means limited to these. As we will see, there may be several ways to estimate a parameter.

5-2 DESIRABLE PROPERTIES OF ESTIMATORS

Since any given statistic (parameter estimate) is a function of the data, then it too is a random variable whose behavior is described by a PDF. That is, given an estimator W and a set of data $\{X_1, X_2, \ldots, X_I\}$ taken from a population with parameter value θ, W produces an estimate $\hat{\theta}$ based on the data: $\hat{\theta} = W(X_1, X_2, \ldots, X_I)$. Knowing this, there are several features that a good estimator will have: small bias, good efficiency, robustness, and consistency. In addition, it should produce physically meaningful results.

5-2.1 Estimator Bias

The PDF of an estimate $\hat{\theta}$ should be centered about the population parameter θ we wish to estimate (Fig. 5-1). If this is not true, the estimator *W* will produce estimates that tend to over- or underestimate θ, and the estimator is said to be *biased*. Bias is given by the expression

$$b = E(\hat{\theta}) - \theta$$

Bias is generally undesirable, but sometimes an estimator can be corrected for it. Bias has two principal sources, measurement resolution and sampling, as the following example illustrates.

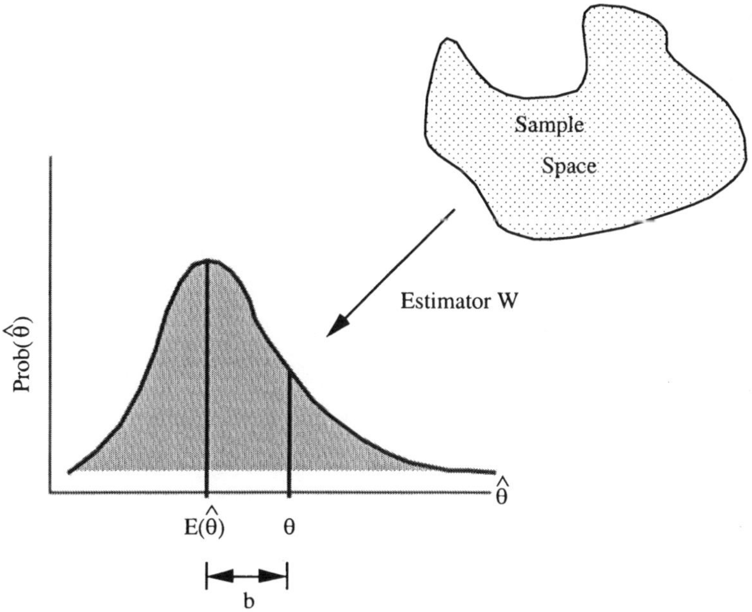

Figure 5-1. Estimator bias (b) depends upon the PDF of $\hat{\theta}$.

Example 1 - Biased Sampling. Data from core-plug samples are sometimes misleading because of preferential sampling. The samples may not be representative for a variety of reasons: incomplete core recovery (nonexistent samples), plug breakage in friable or poorly consolidated rock (failed samples), and operator error (selective sampling).

An interval of core has two lithologies, siltstone and sandstone, in the proportions 30% siltstone and 70% sandstone. $\phi_{silt} = 0.10$ and $\phi_{sand} = 0.20$. If the siltstone is not sampled, what is the bias in the average porosity obtained from plug samples?

The true mean porosity is given by

$\phi_T = (0.30)(0.10) + (0.70)(0.20) = 0.17$

All samples are sandstone with one porosity value. Therefore, ignoring measurement error, the apparent formation porosity is given by

$\phi_a = (1.0)(0.20) = 0.20$

The bias is $b_\phi = \phi_a - \phi_T = +0.03$, or about 18% too large.

5-2.2 Estimator Efficiency

The second property of a good estimator is that the variability of its estimates should be as small as possible. An estimator with a small variance is said to be *precise* or *efficient*. Do not mistake the variance of the sample space for the variance of the estimator. The former is a property of the physical variable and the latter is a property of the data and the estimation technique (Fig. 5-2).

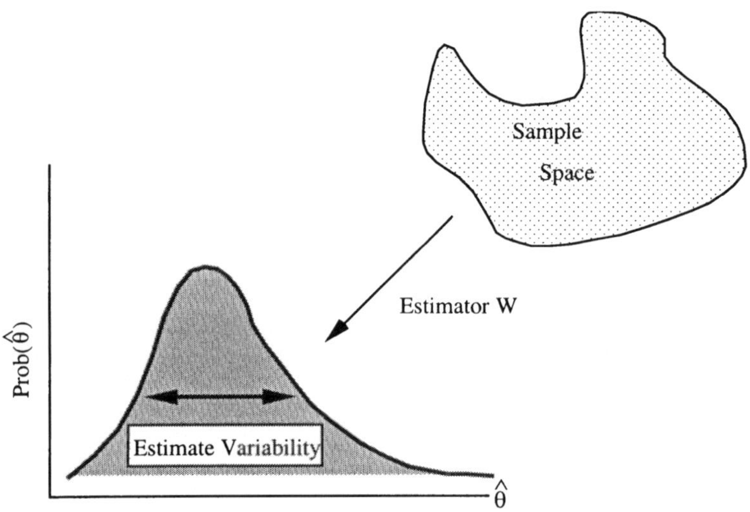

Figure 5-2. Estimator efficiency is based on how variable are the estimates of $\hat{\theta}$.

$[Var(\hat{\theta})]^{1/2}$, the standard deviation of the estimates, is called the *standard error*. There are theoretical statements such as the Cramer-Rao Inequality (Rice, 1988, p. 252) about the minimum variance an estimator can attain. When an estimator has no bias and this minimum variance, it is said to be an MVUE (*minimum variance, unbiased estimator*).

Example 2a - Determining Uniform PDF Endpoints (Method 1). Monte Carlo reserves estimation often assumes a uniform PDF for one or more variables (Chap. 4). How do we obtain the lower and upper limits? Obviously, we must depend on the data to reveal these endpoint values.

In this example, two methods for estimating endpoint values are presented and compared for bias and efficiency. Rohatgi (1984, pp. 495-496) presents a third method. To simplify the discussion, we assume that the upper endpoint is unknown and the lower endpoint is zero. In the context of the previous discussion on estimators, θ is the upper endpoint and the parameter to be estimated (Fig. 5-3).

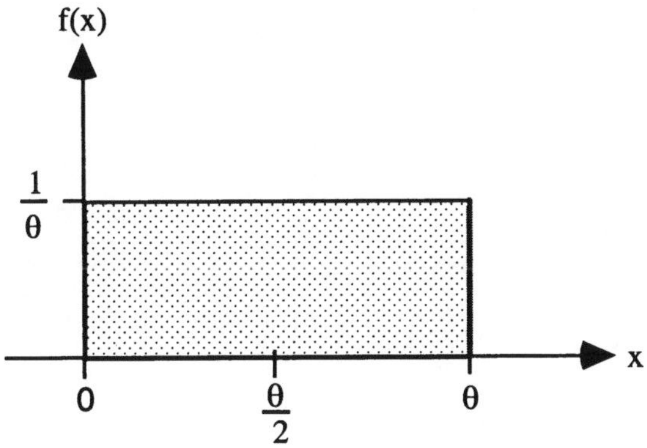

Figure 5-3. Uniform PDF with lower limit 0, upper limit θ, and mean $\theta/2$.

For the uniform PDF, the mean value of the random variable X is given by

$$E(X) = \hat{\theta}_1 = 2\overline{X} = \frac{2}{I}\sum_{i=1}^{I} X_i = \int_{x=0}^{x=\theta} (x/\theta)dx$$

That is, the arithmetic mean is exactly one-half the upper limit θ when the lower limit is zero (Fig. 5-3).

We use this result to provide an estimator for θ. Given I samples of X, X_1, X_2, . . . , X_I, taken from a uniform distribution, the above suggests that twice the arithmetic average should give an estimator for θ:

$$\hat{\theta}_1 = 2\overline{X} = \frac{2}{I}\sum_{i=1}^{I} X_i$$

Let W_1 be this estimator (the subscript 1 on θ, W, and b below refers to the first estimator). The bias of this estimator is as follows:

$$b_1 = E(\hat{b}_1) - \theta = \left(\frac{2}{I}\sum_{i=1}^{I} X_i\right) - \theta = \left(\frac{2}{I}\sum_{i=1}^{I} E(X)\right) - \theta = \frac{2}{I}\frac{\theta I}{2} - \theta$$

or

$b_1 = 0$

Since b_1 is zero, the estimator W_1 is unbiased. The variance of W_1 is

$$Var(\hat{\theta}_1) = E[(\hat{\theta}_1 - \theta)^2] = E(\hat{\theta}_1^2) - \theta^2$$

$$= 4[E(\overline{X}^2) - E(\overline{X})^2] = 4Var(\overline{X})$$

Since (see Eq. (4-4)) $Var(\overline{X}) = Var(X)/I$ for I independent samples X_i,

$$Var(\hat{\theta}_1) = \frac{4}{I}Var(X) = \frac{4}{I}[E(X^2) - E(X)^2]$$

$$= \frac{4}{I}\left[\int_{x=0}^{\theta}(x^2/\theta)dx - \theta^2/4\right] = \theta^2/3I$$

The variability of the estimates $\hat{\theta}_1$ depends upon θ and I, the number of data.

$Var(\hat{\theta}) \propto I^{-1}$ is a very common dependence, i.e., the standard error is halved with a four-fold increase in data-set size. This suggests that overly stringent specifications for estimates can be quite costly. $Var(\hat{\theta}) \propto \theta^2$ is reasonable since, with lower limit fixed at 0, the upper limit θ will control the variability of X.

Example 2b - Determining Uniform PDF Endpoints (Method 2). A more obvious estimator than W_1 for the upper limit of a PDF is to take the maximum value in the data set: $\hat{\theta}_2 = X_{max}$. Let W_2 be defined as $\hat{\theta}_2 = \max\{X_1, X_2, \ldots, X_I\}$.

From theory based on order statistics (Cox and Hinkley, 1974, p. 467),

$$E(\hat{\theta}_2) = [I/(I+1)]\,\theta$$

Hence, W_2 is a biased estimator with $b_2 = -1/(I+1)$. As we might expect, X_{max} never quite reaches θ, no matter how many data there are. We can correct for this bias by redefining W_2 as $[(I+1)/I]\max\{X_1, X_2, \ldots, X_I\}$. This estimator is unbiased because $E(\hat{\theta}_2) = \theta$. It has variance

$$Var(\hat{\theta}_2) = \theta^2/I(I+2)$$

If we compare the variances of W_1 and W_2, we conclude that W_2 is more *efficient* than W_1 because W_2 has a lower variance than W_1. Their relative efficiencies are

$$\frac{Var(\hat{\theta}_1)}{Var(\hat{\theta}_2)} = \frac{I+2}{3} \geq 1 \text{ for } I \geq 1$$

A similar procedure could be used to find an estimator for the lower limit a of a uniform PDF: $\hat{\theta} = [I/(I+1)] \min\{X_1, X_2, \ldots, X_I\}$.

The estimator, W_2, in Example 2b was biased but more efficient. Efficiency is evidently not an intuitively obvious property. W_1 explicitly used all the data in the estimator, whereas W_2 only used the values to order the data and explicitly used only the maximum. It may appear that W_2 made less use of the information contained in the data, but that is not the case.

There are several methods to develop estimators (Rice, 1988, Chap. 8). The methods of moments and of maximum likelihood are two examples. Maximum likelihood requires that the population PDF be known (or assumed). This additional information helps to make likelihood estimators efficient.

Combining the properties of estimator bias and variability introduces an important property known as *consistency*. An estimator W is consistent if, as the number of data, I, becomes large, $\hat{\theta}$ approaches θ (asymptotically unbiased) and the variance of $\hat{\theta}$ approaches zero.

5-2.3 Estimator Robustness

Estimator robustness is the ability of the estimator to be relatively uninfluenced by errors in a small proportion of the data set. We appreciate that errors can occur in the measurement and recording of data; a decimal point could get moved or a specimen could be improperly tested. If the unrepresentative values have a large influence on the estimate produced, then the estimator is not robust. We will always have problems trying to find an estimator that is responsive to meaningful data values but is insensitive to wild, unrepresentative data.

Consider, for example, the following set of porosity data: 8.3, 9.6, 7.4, 8.9, and 9.1. For this set, $\overline{X} = 8.7$ and $\hat{X}_{0.50} = 8.9$, which are in good agreement (within 2%). What happens if one of the data values, 9.1, were misrecorded as 19.1? In that case, $\overline{X} = 10.7$ while $\hat{X}_{0.50} = 8.9$, a disagreement of 18%, which can be significant.

5-2.3.1 The Median

Substantial work has been done concerning robust estimators. Barnett and Lewis (1984, Chap. 3) discuss quantitative assessments for robustness. A simple example is using the median to estimate the mean. The sample median ($\hat{X}_{0.50}$) is the middle value in a set of I data:

$$\hat{X}_{0.50} = \begin{cases} X_{(I+1)/2} & \text{for I odd} \\ \frac{1}{2}\left[X_{(I)/2} + X_{(I+2)/2}\right] & \text{for I even} \end{cases}$$

where $X_1 \leq X_2 \leq X_3 \leq \cdots \leq X_I$ is an ordered data set. If the data come from a normally distributed population with mean μ, $\hat{X}_{0.50}$ is an unbiased estimate of μ, but it is about 25% less efficient than the arithmetic average, $\bar{X}$ (Kendall and Stuart, 1977, pp. 349-350). Frequently, the most efficient estimator is less robust than other estimators. The most efficient estimators use all information, including knowledge of the population PDF, but they can deteriorate quickly with small deviations from the underlying assumptions.

5-2.3.2 *L*-Moments

L-moments are a particularly attractive set of robust estimators. Just like ordinary moments, such as the mean and variance, *L*-moments provide measures of location, variability, skewness, kurtosis, and other, higher-order features. To describe *L*-moments, we begin by recalling that the *r*th moment of a random variable *X* is given by (see Sec. 4-1)

$$\mu_r = \int_{-\infty}^{+\infty} x^r f(x) dx$$

Instead of integrating over all possible values of *X*, we can obtain the same result by integrating over all probabilities

$$\mu_r = \int_0^1 x^r dF$$

where *F* is the CDF of *X*. An alternative to raising each value of *X* to some power *r* is to raise each value of the probability to the power *r*

$$\beta_r = \int_0^1 x F^r dF$$

Thus, β represents a probability-weighted sum of all values of *X*. β_r for $r > 0$ contains information about the shape of the PDF of *X*, just as the conventional moment μ_r does. Finally, just as we combined μ_r's to obtain centered moments (μ_r'), the *L*-moment of order *r* is expressed in terms of the β_r's. The first four *L*-moments are defined as

$$\lambda_1 = \beta_0$$

$$\lambda_2 = 2\beta_1 - \beta_0$$

$$\lambda_3 = 6\beta_2 - 6\beta_1 + \beta_0$$

$$\lambda_4 = 20\beta_3 - 30\beta_2 + 12\beta_1 - \beta_0$$

Definitions for $r > 4$ can be found in Hosking and Wallace (1997). λ_1 is our old friend, the mean value, $E(X)$, and is a measure of location. λ_2 is a measure of spread, like the variance, but is smaller in value since it puts less weight on the extremes of X. λ_3 measures skewness, and λ_4 reflects peakedness. Consequently, for each value of r, λ_r and μ_r measure a similar PDF property.

Because L-moments put less weight on the extremes than conventional moments, however, their estimation is easier. The first two sample moments are defined as follows, assuming $X_1 \leq X_2 \leq X_3 \leq \cdots \leq X_I$ is an ordered data set.

$$\hat{\lambda}_1 = I^{-1} \sum_{i=1}^{I} X_i$$

$$\hat{\lambda}_2 = 2I^{-1} \left(\sum_{i=2}^{I} \frac{i-1}{I-1} X_i \right) - \hat{\lambda}_1$$

The third and fourth L-moment estimates, L-skewness and L-kurtosis, are much less variable than their ordinary counterparts. Therefore, while we would need a very large data set to use skewness and kurtosis to decide which PDF the samples are coming from, we can use much smaller datasets for L-skewness and L-kurtosis.

A lucid discussion of L-moments and hydrological applications is given by Hosking and Wallace (1997). For petroleum reservoir characterization, the use of L-moments is just beginning. We expect that their use will grow considerably as their advantages become more widely known.

We will not dwell further on robust estimators because there are excellent treatments elsewhere (e.g., Hoaglin et al., 1983; Barnett and Lewis, 1984). We will, however, sometimes refer to an estimator as robust or nonrobust. In reservoir description, we must examine the robustness issue from two perspectives: choosing a robust estimator when we have a choice and, when we do not have a choice, assessing the estimator sensitivity.

5-3 SENSITIVITY ASSESSMENT

Quantification of estimator robustness is often difficult. We can, however, make some progress using perturbation analysis. Two such methods are outlined here. Both approaches use perturbation analysis to assess the effects of variability on estimates. They differ in that the first method sums the errors of each component while the second method sums the variances of each component.

5-3.1 Sum-of-Errors Sensitivity

If we have an analytic expression for the estimator W, then we can calculate the sensitivities based on the relation $\hat{\theta} = W(X_1, X_2, \ldots, X_I)$:

$$\Delta\hat{\theta}_i = \frac{\partial W}{\partial X_i}\Delta X_i$$

where $\Delta\hat{\theta}_i$ is the error in the estimate arising from the error ΔX_i in the measurement of the ith datum. This expression comes from the truncated Taylor series expansion of W about the true data values $X_1, X_2, \ldots, X_I$, so it may only be useful for small errors (less than 20% or so) in the X's. The expression also assumes the ΔX_i's are independent. Even with these limitations, it is often useful to examine the estimator sensitivities to measurement errors. Knowing these sensitivities, resources of data acquisition can be better applied to give the least erroneous estimates possible.

Example 3 - Archie's Law Sensitivity Study. We want to estimate hydrocarbon saturation in a formation by measuring formation porosity and resistivity with wireline logs and cementation and saturation exponents in the laboratory. We use a model, Archie's law,

$$S_w^n = \frac{aR_w}{\phi^m R_t}$$

to predict water saturation, where a is a constant, m and n are the cementation and saturation exponents, respectively, and R_w and R_t are the water and formation resistivities (Archie, 1943). Hydrocarbon saturation is one minus the water saturation. In this case, the estimator–Archie's law–is fixed, but we want to know whether to pay for the expensive logging suite (accurate to ±5%) or spend the money acquiring better laboratory data (measuring m and n more accurately).

We now assess the impact of measurement errors in a, ϕ, m, n, R_w, and R_t upon the water saturation S_w. Taking partial derivatives of sums and differences is easier than differentiating products and quotients. Since Archie's law consists only of products, quotients, and exponents, we can simplify things by taking logarithms of both sides before calculating the partial derivatives:

$$\ln(S_w) = \frac{1}{n}[\ln(a) + \ln(R_w) - m\ln(\phi) - \ln(R_t)]$$

Hence, in the case of the saturation exponent n, for example, we have, assuming that a, ϕ, m, n, R_w, and R_t are independent,

$$\frac{\partial \ln(S_w)}{\partial n} = -\frac{1}{n^2}[\ln(a) + \ln(R_w) - m\ln(\phi) - \ln(R_t)] = -\frac{1}{n}\ln(S_w)$$

or

$$\frac{\partial S_w}{\partial n} = -\frac{S_w}{n}\ln(S_w)$$

Hence,

$$\Delta S_{\mathrm{w}} = -\frac{S_{\mathrm{w}}}{n} \ln(S_{\mathrm{w}})\Delta n$$

A similar approach for the other variables gives the following expression for the total error in S_w arising from errors in all the measured variables:

$$\Delta S_{\mathrm{w}} = \frac{S_{\mathrm{w}}}{n}\left[-\ln(S_{\mathrm{w}})\Delta n + \frac{\Delta a}{a} + \frac{\Delta R_{\mathrm{w}}}{R_{\mathrm{w}}} - m\frac{\Delta \phi}{\phi} - \ln(\phi)\Delta m - \frac{\Delta R_{\mathrm{t}}}{R_{\mathrm{t}}}\right]\Delta S_W$$

The proportional or relative change in water saturation is

$$\frac{\Delta S_{\mathrm{w}}}{S_{\mathrm{w}}} = \frac{1}{n}\left[-\ln(S_{\mathrm{w}})\Delta n + \frac{\Delta a}{a} + \frac{\Delta R_{\mathrm{w}}}{R_{\mathrm{w}}} - m\frac{\Delta \phi}{\phi} - \ln(\phi)\Delta m - \frac{\Delta R_{\mathrm{t}}}{R_{\mathrm{t}}}\right]$$

Thus, R_{w}, R_{t}, a, and ϕ all contribute proportionately to the total error ΔS_{w}. If $m = 2$, porosity errors are twice as significant as resistivity errors. On the other hand, m and n contribute directly according to the porosity and water saturation of the material under consideration. Errors in n could lead to significant errors in S_{w} at small water saturations, while errors in m can be important for low-porosity media. Thus, spending money to make careful laboratory measurements may be worthwhile.

Unlike Archie's law, which deals with static quantities, there are also instances where measurements can influence the predicted flow properties of the reservoir. Relative permeability measurements are a good example of this situation.

Relative permeability functions are required to model immiscible multiphase flow. They are a laboratory-determined property and are usually based on results from a few, small rock samples. Thus, these data are susceptible to errors and those errors may be important, as described in the next example.

Example 4 - Assessing the Effects of Relative Permeability Errors. In a common case of one spatial dimension and two phases (oil and water), the frontal advance solution (Willhite, 1986, pp. 59-64) of Buckley and Leverett shows that the fractional flow is the important quantity that relates the relative permeabilities to flooding performance. It is not the relative permeability measurement errors *per se* that matter, but the effects of these errors upon the fractional flow. If we ignore gravitational effects and assume negligible capillary pressure, the water fractional flow is

$$f_{\mathrm{w}} = \frac{1}{1 + \dfrac{k_{\mathrm{ro}}}{k_{\mathrm{rw}}}\dfrac{\mu_{\mathrm{w}}}{\mu_{\mathrm{o}}}}$$

where k_r is the relative permeability, μ is the fluid viscosity, and the subscripts o and w refer to oil and water. By definition $f_w \leq 1$ for this case. Ignoring errors in the viscosities, an analysis similar to that of Example 3 shows that

$$\frac{\Delta f_w}{f_w} = (1 - f_w)\left(\frac{\Delta k_{rw}}{k_{rw}} - \frac{\Delta k_{ro}}{k_{ro}}\right)$$

Thus, errors in f_w are less than errors in the relative permeabilities, since $0 < (1-f_w) < 1$. If a shock front develops, only f_w values exceeding the breakthrough value, f_{wbt}, exist (see Lake, 1989, Chap. 5). This further mitigates the impact of relative permeability errors. The errors in k_{ro}, which can be large when k_{ro} is small, produce smaller changes in f_w.

The approach above offers great flexibility in the form of an estimator. Errors in exponents can be treated, for example, as in Example 3. All that is required is that the derivatives exist. Its weakness, however, is that it is limited to small errors because of the truncated expansion. The following method is more robust to error size because, while it also depends on a first-order expansion, it uses logarithms of the variables and sums their variances.

5-3.2 Sum-of-Variance Sensitivity

The sensitivity analysis approach just presented can be developed further using the properties of the logarithm and the variance operator. This can lead to some interesting insights into the sources of variability and data error. This process is sometimes called *first-order* or *linearized error analysis*.

Suppose we have an estimator of the form

$$Y = Y(X_1, X_2, \ldots X_I)$$

where Y is a quantity to be estimated that depends on the values of the $X_1, X_2, \ldots, X_I$, all independent quantities. We treat the X_i as random variables; even though the relationship is deterministic, this makes Y a random variable also. In general, the function Y is known (usually it is a physical law) and is nonlinear. As discussed in Chap. 1, statistics cannot help in determining the form of Y, but, once it is determined, statistics can determine the sources of its variability. We could actually treat a system of equations of the above form but will not for mathematical brevity.

As in Section 5-3.1, we linearize Y by first writing its differential expansion:

$$dY = \frac{\partial Y}{\partial X_1} dX_1 + \frac{\partial Y}{\partial X_2} dX_2 + \ldots + \frac{\partial Y}{\partial X_I} dX_I$$

Upon multiplying and dividing each term on the right side by the respective X_i, multiplying the entire expression by Y, and recognizing the differential properties of natural logarithms, this equation becomes

$$d\ln(Y) = a_1 d\ln(X_1) + a_2 d\ln(X_2) + \ldots + a_I d\ln(X_I) \tag{5-1}$$

where the coefficients a_i are *sensitivity coefficients* defined as

$$a_i = \frac{\partial \ln(Y)}{\partial \ln(X_i)}$$

These coefficients are not, at this point, constants since they depend on the values of the entire set of X_i. The a_i are the relative change in Y caused by a relative change in the respective X_i. The *relative* part of this statement, which arises in the derivation because of the use of the logarithms, is important because the X_i can be different from each other by several factors of ten. Of course, since Y is known, the sensitivity coefficients can be easily calculated.

Equation (5-1) is thus far without approximation. We now replace the differential changes by discrete changes away from some base set of values (denoted by ^{o}):

$$d\ln(X_1) = \ln(X_1) - \ln(X_1^o) = \ln(X_1 / X_1^o)$$

The resulting equation is now an approximation to the original model, being dependent on the selection of the base values. It is a good approximation only in some neighborhood of the base values. If we, in addition, evaluate the a_i at the base values, the estimator becomes linear with constant coefficients.

$$\ln(Y/Y^o) = \sum_{i=1}^{I} a_i \ln(X_i / X_i^o)$$

This linearized estimator is now in a suitable form for application of the variance operator,

$$Var(Y) = \sum_{i=1}^{I} a_i^2 Var(X_i)$$

from the properties $Var(aX) = a^2 Var(X)$, $Var(X + \text{constant}) = Var(X)$, and $Var(\text{constant}) = 0$. The signs of the a_i no longer matter since they appear squared. This expression assumes that the X_i are independent.

Example 5 - Sensitivities in the Carman-Kozeny Equation. We can use the above procedure on a fairly simple equation to make inferences about the origin of permeability heterogeneity. The application also illustrates other ways to linearize estimators.

Let us view a reservoir as consisting of equal-sized patches, each composed of spherical particles of constant diameter. (The facies-driven explanation of heterogeneity discussed in Chap. 11 suggests that this picture is not too far from reality, although we invoke it here mainly for mathematical convenience.) The Carman-Kozeny (CK) equation

$$k = \frac{1}{72\tau} \frac{\phi^3}{(1 - \phi)^2} D_p^2$$

now gives the permeability within each patch. D_p is the particle diameter, ϕ the porosity, and τ the tortuousity in this equation.

The CK equation is a mix of multiplications and subtractions. It can be partially linearized simply by taking logarithms.

$$\ln(k) = \ln\left[\frac{\phi^3}{(1-\phi)^2}\right] + 2\ln(D_p) - \ln(\tau) - \ln(72)$$

From inspection, the sensitivity coefficients for D_p and τ are $a_{D_p} = 2$ and $a_\tau = -1$. These coefficients are constant because of the logarithms. By differentiating this expression, the sensitivity coefficient for porosity is

$$a_\phi = \frac{3-\phi}{1-\phi}$$

This requires a base value, which we take to be $\phi = 0.2$ from which $a_\phi = 3.5$. Note that the CK permeability estimate is most sensitive to porosity, followed by particle size, then tortuousity ($|a_\phi| > |a_{D_p}| > |a_\tau|$).

The variance of ln(k) is

$$Var(\ln(k)) = a_\phi^2 Var[\ln(\phi)] + a_{D_p}^2 Var[\ln(D_p)] + a_\tau^2 Var[\ln(\tau)]$$

Estimating this variability requires estimates of the variabilities of ln(ϕ), ln(D_p), and ln(τ). We note in passing that porosity, particle size, and tortuousity are presumed independent in this treatment.

Based on typical core data, reasonable values for the variances are $Var[\ln(\phi)] = 0.26$, $Var[\ln(D_p)] = 2.3$, and $Var[\ln(\tau)] = 0.69$. The variance of ln(k) is now 13.08, but perhaps the most insight comes from apportioning this variance among the terms. From this we find that 24% of $Var[\ln(k)]$ comes from ln(ϕ), 70% from ln(D_p) and 6% from ln(τ). An analysis of the CK equation, using the sum-of-errors approach, yields similar results to these. Either way, substantially more of the variability in permeability comes from variability in particle size, even though the estimate is more sensitive to porosity. Additional calculations of this sort, that do not regard ϕ, D_p, and τ as independent, are to be found in Panda (1994).

Although simple, Example 5 illustrates some important truths about the origin of reservoir heterogeneity–namely that most heterogeneity arises because of particle size variations. This observation accounts, in part, for the commonly poor quality of permeability-porosity correlations. Similar analysis on other estimators could infer the source of measurement errors. We conclude with the following three points.

1. The procedure outlined above is the most general with respect to estimator linearization. However, as the example shows, there are other means of linearization, many of which do not involve approximation. These should be used if the form of the estimator allows it. For example, logarithms are not needed if the estimator is originally linear. In such cases, the entire development should be done with variables scaled to a standard normal distribution.

2. The restriction to small changes away from a set of base values is unnecessarily limiting in many cases. When the small change approximation is not acceptable, the general nonlinear estimator $Y = W(X_1, X_2, \ldots, X_I)$ can be used in Monte Carlo algorithms that do not require the estimator to be linear (e.g., Tarantola, 1987, and Chap. 3). Many such algorithms also allow for the X_i to be dependent.

3. Perhaps most important is the fact that the procedure allows a distinction between sensitivity and variability (or error). Sensitivity, as manifest through the sensitivity coefficients, is merely a function of estimator form; variability is a combination of sensitivity and the variability of quantities constituting the estimator. In many engineering applications, variability is the most interesting property.

5-4 CONFIDENCE INTERVALS

During estimation, the information conveyed by the data is "boiled down" by the estimator W to produce the estimate $\hat{\theta}$. There usually is information in the data that is unused. Since $\hat{\theta}$ is a random variable (e.g., Fig. 5-2), the additional information can be used to estimate $\hat{\theta}$'s variability. This variability assessment, $s_{\hat{\theta}} = [\mathrm{Var}(\hat{\theta})]^{1/2}$, provides a confidence interval for $\hat{\theta}$.

Example 2 showed that estimator W_1 produced unbiased estimates with $Var(\hat{\theta}_1) = \theta^2/3I$ while W_2 gave bias-corrected estimates with $\mathrm{Var}(\hat{\theta}_2) = \theta^2/I(I+2)$ for I data. If we assume that the $\hat{\theta}$'s are normally distributed, then $\hat{\theta}_1 \sim N[\theta, \theta^2/3I]$ and $\hat{\theta}_2 \sim N[\theta, \theta^2/3(I+2)]$. These results permit us to say, at any given level of probability, how close our estimates $\hat{\theta}_1$ and $\hat{\theta}_2$ are to the true value θ (Fig. 5-4). For any number of data and value of θ, W_2 will give estimates more closely centered about θ.

An alternative to showing estimated PDF's is to give the standard error $s_{\hat{\theta}}$ along with the estimate, usually written as $\hat{\theta} \pm s_{\hat{\theta}}$. $s_{\hat{\theta}}$ is an estimate of $[Var(\hat{\theta})]^{1/2}$ because we (again) have to depend on the data. For example, in the case of W_1 in Example 2a, we have $\hat{\theta}_1 \pm (\theta^2/3I)^{1/2}$. Clearly, if we are estimating θ, we do not know the value of θ for the standard error of $\hat{\theta}_1$ either. We do the next best thing, however, and use the estimate $\hat{\theta}_1$ again: $\hat{\theta}_1 \pm (\hat{\theta}_1^2/3I)^{1/2}$. In essence, we are giving the best estimate of θ along with our best estimate of its standard deviation.

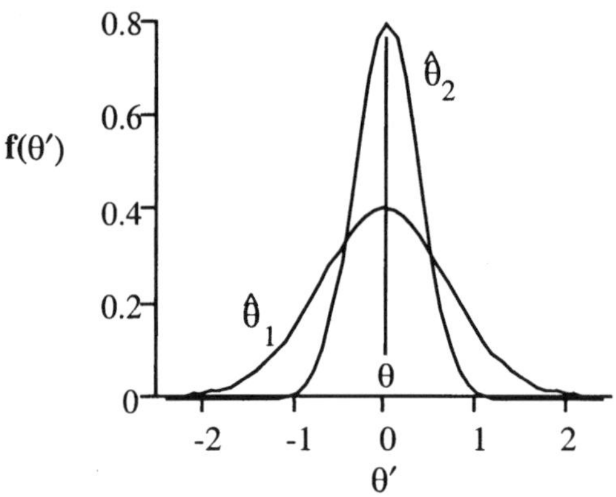

Figure 5-4. PDF's of $\hat{\theta}_1$ and $\hat{\theta}_2$ with the x axis scaled in units of $(\theta/3I)^{1/2}$, centered at θ.

We often assume that unbiased estimates are normally distributed. This is not always true, particularly when I is small (e.g., less than 20), and theory exists for standard errors of some estimators (e.g., the arithmetic average and the correlation coefficient). However, as a working practice for $I \geq 20$, it is often safe to assume $\hat{\theta}$ is normally distributed. We can then use the properties of the normal PDF to state how often θ will be within a given range of $\hat{\theta}$ (see Sec. 4-4.3). Hence, there is a 68% chance that $(\hat{\theta} - s_{\hat{\theta}}) < \hat{\theta} < (\hat{\theta} + s_{\hat{\theta}})$ and a 95% chance that $(\hat{\theta} - 2s_{\hat{\theta}}) < \hat{\theta} < (\hat{\theta} + 2s_{\hat{\theta}})$ (Fig. 5-5).

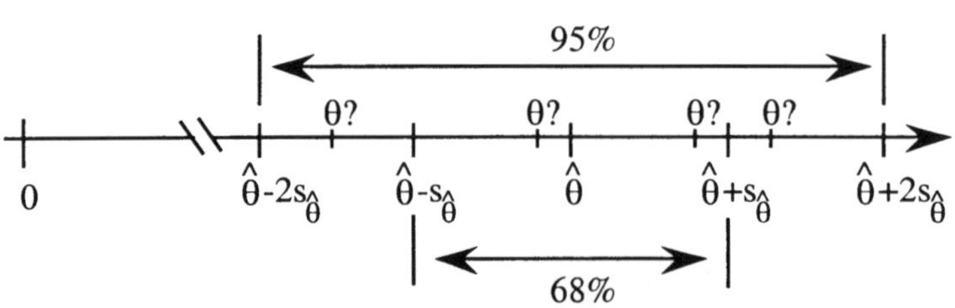

Figure 5-5. The true value for θ is not known. The size of interval where it may be located depends on the probability assignment.

5-4.1 Confidence Intervals and Additivity

To use standard errors to define and interpret confidence intervals requires, strictly speaking, that the variable be additive (ratiometric, Chap. 3). For this type of variable, we can add and subtract multiples of the standard error without difficulty. We do reach a limit, however, because we might, for example, produce a reservoir thickness such as $h = 22$ m $\pm$ 15 m. The 95% confidence limits would be -8 m $< h <$ 52 m, and we know that lower extreme of -8 m cannot be correct. This occurs because the assumption of a normal PDF is no longer valid and we are seeing it break down. But there is also the more difficult area where we want to use standard errors with interval-type variables, which may not be additive. These are very common in reservoir characterization (e.g., permeability, density, and resistivity).

The role of additivity in statistical analysis requires further research. Here, we will take a pragmatic approach that does not demand unconditional additivity of the variables to be analyzed. That is, if X is the reservoir property being considered, we will not require that $\sum \alpha_i X_i$ be a physically meaningful quantity for all α's and X's. We will only require "local" additivity, so that $X \pm \Delta X$, where ΔX is a "small" change in the value X (say, $|\Delta X| \leq 0.20X$), is still physically meaningful. By assuming local additivity, we can continue to interpret standard errors in the conventional way. Large changes, however, may lead to nonphysical values. For example, porosity is clearly nonadditive because it cannot exceed unity. It is, however, a ratio of additive quantities: void volume $V_V \div$ total volume V_T, where $V_V \leq V_T$. For V_T constant, a V_V perturbation, ΔV_V, is directly proportional to the change in porosity: $\Delta\phi = \Delta V_V / V_T$. Hence, as long as $V_V + \Delta V_V \leq V_T$ and V_T is constant, porosity is locally additive. If porosities with different measurement volumes are being added, they can be put on a comparable basis by adjusting for V_T.

The traditional definition of confidence intervals is tied to $Var(\hat{\theta})$. Confidence intervals are also just one type of statistical interval that could be considered. Others exist and are discussed at length in Hahn and Meeker (1991). We will confine discussion here to the traditional usage.

5-4.2 Numerical Estimation of Confidence Intervals

While a very useful tool for conveying uncertainty, in many situations confidence intervals may be difficult to obtain or inaccurate, depending upon the nature of the problem. In particular, small data sets from non-normal populations can give rise to biased estimates and produce erroneous confidence regions if the usual normal-theory assessments are used. Two procedures, called the *jackknife* and the *bootstrap*, address these problems. The idea of these methods is to assess the variability of estimates using incomplete data sets. That is, if we have I samples in a data set, we can generate a number of smaller-size data sets from it. The jackknife and bootstrap methods then use these subsets to assess the estimate variability.

Consider, for example, the jackknife technique on a data set of size I using an estimator W. It is possible to generate I subsets of size I-1 by dropping a different datum for each subset.

Then, using W, we can produce an estimate for each of the I subsets. Let $\hat{\theta}_i$ represent the estimate based on the ith subset, $i = 1, 2, \ldots, I$ and let $\hat{\theta}$ represent the estimate based on the full set of size I. An unbiased estimate of θ is given by

$$\theta^* = I\hat{\theta} - \frac{I-1}{I}\sum_{i=1}^{I}\hat{\theta}_i$$

A variance estimate for θ^*, which we need to obtain a confidence interval, is given by

$$s_{\theta^*}^2 = \frac{I-1}{I}\sum_{i=1}^{I}(\hat{\theta}_i - \frac{1}{I}\sum_{j=1}^{I}\hat{\theta}_j)^2$$

We can then apply the usual confidence interval techniques previously discussed to indicate the variability in θ^*. Clearly, if I exceeds 5 or 10, the jackknife becomes cumbersome, making it a perfect task for a computer. Some of the newer data-analysis packages have this feature available.

The bootstrap is even more computationally intensive than the jackknife but similar in approach. The bootstrap generates subsets allowing replication of data, however. Hence, a much larger number of subsets can be generated. As for the jackknife, W is used to produce an estimate $\hat{\theta}_j$ for $j = 1, 2, \ldots, J$, where J usually exceeds I. The estimate $\hat{\theta}$ for the complete data set is given by

$$\hat{\theta} = \frac{1}{J}\sum_{j=1}^{J}\hat{\theta}_j$$

with a variance of

$$s_{\hat{\theta}}^2 = \frac{1}{J-1}\sum_{j=1}^{J}(\hat{\theta}_j - \hat{\theta})^2$$

The jackknife and bootstrap are useful not only for obtaining confidence intervals of $\bar{X}$. For example, these methods may be very useful for analyzing output from stochastic simulations, where each run can be very costly and, hence, there are very few runs to analyze. Many of the results are likely to come from non-normal distributions (e.g., produced quantities, time to breakthrough, and water cut). Hence, the "average" performance may be hard to assess without the jackknife or bootstrap. Similar arguments can apply to geologic analyses, too. For example, calculation of R^2 estimates from small numbers of core-plug porosity and permeability data may be analyzed using the jackknife or bootstrap. The jackknife, however, is not suitable for order statistics, such as the median. The reader is referred to Schiffelbein (1987) and Lewis and Orav (1989) for further details and references.

Example 6 - Jackknifing the Geometric Average. We compute the jackknife estimate of the geometric mean, a measure of the effective permeability for disordered heterogeneous systems (see Sec. 5-7 for details), from ten permeability measurements, $X_1, \ldots, X_{10}$, shown in Table 5-1. Using all values, the geometric average is given by $\bar{X}_G = [(X_1)(X_2)(X_3)(X_4)\ldots(X_{10})]^{1/10} = 95.6$ mD. $\bar{X}_{G,i}$ is the sample geometric mean (i.e., geometric average) of the data set without the ith point in it.

Table 5-1. Example 5 jackknife analysis of the geometric average.

i	X_i mD	$\bar{X}_{G,i}$ mD
1	15	117
2	25	111
3	40	105
4	60	101
5	100	95
6	115	94
7	152	91
8	210	88
9	350	83
10	550	79

Figure 5-6 is a probability plot of the data, suggesting they appear to come from a log-normal PDF. Because the data appear to have a log-normal PDF, there are theoretical results for the bias and variability of $\bar{X}_G$ (Jensen, 1998) which we can use to compare with the jackknife results. First, we calculate an unbiased estimate of $\bar{X}_G$ using the jackknife.

$$\bar{X}_G^{\bullet} = I\bar{X}_G - \frac{I-1}{I}\sum_{i=1}^{I}\bar{X}_{G,i} = 10(95.6) - \frac{9}{10}(117 + 111 + \ldots + 83 + 79)$$

$$= 956 - 9(963)/10 = 956 - 867 = 89 \text{ mD}$$

The theoretical results suggest $\bar{X}_G$ will be 6.8% larger than X_G. The jackknife results indicates $\bar{X}_G$ is too large by (96 – 89)/89 = 7.8%.

The jackknife also provides a confidence interval estimate of the sample geometric mean:

$$s^2_{\bar{X}_G^{\bullet}} = \frac{I-1}{I}\sum_{i=1}^{I}\left(\bar{X}_{G,i} - \frac{1}{I}\sum_{j=1}^{I}\bar{X}_{G,j}\right)^2$$

$$= \frac{9}{10}\left[(117 - 96.3)^2 + \ldots + (79 - 96.3)^2\right] = 1{,}230 \text{ mD}^2$$

or

$s_{\overline{X}_G^*} = 35$ mD. The theoretical estimate is $s_{\overline{X}_G^*} = 38$ mD.

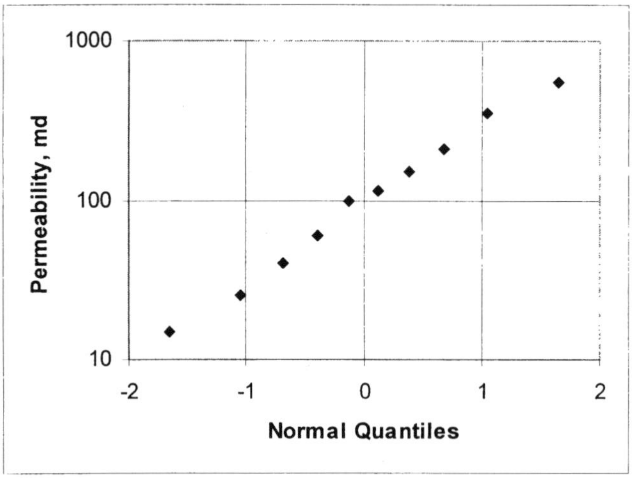

Figure 5-6. Permeability data probability plot showing approximately log-normal PDF.

Thus, at 95% probability, the confidence interval for $\overline{X}_G^*$ is from ($\overline{X}_G^*$ - 2$s_{\overline{X}_G^*}$) to ($\overline{X}_G^*$+ 2$s_{\overline{X}_G^*}$), or 19 mD < $\overline{X}_G^*$ < 159 mD. Clearly, there is a large possible range for $\overline{X}_G^*$.

How well the jackknife performs depends on three items: the estimator, the number of data involved, and the characteristics of these data. When theoretical results are available for comparison, the agreement between theoretical and jackknife results typically improves as the number of data increases. In the above example, the geometric average is a mild, continuous function of its arguments. Thus, agreement is quite good for only 10 data. More sensitive estimators, e.g., the harmonic average, may require more data to perform well.

The jackknife procedure is not restricted to subdividing I data into subsets of size I-1. The procedure is more general. Lewis and Orav (1989, Chap. 9) have the details.

5-5 PROPERTIES OF SELECTED ESTIMATORS

We now list the bias and error behavior of several common estimators.

5-5.1 Properties of the Sample Mean

The arithmetic average is a very common measure of central tendency. It has been extensively studied and we give some results here. Let X_i be a random sample of I observations from a distribution for which the population mean μ and variance σ^2 exist. The arithmetic average or sample mean is

$$\overline{X} = \frac{1}{I}\sum_{i=1}^{I} X_i$$

From the Central Limit Theorem (Chap. 4), one can show (Cramér, 1946, Chap. 28) that $\overline{X}$ has the following properties:

$$\overline{X} \sim N(\mu, \frac{\sigma^2}{I}) \text{ for large } I$$

$$\overline{X} \sim N(\mu, \frac{\sigma^2}{I}) \text{ exactly, if } X_i = N(\mu, \sigma^2),$$

where $E(X_i) = \mu$ and $Var(X_i) = \sigma^2$. The first result is independent of the PDF of the random variable; it says that $\overline{X}$ is an unbiased estimate of μ or, on average, $\overline{X}$ over all samples is equal to the "true" sample space mean. In some cases, $\overline{X}$ approaches normality for surprisingly small I. Papoulis (1965, pp. 267-268) gives an example of independent, uniformly distributed X and $I = 3$.

One of the interesting features of the arithmetic average is how its variability depends on the number of samples, I. The variability of $\overline{X}$ decreases as I increases, which is what we would expect; the more information we put in, the better the estimate ought to be. In particular, the standard deviation of $\overline{X}$ is proportional to $I^{-1/2}$, a property common to many estimators (e.g., Example 2a).

Before we consider further the sampling properties of $\overline{X}$, we first discuss variance estimation.

5-5.2 Properties of the Sample Variance

The sample variance is a measure of dispersion. It is defined as

$$s^2 = \frac{1}{I}\sum_{i=1}^{I}(X_i - \overline{X})^2$$

s^2 is biased since we can show (Meyer, 1966, p. 254) that

$$E(s^2) = \frac{I-1}{I}\sigma^2$$

However, just as in Example 2b, we can formulate an unbiased estimate of σ^2 by using

$$\hat{s}^2 = \frac{I}{I-1} s^2$$

$$= \frac{1}{I-1} \sum_{i=1}^{I} (X_i - \bar{X})^2$$

This result has a more general concept embedded in the use of $(I - 1)$ instead of I in the denominator. Since the sample mean is required to compute the sample variance, we have effectively reduced the number of independent pieces of information about the sample variance by one, leaving $(I - 1)$ "degrees of freedom." Therefore, $(I - 1)$ is a "natural" denominator for a sum of squared deviations from the sample mean. Using $(I - 1)$ instead of I to estimate the variance is not very important for large samples, but we should always use it for small data sets ($I < 20$).

If $X \sim N(\mu, \sigma^2)$, then the $Var(\hat{s}^2) = \sigma^2/2I$ (Kendall and Stuart, 1977, p. 249). More generally, the variability of $\hat{s}^2$ is approximately

$$Var(\hat{s}^2) \approx \frac{E\{[X - E(X)]^4\} - Var(X)^2}{4(I-1)Var(X)}$$

where $E\{[X - E(X)]^4\}$ is the fourth centered moment of X. This result follows the general pattern that the estimate variability for the rth moment depends upon the $2r^{th}$ moment of the PDF.

To estimate the standard deviation, one may usually take the square root of the variance. For very small data sets ($I \leq 5$), however, another effect comes in. The square root is a nonlinear function and causes the estimated standard deviation to be too small. When $I \leq 5$, multiply $\hat{s}$ by $1 + (1/4I)$ (Johnson and Kotz, 1970, pp. 62-63).

5-5.3 Confidence Intervals for the Sample Mean

The arithmetic average, $\bar{X}$, is often used to estimate $E(X)$. There are some theoretical results for the confidence limits of $\bar{X}$ when the samples come from a normally distributed population. If we know the population variance, $\sigma^2 = Var(X)$, we would know that $(\mu - z\sigma/\sqrt{I}) \leq \bar{X} \leq (\mu + z\sigma/\sqrt{I})$ for a fraction $[1 - P(z)]/2$ of the time. (Recall P is the standard normal CDF.) Usually, however, we do not know σ but can estimate it from the same samples we used to calculate $\bar{X}$. If that is the case, we have to introduce a different term that acknowledges that $\hat{s}$ is also subject to statistical variation:

$$\mu = \bar{X} \pm t(\alpha/2, df)\ \hat{s}/\sqrt{I}$$

where $t(\alpha/2, df)$ is the "t value" from Student's distribution (Fig. 5-7) with confidence level α, $df = I - 1$ degrees of freedom, and $\hat{s}$ is the sample standard deviation.

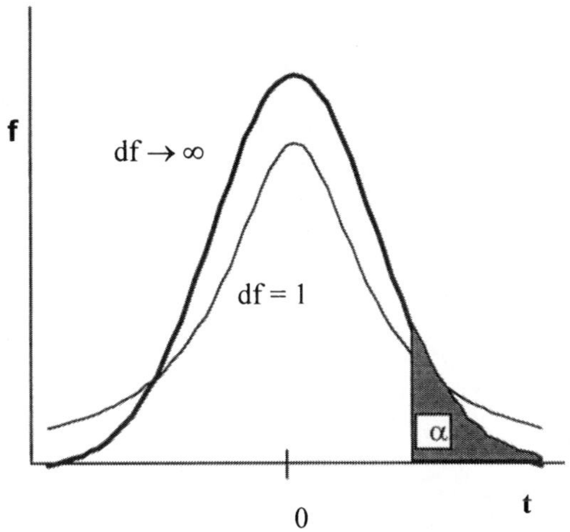

Figure 5-7. The Student's distributions change shape with the degrees of freedom (*df*). *t* is the value for which the tail area is α at the stipulated *df* value.

Values of the function $t(\alpha/2, df)$ are widely tabulated (e.g., Table 5-2) and appear in spreadsheet libraries. α is the complement of the fractional confidence limit (e.g., if we want 95% confidence limits, then $\alpha = 0.05$). As *df* (i.e., the sample size) becomes large, the *t* value approaches the normal PDF value for the same confidence level.

Table 5-2. *t* values.

	Tail area probability, α		
df	**0.05**	**0.025**	**0.005**
1	6.314	12.706	63.657
2	2.920	4.303	9.925
3	2.353	3.182	5.841
4	2.132	2.776	4.604
5	2.015	2.571	4.032
6	1.943	2.447	3.707
7	1.895	2.365	3.499
8	1.860	2.306	3.355
9	1.833	2.262	3.250
10	1.812	2.228	3.169
11	1.796	2.201	3.106
12	1.782	2.179	3.055
13	1.771	2.160	3.012
14	1.761	2.145	2.977
15	1.753	2.131	2.947
16	1.746	2.120	2.921
18	1.734	2.101	2.878
20	1.725	2.086	2.845
22	1.717	2.074	2.819
24	1.711	2.064	2.797
26	1.706	2.056	2.779
28	1.701	2.048	2.763
30	1.697	2.042	2.750
40	1.684	2.021	2.704
60	1.671	2.000	2.660
∞	1.645	1.960	2.576

For a given value of $t = t_0$, the symmetrical form of Student's distribution implies that any random variable T with this PDF is just as likely to exceed some value t_0 as it is to be less than $-t_0$. Consequently, when we set a confidence level α for T, we usually want $t(\alpha/2, df)$ because T can be either positive or negative with equal probability. For example, if our confidence level is $(1 - \alpha) = 0.95$, we find $t(0.025, df)$. This situation is called a *double-sided* confidence interval. If we knew, however, that T could only have one sign or we were interested only in excursions in one direction, then we would use $t(\alpha, df)$ for a *single-sided* confidence interval.

Example 7 - Estimating the Mean Value. Suppose we wish to specify the 95% (α = 0.05) confidence interval for the population mean from a sample of 25 data points (df = 24) with $\bar{X}$ = 30 and $\hat{s}$ = 3. The t value from the above table is 2.064 and the above formula yields

$$\mu = 30 \pm (2.064)\ 3/\sqrt{25}$$

or

$$\mu \in \{28.76, 31.23\}$$

Thus, we can be 95% certain that the population mean is within the stated limits (Fig. 5-8).

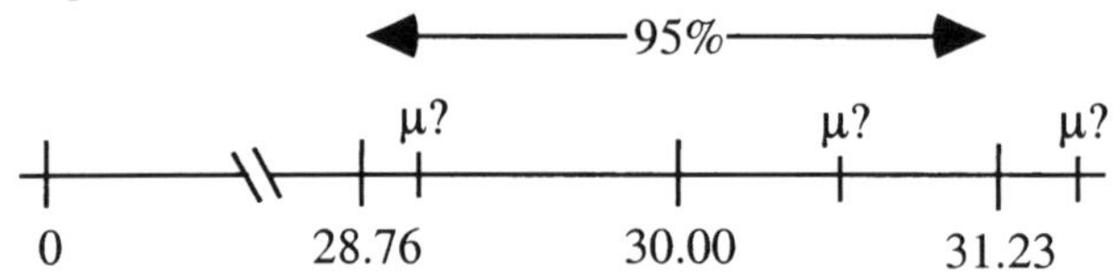

Figure 5-8. Confidence limits for $\bar{X}$. There is a 2.5% probability that μ > 31.23 and a similar chance that μ < 28.76.

Had we used the z value instead of the t value (i.e., if we ignored the fact that we have estimated the standard deviation from the data as well as the mean), the factor would be 1.960 (Chap. 4) instead of 2.064.

Despite its considerable attractions, there are some problems with the sample mean. It can be inefficient when the X_i are coming from skewed populations, such as the log-normal PDF. Consequently, we can expect that the arithmetic averages of quantities like permeability and grain size may be inaccurate. Agterberg (1974, pp. 235 ff) discusses an efficient method for estimating the mean for log-normal populations.

5-5.4 Properties of the Sample Median

Besides the sample mean, the sample median $\hat{X}_{0.50}$ is another measure of central tendency. We have already observed that $\hat{X}_{0.50}$ is more robust than $\bar{X}$. $\hat{X}_{0.50}$, however, does not have the nice distributional properties obtained through the central limit effect that X has.

The exact PDF and properties of $\hat{X}_{0.50}$ are discussed by Kendall and Stuart (1977, pp. 252, 348-351). When $X \sim N(\mu, \sigma^2)$, $\hat{X}_{0.50}$ is unbiased and its standard error is approximated by

$$\sqrt{Var(\hat{X}_{0.50})} = \sigma\sqrt{\pi / 2I}$$

for I data. Since $\text{Var}(\overline{X}) = \sigma^2 / I$ we find that, for large I,

$$\frac{Var(\overline{X})}{Var(\hat{X}_{0.50})} = \frac{\sigma^2 / I}{\pi\sigma^2 / 2I} = 2/\pi$$

Hence, for the normal PDF, where $X_{0.50} = E(X)$, estimation of the mean using $\hat{X}_{0.50}$ is less efficient (i.e., needs about 60% more data) than using $\overline{X}$.

5-5.5 Properties of the Interquartile Range

The *interquartile range* (IQR) is the difference $(X_{0.75} - X_{0.25})$ and is another measure of dispersion. It has been used as a robust substitute for the standard deviation, $\sqrt{Var(X)}$. Apart from robustness, the sample IQR, $(\hat{X}_{0.75} - \hat{X}_{0.25})$, has several other features in common with the median. Both are based on *order statistics*, which are sample quantiles.

For $X \sim N(\mu, \sigma^2)$, $(X_{0.75} - X_{0.25}) = 1.35\sigma$. Hence, besides the sample variance, an estimator for σ is $\hat{\sigma}_{IQR} = (\hat{X}_{0.75} - \hat{X}_{0.25})/1.35$ when X is normally distributed. For large I, the sample IQR is unbiased and has the sampling variability $Var(\hat{X}_{0.75} - \hat{X}_{0.25}) = 3\sigma^2/2I$ (Kendall and Stuart, 1977, p. 350). $\hat{\sigma}_{IQR}$ is less efficient than $\hat{s}^2$ at estimating σ^2. Depending upon I, $Var(\hat{\sigma}_{IQR})$ is between 30% ($I = 6$) and 65% (I large) larger than $Var(\hat{s}^2)$. This puts $\hat{\sigma}_{IQR}$ at about the same efficiency as the sample median for estimating their respective quantities when X is normally distributed.

5-6 ESTIMATORS FOR NONARITHMETIC MEANS

There are other means besides the arithmetic mean, $X_A = E(X)$, that arise in reservoir evaluation. The geometric and harmonic means of X are defined as $X_G = \exp\{E[\ln(X)]\}$ and $X_H = [E(X^{-1})]^{-1}$, respectively. All three means are specific instances of a more general situation (Kendall and Stuart, 1977, pp. 37-38), the *power mean*:

$$X_p = \begin{cases} [E(X^p)]^{1/p} & p \neq 0 \\ \exp\{E[\ln(X)]\} & p = 0 \end{cases}$$

X_H is the case when $p = -1$, X_G when $p = 0$, and X_A corresponds to $p = 1$. X_p can be viewed as being the arithmetic mean of X^p raised to the $1/p$ power. As long as the reservoir property X is nonnegative, $X_{p1} \leq X_{p2}$ if $p_1 \leq p_2$ (Kendall and Stuart, 1977, p. 38). Hence, $X_H \leq X_G \leq X_A$ for many reservoir properties.

We have already considered estimation of X_A. Compared to $\overline{X}_A$, the properties of X_G and X_H estimators have received little attention. Two common estimators for X_G and X_H are

$$X_G = exp\left[\frac{1}{I}\sum_{i=1}^{I} \ln(X_i)\right] \qquad \text{(geometric average)}$$

and

$$X_H = I\left[\sum_{i=1}^{I} (X_i)^{-1}\right]^{-1} \qquad \text{(harmonic average)}$$

Neither is unbiased. For example, $\overline{X}_G$ overestimates X_G if $\ln(X) \sim N(\mu, \sigma^2)$: $E(\overline{X}_G) = X_G \exp(\sigma^2/2I)$, so $\overline{X}_G$ is consistent. Under similar conditions, $E(\overline{X}_H) = X_H\{1 + [\exp(\sigma^2) - 1]/I\}$. $\overline{X}_G$ and $\overline{X}_H$ are not especially robust, either. Their standard errors for the log-normal case are discussed in Jensen (1998) and Limbrunner et al. (2000). Jensen (1998) shows that $\overline{X}_G$ has less variability than either $\overline{X}_A$ or $\overline{X}_H$. Limbrunner et al. (2000) discuss the case of X_H estimation when the data are extremely variable with C_V's in excess of 10.

$\overline{X}_G$ and $\overline{X}_H$ are also sensitive to small, uncertain data values. One way to avoid this problem is to use the sample median; if $Y = X^p$, choose $\hat{X}_p = [\hat{Y}_{0.50}]^{1/p}$. Another approach is to use a trimmed or Winsorized mean of Y, which is discussed below.

5-7 FLOW ASSOCIATIONS FOR MEAN VALUES

As we pointed out at the beginning of this chapter, an important element of the estimation procedure is to decide on the quantity to be estimated, θ. Sometimes the decision is fairly obvious. For example, $E(\phi)$ is often a useful measure for porosity in many applications because it represents a central value for a physically interpretable quantity. Permeability, on the other hand, is usually a more difficult property for which to find a representative value, and only a few systematic guidelines are available.

Why is permeability so troublesome? Since it pertains to flow, permeability is clearly an important property. But, permeability is an intensive variable (Chap. 3). Depending upon the situation, it may not be additive. The flow transmitted by any given region depends upon the permeabilities of surrounding regions. For example, consider a highly permeable region of a reservoir encased in shale (e.g., lenticular bedding). No matter how permeable is the center, the outer "shell" of low-permeability material prevents flow. This is not the case for porosity, for example, where the pore volume of fluids in a lens contributes to the total amount of fluid in a region without regard to whether the fluids can move or not. There are a few cases, however, where it is clear what the additive properties of permeability are.

In the case of linear flow parallel to a stratified medium, typical of shallow marine sheet sands, of I layers (Fig. 5-9), the aggregate permeability (k_t) of the region is the expected value of the layer permeabilities. To see this, we take the expression for k_t and rearrange it:

$$k_t = \frac{\sum_{i=1}^{I} h_i k_i}{\sum_{i=1}^{I} h_i} = \Sigma k_i \frac{h_i}{\sum_{i=1}^{I} h_i} = \sum_{i=1}^{I} k_i p_i = E(k)$$

since p_i is the probability that the permeability is k_i. Clearly, permeability is additive in this case.

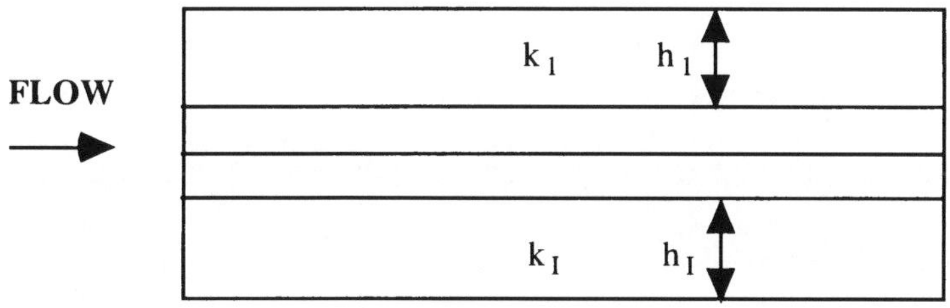

Figure 5-9. Effective permeability for layer-parallel flow.

When linear flow is orthogonal to the layers (e.g., dune crossbeds), permeability is no longer additive. Its *inverse*, however, is additive. The aggregate *resistance* to flow, k_t^{-1}, is given by the expected value of the layer resistances to flow (Fig. 5-10):

$$k_t^{-1} = \frac{\sum_{i=1}^{I} h_i / k_i}{\sum_{i=1}^{I} h_i} = \Sigma k_i^{-1} h_i = \sum_{i=1}^{I} k_i^{-1} p_i = E(k^{-1})$$

Consequently, here are two situations for which we know the quantity to be estimated, θ, when the aggregate permeability is desired:

$k_t = E(\mathrm{k})$ for linear flow parallel to layers

$k_t^{-1} = E(k^{-1})$ for linear flow perpendicular to layers

Many geological systems, while containing some degree of ordered permeability variation, do not meet the layered situations described above (e.g., turbidites, braided deposits). What are the expressions for k_t in those cases? The best general answer comes from Matheron (1967): $E(k) \geq k_t \geq [E(k^{-1})]^{-1}$. In other words, k_t appears to be the power mean of permeability for some exponent p, $-1 \leq p \leq 1$. The work of Desbarats (1987) has supported this result, but it is not clear how to select *p a priori*.

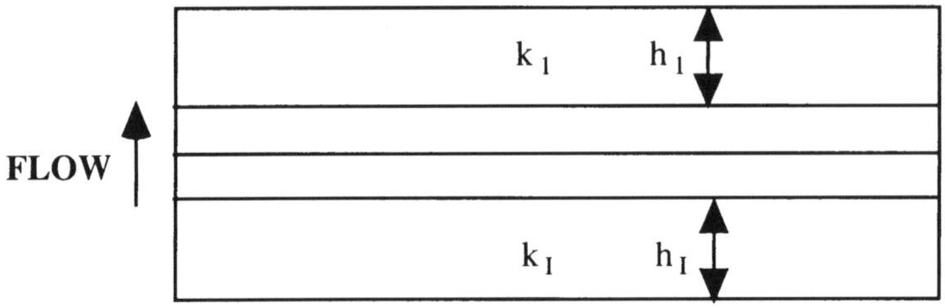

Figure 5-10. Effective permeability for layer-transverse flow.

There are two other situations for which theoretical results exist. They both require that $\ln(k) \sim N(\mu, \sigma^2)$ and σ is small. When the flow is two-dimensional (2D), $\ln(k_t) = E[\ln(k)]$ (Matheron, 1967). Hence, the geometric mean is appropriate. For 3D flow, $(k_t)^{-1/3} = E[(k)^{1/3}]$ (Noetinger, 1994). Various simulation studies (e.g., Warren and Price, 1961) have extended the former result to less-restrictive situations with some success. These results do not require that the system be totally disordered. As long as the scale of the ordered or structured element is small compared to the region for which k_t is sought, given that k is log-normally distributed and exhibits only moderate variation.

In summary, then, there are three definite results–all assuming linear 2D flow–for determining k_t.

1. Layered system, flow parallel to layers: arithmetic mean ($p = 1$).

2. Layered system, flow normal to layers: harmonic mean ($p = -1$).

3. Random system, log-normal k, small variation: geometric mean ($p = 0$).

For 3D systems, $p = {}^1/_3$ applies for the same conditions as item 3 above. Otherwise, $E(k) \geq k_t \geq [E(k^{-1})]^{-1}$ is the best that can be said. The three definitive results suggest that some function of permeability may always be additive, but the precise form of the function changes with the flow geometry.

5-8 PHYSICAL INTERPRETABILITY IN ESTIMATION

Any estimator will manipulate data in some way to produce an estimate. Despite the large amount written on estimation, it appears very few have considered whether the estimator combines variable values in a physically meaningful manner. For example, to obtain an arithmetic average, sample values that are physically meaningful have to be added together. To add values and produce something that is still physically meaningful requires that the variable be ratiometric (Chap. 3), i.e., additive. We have just observed that permeability may not be additive, so calculating the average permeability may give a nonsensical result.

A still more questionable estimator is the sample variance. Differences of sample values are taken, squared, and summed. This suggests that the variable *and* its square are both additive, a situation that would apply, for example, with length because area is also additive.

In this light, the sample IQR might be preferable because it does not take squares. Unfortunately, we cannot simply take the local additivity approach with estimators that we took with their standard errors. Some estimators, such as the arithmetic average or sample variance, add or subtract quantities of similar magnitude and manipulate them further. It is not clear what effect the lack of additivity has on results and further research is needed.

In this respect, PDF's could play an interesting role. It is well-known that two variables with normal PDF's give a normally distributed sum (Cramér, 1946, p. 212). The squares of normal PDF variables are also additive, giving the χ-squared PDF (Cramér, 1946, p. 233). Hence, statistical additivity (preservation of the PDF when variables are added or subtracted) is satisfied for the average or variance when normal PDF variables are involved. It may be that the effects of physical nonadditivity are mitigated when variables have normal PDF's prior to use in estimators. For example, for sound statistical reasons, $X_{0.50} \neq X_A$ when X is log-normally distributed (Chap. 4).

If $\ln(X) \sim N(\mu, \sigma^2)$ and $\sigma > 0$, then $X_{0.50} = \exp(\mu)$ compared to $E(X) = X_A = \exp(\mu + \sigma^2/2)$. $X_{0.50}$ does not involve a summation of X (recall that $X_{0.50}$ is defined as $0.5 = \int f(x)dx$), but only summation of X's PDF, $f(x)$. Compare this with X_A that does involve a summation of all possible values of X: $X_A = \int xf(x)dx$. Could it be that the difference between X_A and $X_{0.50}$ is also a reflection of X's nonadditivity?

Until this issue is resolved, it would appear that two approaches might be taken: (1) use the median and IQR instead of the mean and standard deviation for central tendency, dispersion, and variability measures, or (2) continue to use both the arithmetic average and standard deviation, but also calculate the sample median and IQR to compare and diagnose nonadditivity. For example, use $(\overline{X}_A - \hat{X}_{0.50}) \neq 0$ as a diagnostic tool. When variable PDF's are near-normal, the median and mean are similar in value; they can differ substantially for non-normal PDF's. In the latter case, transform the variable to normal (see Chap. 4) before applying estimators or re-express (if possible) the variable in terms of quantities that are additive.

5-9 WEIGHTED ESTIMATORS

Every estimator implicitly weights each data point upon which it operates. $\overline{X}_A$, for example, weights all the data equally. Many other estimators give equal influence or weight to the data points. This need not necessarily be the case, however.

There are several reasons for weighting some data more heavily than others. First, some data may be more prone to errors than others. Particularly small or large values may be poorly measured, and we especially do not want inaccurate data to have as much influence as the more precisely measured values. Second, all the data may not be equally representative. In Fig. 5-11, for example, a representative reservoir porosity would probably not be given by $(\phi_1 + \phi_2 + \phi_3 + \phi_4)/4$ because ϕ_1, ϕ_2, and ϕ_3 represent similar portions of the reservoir, the crest, while ϕ_4 represents only a portion of the flank-region porosity. In other words, the four measurements do not represent equal amounts of information concerning the large-scale reservoir porosity. Sampling points are seldom chosen at random. Third, the data may represent different portions of the whole. Measurements may be made on different volumes of rock and, therefore, need a weight that reflects the amount of material sampled.

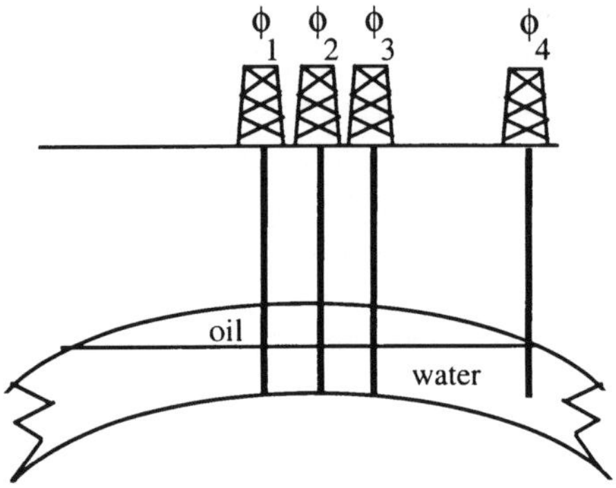

Figure 5-11. Porosity measurements from four locations in a reservoir.

If $W(X_1, X_2, \ldots, X_I)$ is an estimator, then $W(X_1, X_2, \ldots, X_I, \alpha_1, \alpha_2, \ldots, \alpha_I)$ is the weighted estimator, where the α's are the weights and

$$\sum_{i=1}^{I} \alpha_I = 1$$

is a condition to ensure unbiasedness. For example, if $\overline{X}_A = W$, then $\overline{X}_A^W$ is the weighted sample mean given by (Hald, 1952, pp. 243-245)

$$\overline{X}_A^w = \frac{1}{I}\sum_{i=1}^{I} \alpha_i X_i$$

Weighted averages occur in the Kriging estimator to be discussed in Chap. 12. The weighted sample variance is

$$\hat{s}_w^2 = \frac{1}{I-1}\sum_{i=1}^{I} \alpha_i (X_i - \overline{X})$$

When $\alpha_1 = \alpha_2 = \cdots = \alpha_I = 1$, we have the unweighted version. If we further assume that $X_1 \leq X_2 \leq \cdots \leq X_I$, then letting $\alpha_1 = \alpha_2 = \cdots = \alpha_r = \alpha_{(I-r+1)} = \alpha_{(I-r+2)} = \cdots = \alpha_I = 0$ and $\alpha_{(r+1)} = \alpha_{(r+2)} = \cdots = \alpha_{(I-r)} = I/(I - 2r)$ gives the *r-fold symmetrically trimmed* estimator. In other words, the trimmed estimator ignores the r lowest and r highest values in the data set. A variation on trimming is to *Winsorize* the data set: $\alpha_1 = \alpha_2 = \cdots = \alpha_r = \alpha_{(I-r+1)} = \alpha_{(I-r+2)} = \cdot$

$\cdots = \alpha_I = 0$ (as with trimming), but $\alpha_{(r+1)} = \alpha_{(I-r)} = r + 1$ and $\alpha_{(r+2)} = \cdots = \alpha_{(I-r-1)} = 1$. Winsorizing still ignores the r lowest and highest data, but it replaces them with r values of X_r and $X_{(I-r)}$. Asymmetrical trimming is also possible, but it is usually reserved for cases where the PDF is asymmetrical. Barnett and Lewis (1984, Chap. 3) discuss these methods further.

Trimming or Winsorizing is useful when data sets are censored and the proportions of lost data are known or when some data at the extremes are particularly corrupted. If the variances of the corrupting factors are known, the weights can be modified so that all data are included. For example, suppose data points 1, 2, . . . , r are obtained using an instrument with normal measurement errors of variance σ_1^2 while r+1, r+2, . . . , n have variability σ_2^2. A set of weights is

$$\alpha_1 = \alpha_2 = \cdots = \alpha_r = \frac{n\sigma_2^2}{r\sigma_2^2 + (n-r)\sigma_1^2}$$

and

$$\alpha_{(r+1)} = \alpha_{(r+2)} = \cdots = \alpha_I = \frac{n\sigma_1^2}{r\sigma_2^2 + (n-r)\sigma_1^2}$$

More sophisticated approaches are available, depending upon the information available about the measurement properties (Fuller, 1987).

5-10 SUMMARY REMARKS

Estimation is an important aspect of data analysis. It is important to understand what the estimator properties are so that the most appropriate estimator is used for the analysis at hand. The uncertainties of estimates can be assessed using confidence intervals.

Confidence-interval size is a strong function of data variability; a highly variable material produces estimates with large confidence intervals. Since the geological medium has variability that cannot be changed, the only methods for reducing the size of confidence intervals is to collect more data or apply an estimator that uses the data more efficiently. Chapter 6 presents a method for estimating the number of samples needed, based on the geological characterization and heterogeneity of the rock.

CHAPTER 6

MEASURES OF HETEROGENEITY

The most common statistics of a random variable are measures of central tendency, dispersion, and correlation. Chapter 5 dealt with measures of central tendency and their associated uses, along with introducing some simple measures of dispersion. This chapter further discusses measures of dispersion or variation with a special emphasis on measures used in assessing their impact on flow performance.

A measure of variability can, of course, be applied to any reservoir property. However, permeability varies far more than other properties that affect flow and displacement. Hence, in the petroleum sciences, measures of heterogeneity are almost exclusively applied to permeability data.

Heterogeneity measures are useful for a number of purposes. Since heterogeneity influences the performance of many flow processes, it is helpful to have a single statistic that will convey the permeability variation. Variabilities can be compared for geologically similar units and sampling schemes can be adjusted for the variability present. Performance models have been developed that show how permeability heterogeneity will influence a particular recovery process (e.g., Lake, 1989, pp. 411-416). Heterogeneity measures are also helpful when comparing performance for two or more fields. It should be kept in mind, however, that in summarizing variability, a considerable amount of information is lost.

Heterogeneity measures are not a substitute for detailed geological study, measurements, and reservoir analysis. They are simply one way of beginning to assess a formation unit. Most measures do not include any information about spatial arrangement and, even when they do, they tend to ignore most of the structure present.

6-1 DEFINITION OF HETEROGENEITY

Informally, we use heterogeneity and variability interchangeably. In reservoir characterization, however, heterogeneity specifically applies to variability that affects flow. Consider a high-rate displacement using matched mobility and density and chemically inert, miscible fluids. *Heterogeneity* is the property of the medium that causes the flood front, the boundary between the displacing and displaced fluids, to distort and spread as the displacement proceeds. Permeability variation is usually the prime cause of flood-front spread and distortion; for a displacement in a hypothetical homogeneous medium, the rate of distortion and spreading is zero. As the permeability variability increases, both distortion and spreading increase. Of course, the arrangement of the permeability is also important since this governs the number and size of interwell pathways. We discuss ways to measure spatial arrangement in Chap. 11.

Whatever the reservoir properties involved, heterogeneity measures can be classified into two groups, static and dynamic. *Static measures* are based on measured samples from the formation and require some flow model to be used to interpret the effect of variability on flow. *Dynamic measures* use a flow experiment and are, therefore, a direct measure of how the heterogeneity affects the flow.

Each measure type has advantages and disadvantages. For example, an advantage to dynamic measures is that, if the process used during the flow experiment closely parallels the process that is expected to be applied to the reservoir, the results are most directly applicable with a minimum of interpretation. Disadvantages include the cost, the complexity, and the selection of "representative" elements of the reservoir for conducting the flow experiments at the appropriate scale.

6-2 STATIC MEASURES

We discuss four types of static heterogeneity measures and a few of their properties.

6-2.1 The Coefficient of Variation

A static measure often used in describing the amount of variation in a population is the *coefficient of variation*,

$$C_V = \frac{\sqrt{Var(k)}}{E(k)}$$

This dimensionless measure of sample variability or dispersion, introduced in Chap. 4, expresses the standard deviation as a fraction of the mean. For data from different populations or sources, the mean and standard deviation often tend to change together such that C_V remains relatively constant. Any large changes in C_V between two samples would indicate a dramatic difference in the populations associated with those samples.

A C_V estimator, based on the sample mean and standard deviation, is $\hat{C}_V = \hat{s}/\bar{k}$. The statistical properties of $\hat{C}_V$ are not easily determined in general. Hald (1952, pp. 301- 302) gives results for samples from a normal population and Koopmans et al. (1964) give results for the log-normal case. Figure 6-1 shows these results for a sample size $I = 25$. For $C_v \leq 0.5$, the sampling variabilities from the two distributions are quite similar. When $C_v > 0.5$, where a normal PDF is not possible for a nonnegative variable, the lower and upper limits become increasingly asymmetrical about the line $\hat{C}_V = C_V$. For I samples, $\hat{C}_V \leq \sqrt{I-1}$ (Kendall and Stuart, 1977, p. 48), which is achieved when (I-1) samples are one value and one sample is another value.

Estimators other than $\hat{C}_V = \hat{s}/\bar{k}$ may be used if the PDF is known, with some increase in efficiency. For example, if k is log-normally distributed with mean μ and variance σ^2 ,

$k_A / k_H = \exp(\sigma^2)$, where k_A and k_H are the arithmetic and harmonic means (Johnson and Kotz, 1970, p. 115). Hence,

$$\hat{C}_V = \left[(k_A / k_H) - 1\right]^{1/2}$$

The coefficient of variation is being increasingly applied in geological and engineering studies as an assessment of permeability heterogeneity. C_V has been used in a study of the effects of heterogeneity and structure upon unstable miscible displacements (Moissis and Wheeler, 1990). It is also useful when comparing variabilities of different facies, particularly when there can be competing causes for permeability variation. Corbett and Jensen (1991) for example, used C_V to assess the relative effects of grain-size variation and mica content upon permeability variation. Comparisons of geologically similar elements in outcrop and subsurface showed that, despite large changes in the average permeability, the C_V's remained similar (Goggin et al., 1992; Kittridge et al., 1990). These and other studies (e.g., Corbett and Jensen, 1992a) suggest that C_V's may be transportable for elements with similar geologies. C_V's and scale may also be linked as shown in the following example.

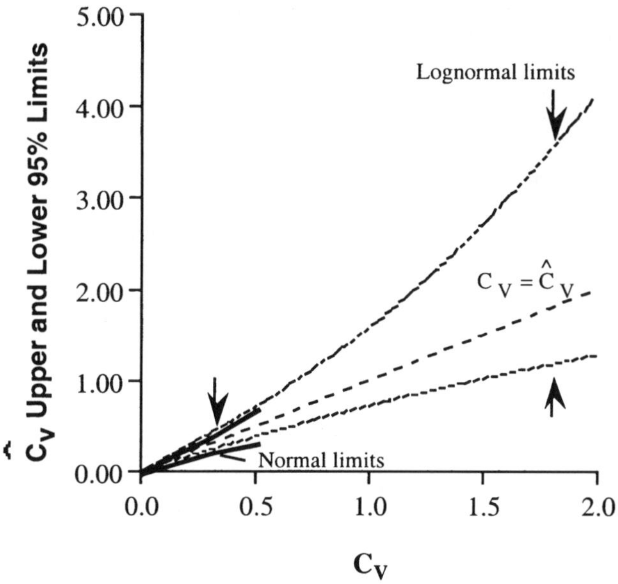

Figure 6-1. $\hat{C}_V$ variability (I = 25 samples) for normal and log-normal populations.

Example 1 - *Geological Variation and Scale.* C_V, as a measure of permeability variation, can be used to compare and contrast facies and formations. The smaller stratal elements appear to have the most uniform properties. The larger the element, the more opportunity for variability (Fig. 6-2).

In Fig. 6-2, we have not highlighted the scale (i.e., grid size) associated with each measurement. We prefer to note lamina or subfacies (e.g., grainfall, grainflow), bedform or facies (e.g., trough crossbedding), or formation (e.g., Etive, Rannoch, Rotliegendes) terminology in preference to specific dimensions. If a bedform consists of a single lamina type, or a formation consists of a single bedform type, then it is reasonable to expect similar variabilities with increasing scale (e.g., low-contrast lamination and SCS, HCS, and Rannoch). The occurrence of a rock type with a certain C_V (e.g., fluvial trough crossbeds) does not mean that they always occur with this level of heterogeneity, as has been pointed out by Martinus and Nieuwenhuijs (1995). An interesting point to note in Fig. 6-2 is that the two samples of fluvial trough crossbeds (from the same formation) have similar C_V's and, measured by different geologists, this suggests a certain transportability.

With the combination of several subfacies or bedforms, the variability increases. For example, bedforms are composed of numerous laminations. In the aeolian environment, dune bedforms are composed of wind-ripple and grainflow components, each of which has an intrinsic variability. Goggin et al. (1988) found that $C_V = 0.26$ for wind ripple (unimodal PDF) and 0.21 for grainflow (unimodal PDF) in a unit of the Page Sandstone. The combination for the dune (bimodal PDF) produced $C_V = 0.45$, approximately the sum of the grainflow and wind-ripple C_V's.

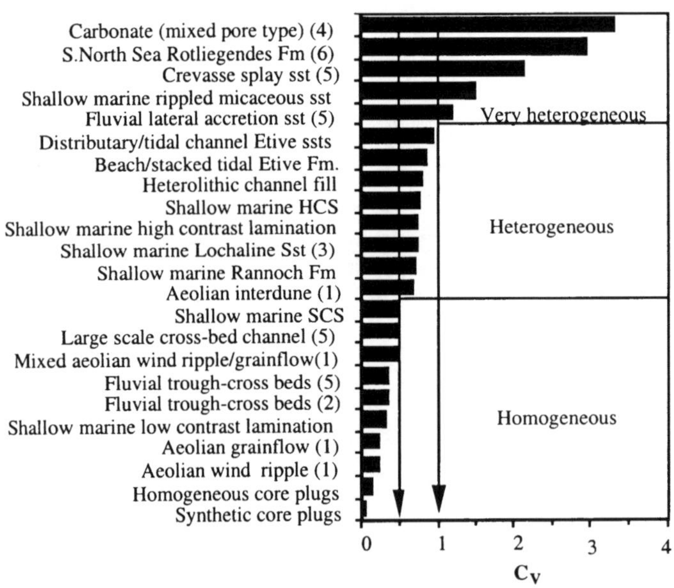

Figure 6-2. Variation of C_V with scale and depositional environment. SCS = swaley cross-stratification and HCS = hummocky cross-stratification. Homogeneous region is taken as $C_V \leq 0.5$; $0.5 < C_V \leq 1$ is heterogeneous; and $C_V > 1$ is very heterogeneous. Sources of data for this plot are (1) Goggin et al. (1988); (2) Dreyer et al. (1990); (3) Lewis and Lowden (1990); (4) Kittridge et al. (1990); (5) Jacobsen and Rendall (1991); and (6) Rosvoll, personal communication (1991).

Corbett and Jensen (1993b) observed that an HCS bedform in the shallow marine environment has the following facies variabilities: C_V = 0.2 to 0.7 for the low-contrast (low-mica) laminations; C_V = 1.0 to 1.3 for the high-contrast (high-mica) laminations; and C_V = 1.0 to 1.3 for the rippled laminations. The HCS bedform is approximately composed of 60% low-contrast rock, 35% high-contrast rock, and 5% rippled material. The weighted sum of the component C_V's gives 0.60 (0.6) + 0.35 (1.2) + 0.05 (1.2) = 0.84. This agrees well with the overall HCS C_V of 0.86.

Despite the evidence of this example, C_V's are not, in general, additive. For example, consider a formation unit composed of two elements (a mixture), each having independent random permeabilities, k_1 and k_2, and present in proportions p and $1-p$, respectively. For this mixture of random variables, let $E(k_1) = \mu_1$, $Var(k_1) = \sigma_1^2$, $E(k_2) = \mu_2$, and $Var(k_2) = \sigma_2^2$. The C_V of the unit is given by (Chap. 4, Example 4)

$$C_{V\text{total}} = \frac{\sqrt{p\sigma_1^2 + (1-p)\sigma_2^2 + p(1-p)(\mu_1 - \mu_2)^2}}{p\mu_1 + (1-p)\mu_2}$$

which is not the sum of the constituent C_V's. If we let $\mu_2 = r\mu_1$ for some factor r, $C_{V_{\text{total}}}$ can be expressed as

$$C_{V\text{total}} = \frac{\sqrt{pC_{V1}^2 + (1-p)r^2C_{V1}^2 + p(1-p)(1-r)^2}}{p + r(1-p)}$$

When the component means are identical, $r = 1$, and

$$C_{V\text{total}} = \sqrt{pC_{V1}^2 + (1-p)C_{V1}^2}$$

so the total is the root mean square of the weighted individual component variabilities. Thus, $C_{V_{\text{total}}}$ is guaranteed only to be the minimum of C_{V_1} and C_{V_2}. In most geological situations, $r \neq 1$ because the elements have differences in the energies, source materials, and other factors that produced them. When $r >> 1$ and $p < 1$,

$$C_{V\text{total}} = \sqrt{\frac{C_{V1}^2 + p}{1-p}}$$

so the variability of the larger mean population dominates and is amplified by $(1-p)^{-1/2}$. For these cases, $C_{V_{\text{total}}}$ may be near $C_{V_1} + C_{V_2}$.

One case where C_V's are additive concerns the sensitivity analysis of estimators to errors in the data (Chap. 5). We saw that random variables that are products of other random variables give an especially simple form to the sensitivity equation. For example, if m and n in Archie's law (Chap. 5) are constant, the sensitivity of S_w to errors in R_w, R_t, a, and ϕ is

$$\frac{\Delta S_w}{S_w} = \frac{1}{n}\left(\frac{\Delta R_w}{R_w} - \frac{\Delta R_t}{R_t} + \frac{\Delta a}{a} - m\frac{\Delta \phi}{\phi}\right)$$

If we interpret the Δ's as being the standard deviations of these variables, we can recast the above as

$$\frac{s_{S_w}}{S_w} = \frac{1}{n}\left(\frac{s_{R_w}}{R_w} - \frac{s_{R_t}}{R_t} + \frac{s_a}{a} - m\frac{s_\phi}{\phi}\right)$$

Each term has the form of a variability over a true value, giving it the same form as a coefficient of variation. Hence, we can express the variability of S_w in the following manner:

$$C_{V S_w} = \frac{1}{n}\left(C_{VR_w} + C_{VR_t} + C_{V_a} + mC_{V_\phi}\right)$$

Consequently, we can see at a glance which measurement(s) will contribute the most to the uncertainty in S_w. A similar treatment can be made for other equations, such as the STOIIP expression discussed in Chap. 3.

The above expression for the total C_V is an approximation that works best for small C_V's (i.e., $C_V < 0.5$). It will show the relative contributions of each component of variability. A better approximation applies if the variables are all log-normally distributed in a product. Recall from Chap. 4 that if $\ln(X) \sim N(\mu_X, \sigma_x^2)$, then $\sigma_x^2 = \ln(1 + C_{V_x}^2)$ Hence, if $\ln(Y) \sim N(\mu_X, \sigma_y^2)$ and $Z = XY$, then

$$\sigma_z^2 = \sigma_x^2 + \sigma_y^2 \text{ and } \ln(1 + C_{V_z}^2) = \ln(1 + C_{V_x}^2) + \ln(1 + C_{V_y}^2)$$

Solving for $C_{V_z}^2$ gives

$$C_{V_z}^2 = (1 + C_{V_x}^2)(1 + C_{V_y}^2) - 1 = C_{V_x}^2 + C_{V_y}^2 + C_{V_x}^2 C_{V_y}^2$$

Neglecting cross terms and extending to products of I terms, we obtain the approximation

$$C_{V_{total}}^2 = \sum_{i=1}^{I} C_{V_i}^2$$

This equation is the basis for the "parametric method" advocated by Smith et al. (1993) for assessing which variables are important in reserves estimation. The C_V of the field reserves can be easily related to the C_V of each of the input variables . In their treatment, C_V^2 is denoted κ^2.

We also note that, because $C_V = [\exp(\sigma^2) - 1]^{1/2}$ for $\ln(k) \sim N(\mu, \sigma^2)$, C_V is an estimator for the standard deviation σ when σ is small (e.g., $\sigma < 0.5$).

C_V can also be used to guide sampling density. The so-called "*N-zero method*," discussed by Hurst and Rosvoll (1991), is based on two results of statistical theory:

1. The Central Limit Theorem states that, if I_S independent samples are drawn from a population (not necessarily normal) with mean μ and standard deviation σ, then the distribution of their arithmetic average will be approximately normal.
2. The sample average will have mean μ and standard error $\sigma/\sqrt{I_S}$.

See Chap. 5 for more details. From these two points, the probability that the sample average ($\bar{k}_S$) of I_S observations lies within a certain range of the population mean (μ) can be determined for a given confidence interval.

For a 95% confidence level, the range of the average is given by $\pm\, t\, SE$, where the standard error (SE) is approximated by $\hat{s}/\sqrt{I_S}$. The larger the sample number I_S, the more confident we can be about estimates of the mean. SE is the standard deviation of the sample mean, drawn from a parent population, and is a measure of the difference between sample and population means.

The Student's t is a measure of the difference between the estimated mean, for a single sample, and the population mean, normalized by the SE. For normal distributions, the t value varies with size of sample and confidence level (Chap. 5). The above statement can be expressed mathematically as

$$\mathrm{Prob}\left[\left(\mu - t\frac{\hat{s}}{\sqrt{I_s}}\right) \le \bar{k}_s \le \left(\mu + t\frac{\hat{s}}{\sqrt{I_s}}\right)\right] = 95\%$$

Consider now another sample of size I_O such that $\bar{k}_O \pm P\%$ tolerance satisfies the predetermined confidence interval:

$$\mathrm{Prob}\left[\left(\mu - t\frac{P\bar{k}_o}{\sqrt{I_s}}\right) \le \bar{k}_o \le \left(\mu + t\frac{P\bar{k}_o}{\sqrt{I_s}}\right)\right] = 95\%$$

This time, we have expressed the permissible error in terms of a percentage of k_o. When both conditions are satisfied, $I_S = I_o$ and

$$\frac{P\bar{k}_o}{100} = t\frac{\hat{s}}{\sqrt{I_o}}$$

Rearranging this gives an expression for the appropriate number of specimens, I_o:

$$I_o = \left(\frac{100t\hat{s}}{P\bar{k}_o}\right)^2$$

where the nearest integer value for the right side is taken for I_o. For $I_S > 30$, $t = 2$ and with a 20% tolerance (i.e., the sample mean will be within ±20% of the parent mean for 95% of all possible samples, which we consider to be an acceptable limitation), this expression reduces to

$$I_o = \left(\frac{200\hat{C}_V}{20}\right)^2 \quad \text{where } \hat{C}_V = \hat{s}/\bar{k}_o$$

or

$$I_o = \left(10\hat{C}_V\right)^2$$

This rule of thumb is a simple way of determining sample sufficiency. Of course, since $\hat{C}_V$ is a random variable, I_o is also random in the above expression. I_o will change because of sampling variability. The above is called the I_o-*sampling approach*.

Although derived for the estimate of the arithmetic mean from uncorrelated samples by normal theory, we have found it useful in designing sample programs in a range of core and outcrop studies. Having determined the appropriate number of samples, the domain length (d) will determine the sample spacing (d_o) as $d_o = d / I_o$.

An initial sample of 25 measurements, evenly spaced over the domain, which can be a lamina, bedform, formation, outcrop, etc., is recommended. If $\hat{C}_V$, estimated from this sample, is less than 0.5, sufficient samples have been collected. If more are required, infilling the original with 1, 2, or J samples will give 50, 75, or $25J$ samples. In this way, sufficient samples can be collected.

Example 2 - Variability and Sample Size. In a 9-ft interval of wavy bedded material (Fig. 6-3), the performance of core plugs is compared with the probe permeameter for the estimate of the mean permeability of the interval. Based on the core plugs,

$\hat{C}_V = 0.74$, giving $I_o = 55$ and $d_o = 2$ in., well below the customary 1-ft sample spacing. The probe data give $I_o = 98$ and $d_o = 1$ in. For such variability, about 100 probe-permeameter measurements are needed for estimates within ±20% but, because of the way core plugs are taken (Chap. 1), plugs are an impractical method for adequately sampling this interval. Nine plugs, taken one per foot over the 9-ft interval, are clearly insufficient even if they are not biased towards the high-permeability intervals.

Comparing the sample means, the plug estimate is 2.3 times the probe value. Why are they so different? Equation (6-1) can be rearranged to solve for P for both data sets. In the case of the nine plug data, $\hat{C}_V = 0.74$ and $t = 2.3$, so that $P = 57\%$. The true mean is within 390 mD ± 57% about 95% of the time. A similar calculation for the probe data shows the mean to be in the range 172 mD ± 12%. Thus, the estimates are not statistically different. That is, the difference between the estimated means can be explained by sampling variability. No preferential sampling by the plugs in the better, higher-permeability material is indicated. The plugs have simply undersampled this highly variable interval.

Corbett and Jensen (1992a) have suggested that C_V's in clastic sediments could have sufficient predictability that sample numbers could be estimated by using the geological description. While further data are needed, there are indications that stratal elements of similar depositional environments have similar variabilities. For example, the two fluvial trough crossbed data sets of Fig. 6-2, obtained at different locations of the same outcrop, show very similar C_V's. Thus, applying the I_o-sampling approach, sample numbers could be predicted on the basis of the geological element (e.g., aeolian dune foreset, fluvial trough crossbed, fluvio-marine tidal channel).

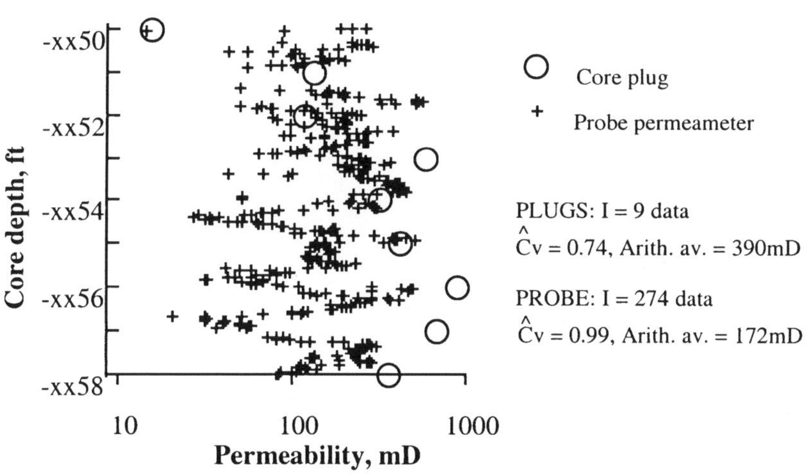

Figure 6-3. Core-plug and probe-permeameter data in a Rannoch Formation interval.

6-2.2 Dykstra-Parsons Coefficient

The most common measure of permeability variation used in the petroleum industry is V_{DP}, the Dykstra-Parsons coefficient (Dykstra and Parsons, 1950):

$$V_{DP} = \frac{k_{0.50} - k_{0.16}}{k_{0.50}} \tag{6-2}$$

where $k_{0.50}$ is the median permeability and $k_{0.16}$ is the permeability one standard deviation below $k_{0.50}$ on a log-probability plot (Fig. 6-4). V_{DP} is zero for homogeneous reservoirs and one for the hypothetical "infinitely" heterogeneous reservoir. The latter is a layered reservoir having one layer of infinite permeability and nonzero thickness. V_{DP} has also been called the *coefficient of permeability variation*, the *variance*, or the *variation*. Other definitions of V_{DP} involving permeability-porosity ratios and/or variable sample sizes are possible (Lake, 1989, p. 196).

The Dykstra-Parsons coefficient is computed from a set of permeability data ordered in increasing value. The probability associated with each data point is the thickness of the interval represented by that data point. Dykstra and Parsons (1950) state that the values to be used in the definition are taken from a "best-fit" line through the data when they are plotted on a log-probability plot (Fig. 6-4). Some later published work, however, does not mention the line and depends directly on the data to estimate $k_{0.16}$ and $k_{0.50}$ (Jensen and Currie, 1990).

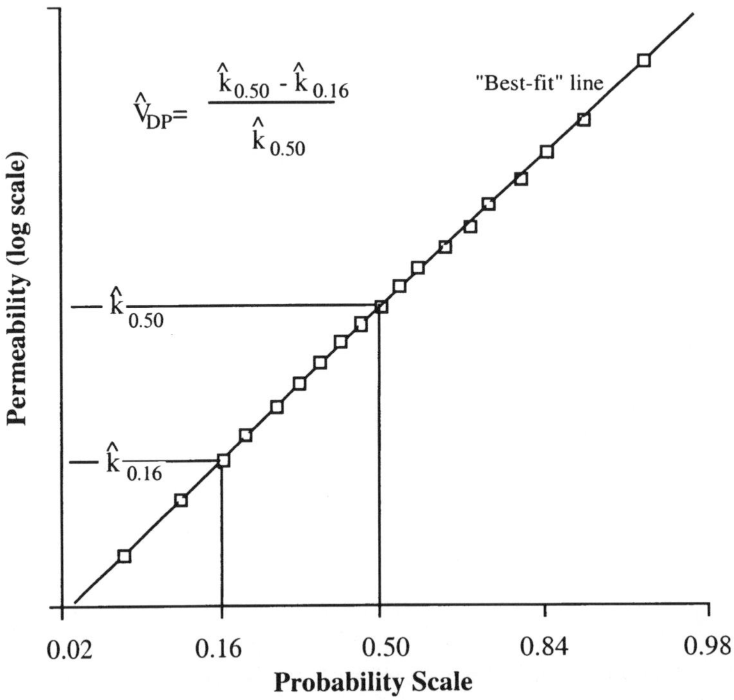

Figure 6-4. Dykstra-Parsons plot.

There are several drawbacks in using the "best fit" approach, especially if the data set is not log-normally distributed. Jensen and Lake (1988) demonstrate the nonunique properties of V_{DP} in light of an entire family of p-normal distributions. Although Dykstra and Parsons were able to correlate V_{DP} with expected waterflood performance, caution should be exercised in the universal application of V_{DP} for data sets that are not log-normal. See, for example, Willhite's (1986) Example 5.6.

From the definition for V_{DP} (Eq. (6-2)), $V_{DP} = 1 - \exp(-\sigma)$ when $\ln(k) \sim N(\mu, \sigma^2)$. This relation between V_{DP} and σ provides an alternative V_{DP} estimator: $\hat{V}^*_{DP} = 1 - \exp(-\hat{s})$, where $\hat{s}$ is the sample standard deviation of $\ln(k)$. $\hat{V}^*_{DP}$ is about twice as efficient as $\hat{V}_{DP}$ (Jensen and Currie, 1990).

A study by Lambert (1981) shows that V_{DP} estimated from vertical wells ranges between 0.65 and 0.99. V_{DP}'s measured areally (from arithmetically averaged well permeabilities) range from 0.12 to 0.93. Both the areal and vertical V_{DP}'s are normally distributed, even though most of the permeability values themselves are not normally distributed. Of greater significance, however, is the observation that V_{DP} did not apparently differentiate between formation type. This observation suggests that either V_{DP} is, by itself, insufficient to characterize the spatial distribution of permeability and/or the estimator itself is suspect. Probably both suggestions are correct; we elaborate on the latter theme in the next few paragraphs.

The simplicity of V_{DP} means that analytical formulas may be derived for the bias and standard error (Jensen and Lake, 1988), assuming $\ln(k)$ is normally distributed. The bias is given by

$$b_V = -0.749[\ln(1 - V_{DP})]^2 (1 - V_{DP}) / I$$

and the standard error is

$$s_V = -1.49[\ln(1 - V_{DP})](1 - V_{DP}) / \sqrt{I}$$

where I is the number of data in the sample. The bias is always negative ($\hat{V}_{DP}$ underestimates the heterogeneity), is inversely proportional to I, and reaches a maximum in absolute value when $V_{DP} = 0.87$. However, the bias is generally small. For example, when $V_{DP} = 0.87$, only $I = 40$ data are required to obtain $b_V = -0.009$.

The variability of $\hat{V}_{DP}$, on the other hand, can be significant (Fig. 6-5). The standard error decreases as the inverse of the square root of I and s_V attains a maximum at $V_{DP} = 0.63$. The number of data needed to keep s_V small is quite large for moderately heterogeneous formations. For example, at $V_{DP} = 0.6$, 120 samples are required to attain a precision of $s_V = 0.05$. From the definition of standard error, this value means that V_{DP} has a 68% (± one standard deviation) chance of being between 0.55 and 0.65, ignoring the small bias.

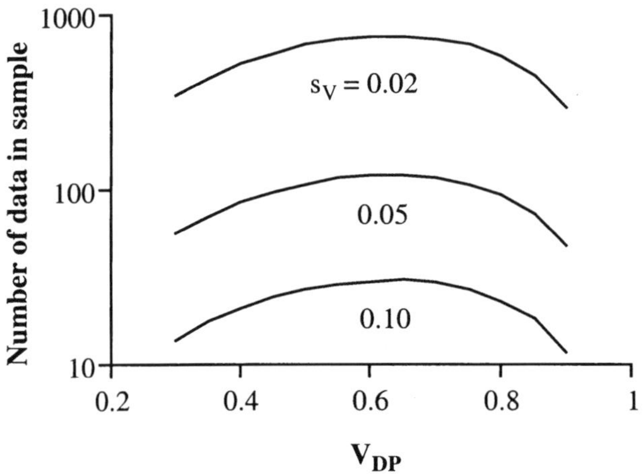

Figure 6-5. Variability performance of V_{DP} estimator. From Jensen and Lake (1988).

Example 3 - Estimating V_{DP} for a Data Set. Estimate V_{DP} for the data set in Example 4, Chap. 3: 900, 591, 381, 450, 430, 1212, 730, 565, 407, 440, 283, 650, 315, 500, 420, 714, and 324. Assume each datum represents the same volume of the reservoir. The ordered data set (X_i) and its logarithm (x_i^*) are shown in the table below, repeated from Example 4.

Figure 6-6 shows the probability plot for x^*. From the "best fit" line established by eye, $\hat{V}_{DP}$ = (493 - 308) / 493 = 0.38. These data represent a fairly homogeneous reservoir. If we assume that $\hat{V}_{DP} = V_{DP}$, then $b_V = -0.006$ and $s_V = 0.11$. While b_V is negligible, s_V is sufficiently large that rounding $\hat{V}_{DP}$ to 0.4 is appropriate.

As we have stated before, statistics such as V_{DP} can gloss over a number of details. Even in this example, with very few data, this is true. The CDF of Fig. 6-6 manifests evidence of two populations; a small-value sample that is even more homogeneous than V_{DP} = 0.4, and a large-value portion that is about as homogeneous as the entire sample.

A number of investigators have used V_{DP} to correlate oil recovery from core-scale waterfloods (see Lake and Jensen, 1991, for examples). A significant feature of many studies is the small sensitivity of models to variations in V_{DP} when $V_{DP} \leq 0.5$ while, for the large-heterogeneity cases, there is a large sensitivity. This behavior reflects, in part, that V_{DP}, with a finite range of 0 to 1, poorly discerns large-heterogeneity situations.

No.	X_i	x_i^*	Prob., p_i
1	283	5.65	0.029
2	315	5.75	0.088
3	324	5.78	0.147
4	381	5.94	0.206
5	407	6.01	0.265
6	420	6.04	0.324
7	430	6.06	0.382
8	440	6.09	0.441
9	450	6.11	0.500

No.	X_i	x_i^*	Prob., p_i
10	500	6.21	0.559
11	565	6.34	0.618
12	591	6.38	0.676
13	650	6.48	0.735
14	714	6.57	0.794
15	730	6.59	0.853
16	900	6.80	0.912
17	1212	7.10	0.971

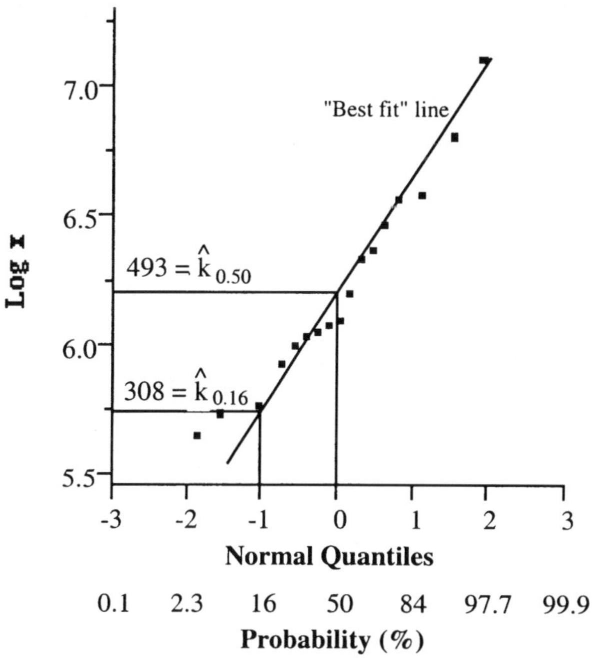

Figure 6-6. Probability plot for $\hat{V}_{DP}$ in Example 3.

6-2.3 Lorenz Coefficient

A less well-known but more general static measure of variability is the *Lorenz coefficient*, L_C. To compute this coefficient, first arrange the permeability values in decreasing order of k/ϕ and then calculate the partial sums.

$$F_J = \sum_{j=1}^{J} k_j h_j \Big/ \sum_{i=1}^{I} k_i h_i \quad \text{and} \quad C_J = \sum_{j=1}^{J} \phi_j h_j \Big/ \sum_{i=1}^{I} \phi_i h_i$$

where $1 \le J \le I$ and there are I data. We then plot F versus C on a linear graph (Fig. 6-7) and connect the points to form the Lorenz curve BCD. The curve must pass through (0, 0) and (1, 1). If A is the area between the curve and the diagonal (shaded region in Fig. 6-7), the Lorenz coefficient is defined as $L_C = 2A$. Using the trapezoidal integration rule, we have for equal-thickness layers (Lake and Jensen, 1991)

$$\hat{L}_C = \frac{1}{2I \sum_{i=1}^{I} \frac{k_i}{\phi_i}} \sum_{i=1}^{I} \sum_{j=1}^{J} \left| \frac{k_i}{\phi_i} - \frac{k_j}{\phi_j} \right|$$

Just as for V_{DP}, L_C is 0 for homogeneous reservoirs and 1 for infinitely heterogeneous reservoirs, and field-measured values of L_C appear to range from 0.6 to 0.9. However, in general, $V_{DP} \neq L_C$. For $\ln(k) \sim N(\mu, \sigma^2)$, L_C, V_{DP}, C_V , and σ are related by the following expressions (if porosity is constant):

$$L_C = \mathrm{erf}\ (\sigma/2) = \mathrm{erf}\ [-\frac{1}{2} \ln(1 - V_{DP})] = \mathrm{erf}\ [\frac{1}{2}\sqrt{\ln(1 + C_V^2)}\ ]$$

where erf() is the error function discussed in Chap. 4. L_C is also known as *Gini's coefficient of concentration* in statistical texts (Kendall and Stuart, 1977, p. 48).

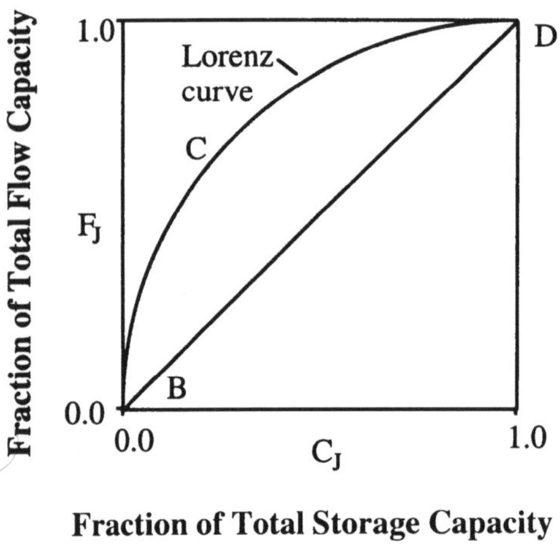

Figure 6-7. Lorenz plot for heterogeneity.

The Lorenz coefficient has several advantages over the Dykstra-Parsons

1. It can be calculated with good accuracy for any distribution. However, for the p-normal distributions, L_C is still not a unique measure of variability.

2. It does not rely on best-fit procedures. In fact, being essentially a numerical integration, there is typically less calculation error in L_C than in V_{DP}.

3. Its evaluation includes porosity heterogeneity and (explicitly) variable-thickness layers.

L_C is, however, somewhat harder to calculate than V_{DP} and has not, as yet, been directly related to oil recovery. For a reservoir consisting of I uniformly stratified elements between wells through which is flowing a single-phase fluid, the F-C curve has a physical interpretation, one of the few such instances of a physical meaning for a CDF. F represents the fraction of the total flow passing a fraction C of the reservoir volume. For example, in Fig. 6-7, approximately 80% ($F = 0.8$) of the flow is passing through 50% ($C = 0.5$) of the reservoir. This curve, therefore, plays the same role, at a large scale, that a water-oil fractional flow curve does on a small scale, which accounts for its use in the same fashion as the Buckley-Leverett theory (Lake, 1989, Chap. 5).

In addition to providing a technique for establishing a numerical coefficient of heterogeneity, the Lorenz plot also displays the petrophysical data in a useful way for reservoir engineering. Consider a standard porosity versus log-permeability crossplot (Fig. 6-8). The plot shows a broad trend of increasing permeability with porosity, but there is a broad scatter of points around this trend. The scatter is due to grain size, grain sorting and clay content variations, that are typical of braided fluvial sandstone reservoirs. A 1 mD net-pay criterion would indicate that a high proportion (75%) of this reservoir is productive pay.

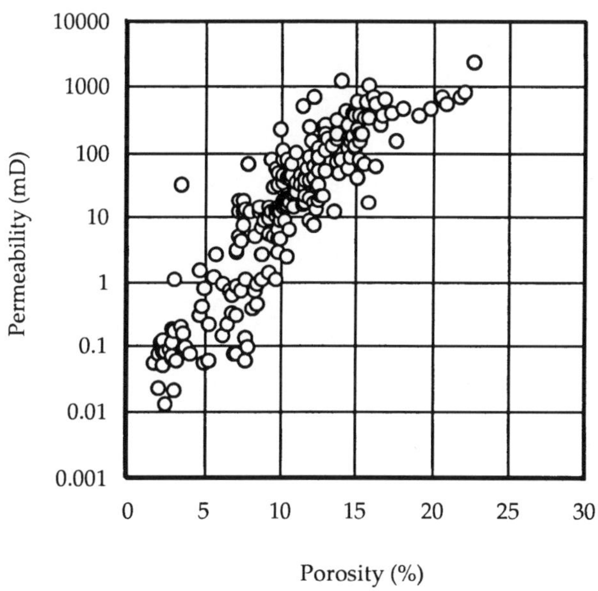

Figure 6-8. Log(permeability) versus porosity for a Triassic Sherwood Sandstone interval.

A Lorenz plot of the interval (Fig. 6-9) shows that, in addition to seeing a large variability ($\hat{L}_C = 0.76$), 80% of the transmissivity (kh) is associated with only 30% of the storativity (ϕh). The properties transmissivity and storativity are important in well test analysis and the identification of flowing intervals will affect the thickness assigned in the determination of well test permeability (Zheng et al., 2000). Net pay from Fig. 6-8 (75%) is greater than the predominant flowing interval indicated in Fig. 6-9 (30%).

The transmissivity - storativity pairs of data may be ordered stratigraphically (i.e., following their unordered depth locations, rather than sorted by permeability) following a technique introduced by Gunter et al. (1997). In the unordered data, the presence of a number of high-transmissivity intervals (thief zones in the plot) and low-transmissivity (barriers) can be seen. These correspond to regions of large and small flow, respectively. The unordered Lorenz plot is no longer a cumulative distribution function, however.

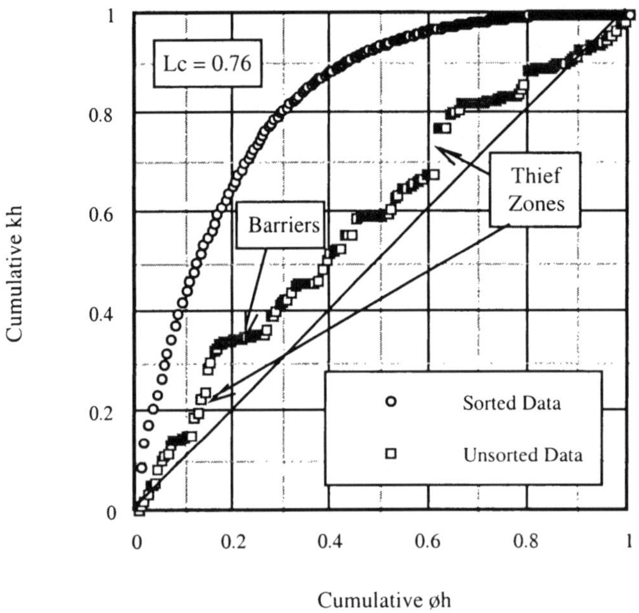

Figure 6-9. Lorenz plot for the same Sherwood Sandstone interval shown in Fig. 6-8.

It is more difficult to develop estimates of bias and precision for $\hat{L}_C$ than for $\hat{V}_{DP}$. Using numerical methods, however, we can present the results graphically, assuming ln(k) is normally distributed. Figure 6-10 presents the bias as a fraction of the true value in L_C. Each curve is for different values of L_C beginning at 0.3 (topmost curve) down to 0.9 (lowest curve). The bias can be pronounced, particularly at high L_C and small sample sizes. For example, for I = 40 data and a true L_C = 0.80, repeated measurements will actually yield an estimated L_C of 0.72 (since the bias is -0.08). $\hat{L}_C$ is, on average, lower than the true value; thus, we are again underestimating the heterogeneity.

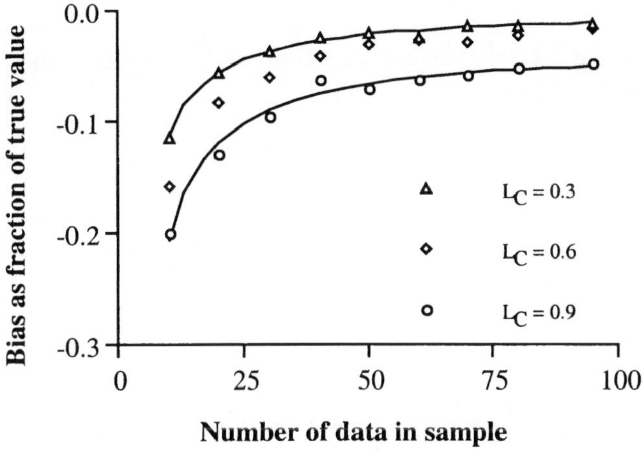

Figure 6-10. Bias performance of the Lorenz coefficient estimator. Lines are best-fit curves for $L_C = 0.3$ and 0.9. From Jensen and Lake (1988).

Figure 6-11 gives the standard error s_L for $\hat{L}_C$. For example, when $L_C = 0.8$ (neglecting the correction for bias), about 140 samples are required to determine L_C to within a standard error of 0.05. Do not try to compare this figure with the similar figure for V_{DP}, because s_L is a measure of the error in the Lorenz estimate in units of L_C and s_V measures error in units of V_{DP}. Since $V_{DP} \neq L_C$ in general, the units are not the same. For an equivalent situation, it turns out that $\hat{L}_C$ usually has a lower error than the $\hat{V}_{DP}$ estimate. See Jensen and Lake (1988) for further details.

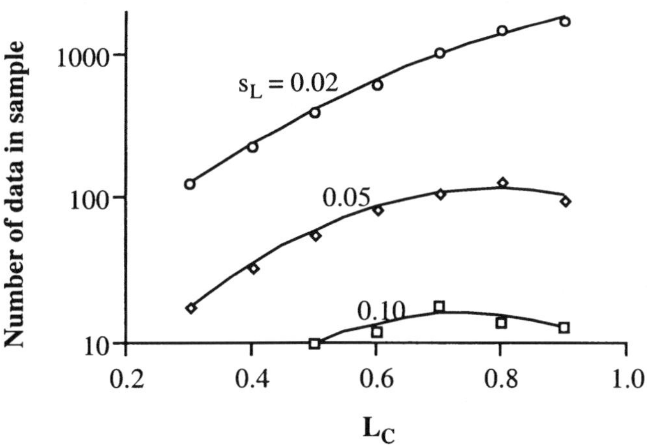

Figure 6-11. Variability of Lorenz coefficient estimates. After Jensen and Lake (1988).

6-2.4 Gelhar-Axness Coefficient

A combined static measure of heterogeneity and spatial correlation is the *Gelhar-Axness coefficient* (Gelhar and Axness, 1983), defined as

$$I_H = \sigma^2_{\ln(k)} \lambda_D$$

where $\sigma^2_{\ln(k)} = Var[\ln(k)]$ and λ_D is the autocorrelation length in the main direction along which flow is taking place, expressed as a fraction of the inlet-to-outlet spacing. λ_D is a measure of the distance over which similar permeability values exist. We discuss λ_D in more detail in Chap. 11.

I_H has the great advantage of combining both heterogeneity and structure in a quite compact form. C_V, V_{DP}, and L_C all ignore how the permeability varies in the rock; regions of high and low permeability could be far apart or quite close and yet have the same C_V, V_{DP}, and L_C. As we have discussed, however, C_V with the geological description does appear to have potential for guiding sampling strategies. I_H begins to incorporate a numerical measure of the geological organization present in the rock. As with V_{DP}, I_H asumes $\ln(k)$ is normally distributed.

Numerical simulation work by Waggoner et al. (1992) and Kempers (1990) indicates that I_H is a much improved indicator of flow performance compared to V_{DP}. However, it cannot be satisfactory for all cases since, in the limit of $\lambda_D \to \infty$ (uniform, continuous layers), the flow must again be governed by V_{DP}. Sorbie et al. (1994) discuss some limitations of I_H, particularly as it relates to prediction of flow regimes for small λ_D.

6-3 DYNAMIC MEASURES

In principle, these measures should be superior to static measures since they most directly characterize flow. In practice, dynamic measures are difficult to infer and, like the static measures, are unclear as to their translation to other systems and scales.

The measures we discuss are based on end-member displacements. A displacement that has an uneven front is called *channeling* and its progress is to be characterized by a Koval factor. If it has an even front it is *dispersive* and is characterized by a dispersion coefficient. See Chap. 13 and Waggoner et al. (1992) for more details.

6-3.1 Koval Factor

The *Koval heterogeneity factor* H_K is used in miscible flooding to empirically incorporate the effect of heterogeneity on viscous fingering (Koval, 1963). It is defined as the reciprocal of the dimensionless breakthrough time in a unit-mobility-ratio displacement:

$H_K = 1/t_{D_{breakthrough}}$

It follows from this that the maximum oil recovery has been obtained when H_K pore volumes of fluid have been injected:

$t_{D_{sweepout}} = H_K$

In these equations, t_D is the volume of displacing fluid injected divided by the pore volume of the medium. *Breakthrough* means when the displacing fluid first arrives at the outlet and *sweepout* means that the originally resident fluid is no longer being produced. For a homogeneous medium, both occur at $t_D = 1$ and $H_K = 1$, but there is no upper limit on H_K. Obviously, large values of H_K are detrimental to recovery.

If we interpret the above response as occurring in a uniformly layered medium with a log-normal permeability distribution, then H_K and V_{DP} are empirically related by

$\log_{10}(H_K) = V_{DP}/(1 - V_{DP})^{0.2}$

which plots as Fig. 6-12.

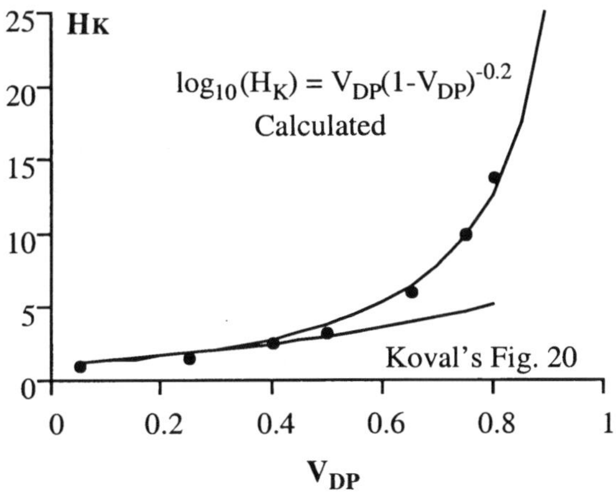

Figure 6-12. Koval factor and Dykstra-Parsons coefficient for a uniformly layered medium. From Paul et al. (1982).

H_K has many of the same problems as V_{DP} and L_C, but it is a far more linear measure of performance, being bounded between zero and infinity. If the medium is not uniformly layered, H_K can be related to I_H, the Gelhar-Axness coefficient, as in Fig. 6-13.

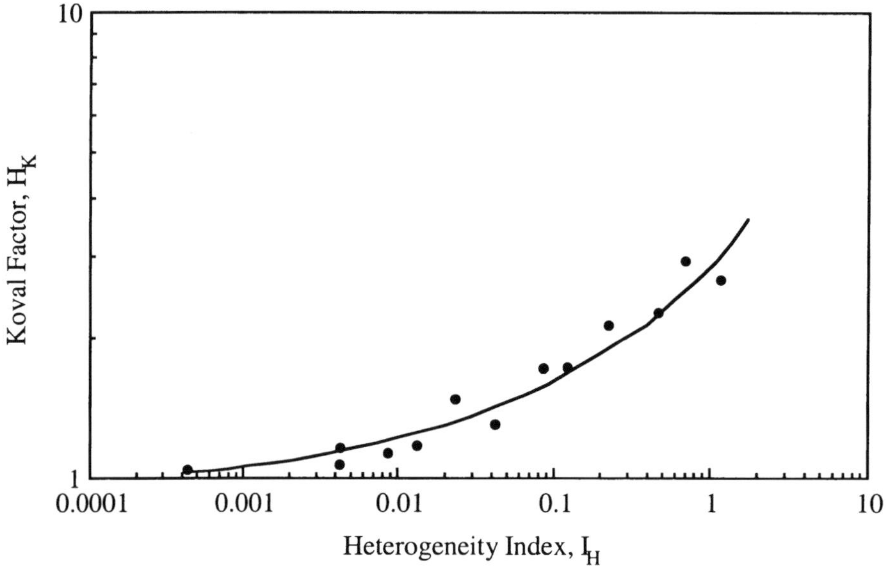

Figure 6-13. Relationship between Koval factor and Gelhar-Axness coefficient. From Datta Gupta et al. (1992).

6-3.2 Dispersion

If a miscible displacement proceeds through a one-dimensional "homogeneous" permeable medium, its concentration at any position x and time t is given by

$$C(x, t) = \frac{1}{2}\left[1 - \text{erf}\left(\frac{x - vt}{2\sqrt{K_l t}}\right)\right]$$

where v is the mean interstitial velocity of the displacement. The most important parameter in this expression is K_l, the *longitudinal dispersion coefficient.* K_l has been found to be proportional to v according to

$$K_l = \alpha_l v$$

where α_l is the *longitudinal dispersivity*, an intrinsic property of the medium. Clearly, the degree of spreading depends on the dispersivity; the larger α_l, the more spreading occurs.

The derivation and implications of the equation for $C(x,t)$ are too numerous to discuss here (see Lake, 1989, pp. 157-168). However, both H_K and α_l are clearly manifestations of heterogeneity in the medium because there would be no spreading of the displacement front if the medium were truly homogeneous. Indeed, several attempts have been made to relate α_l to the statistical properties of the medium. In the limiting case of small autocorrelation and heterogeneity, α_l divided by the length of the medium is proportional to I_H (Arya et al., 1988). This should come as no surprise if we view such dispersion as a series of uncorrelated particle jumps during the displacement. In such circumstances, the average particle position, now being the result of a series of such jumps, should take on a Gaussian character as required by the Central Limit Theorem (Chap. 4) and suggested by the error function in the $C(x,t)$ equation.

Unfortunately, a substantial amount of data collected in the field suggests that α_l depends on the scale of the measurement. Such behavior is called *non-Fickian dispersion*. In such cases, α_l can still be related to the statistical properties of the medium (Dagan, 1989, Chap. 4; Gelhar, 1993), but the connection is much more involved and its utility in subsequent applications is reduced.

Datta Gupta et al. (1992) have shown that the concentration profile of miscible displacements through a correlated medium behaves as a truncated Gaussian distribution, becoming more Gaussian as the scale of correlation decreases with respect to the length of the medium. Adopting this point of view, the dispersion relation (with constant α_l) and Koval approach (with constant H_K) represent the extremes of small and large spatial correlation, respectively. (If we applied the Koval approach to uncorrelated media, we would find that H_K depends on scale.) Nevertheless, both approaches give material properties (α_l or H_K) that are related to the heterogeneity of the medium.

6-4 SUMMARY REMARKS

The progression of this chapter mimics the progress of the entire book in a way. We have begun by discussing heterogeneity measures and their properties. But during the dynamic measure discussion, we could not avoid speaking about autocorrelation or spatial arrangement. This is a subject we will return to later in the book. But we must depart this chapter with a bit of foreshadowing.

The subject of autocorrelation (and its existence in permeable media properties) means that measures of heterogeneity must be scale-dependent. Indeed, laborious measurements in the field have shown this to be the case (Goggin et al., 1988). There are entire statistical tools devoted to such measurements, and these tools frequently combine spatial arrangement and heterogeneity. Chapter 11 contains a discussion of these, along with the implications for the geological character.

Spatial arrangement implies some force or causality behind the observation. Since we are dealing with naturally occurring media, a study and understanding of this causality must deal with the geologic reasons behind the observations. Indeed, we shall find that geologic insight can frequently unravel a statistical conundrum, allow us to fill in missing data, or even indicate that a statistical approach is not necessary. Such a combination is most powerful and forms an underlying theme for this book.

CHAPTER 7

HYPOTHESIS TESTS

Every computation produces an estimate of some property. Given unlimited resources, we could very accurately establish the exact value of that property using the entire volume of material we wish to assess. Since resources are limited, the values we calculate use modest sample numbers and sample volumes, may differ from the exact value because of sampling variability. That is, if we could obtain several sets of measurements, the estimates would vary from set to set because each estimate is a function of the data in that set. An estimate will have an associated error range, the *standard error*, which gives an idea of the precision of the estimate attributable to sampling fluctuation. The standard error is based on the number and variability of the measurements and does not take into account biased sampling procedures or other inadequacies in the sampling program. Chapter 5 described some methods for determining standard errors.

This chapter is about making comparisons involving one or more random variables. A hypothesis test (or a confidence test) is a formal procedure to judge whether an estimate is different, in a statistical sense, from another quantity. That latter quantity can be a second estimate or a number ("truth") obtained by another method. We will use several methods for comparing estimates. Whatever the technique used, the comparison is made to answer the question: can the difference in value between the two quantities be explained by sampling variability? If the answer is yes, then we say the two quantities are not statistically different.

In many cases, hypothesis tests may be unnecessary because they do not answer the appropriate question. Or they only formalize what we already know. For example, suppose we have two laterally extensive facies, f_1 and f_2, with f_1 stratigraphically above f_2. For the purposes of developing a numerical flow model, we want to know whether they can be combined or if they should be modeled separately. Assuming that the arithmetic average is the appropriate average for flow in each of these facies, we calculate the average permeabilities and standard errors from core samples and obtain the results

facies f_1: $\overline{k_1} = 30 \pm 10$ mD
facies f_2: $\overline{k_2} = 100 \pm 20$ mD

Do we need a hypothesis test to establish that f_1 and f_2 are statistically different? No! A simple sketch shows that the 95% confidence intervals do not intersect (Fig. 7-1).

So the statistical significance of the averages is clear: f_1 and f_2 have average permeabilities that are different at the 95% level. What is less evident is whether, under the desired flow process, separating the facies into two strata will improve the predictions made by the numerical model. Answering this issue is beyond the ability of hypothesis testing.

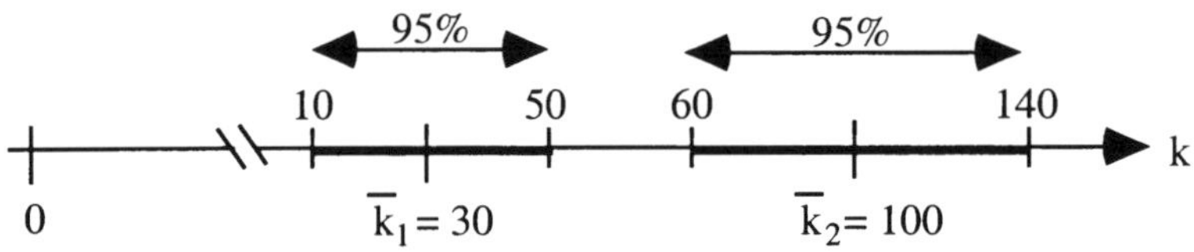

Figure 7-1. Two arithmetic averages and their 95% confidence intervals (twice their standard errors).

Hypothesis tests are also unsuitable for analyzing noncomparable quantities. One indicator of comparability is whether the two quantities have the same units (dimensions), but this is only a partial test. In addition to this, the quantities should be additive (see Chaps. 3 and 5) because hypothesis tests assume that differences in estimates are meaningful. For example, consider the two regions shown in Fig. 7-2. If they have equal areas of the front and back faces for injection or production and no-flow boundaries otherwise, they have identical abilities to transmit a single phase. Thus, while a hypothesis test may indicate that the permeabilities of the two regions are statistically different, their transmissibilities are equal. This problem arises because permeability is an intensive property and, therefore, is not additive. Flow resistance or conductance, however, may be additive (Muskat, 1938, pp. 403-404) so that, with similar boundary conditions, the sample statistics of these variables may be comparable. It is easy to overlook this aspect of hypothesis testing in the heat of the statistical battle.

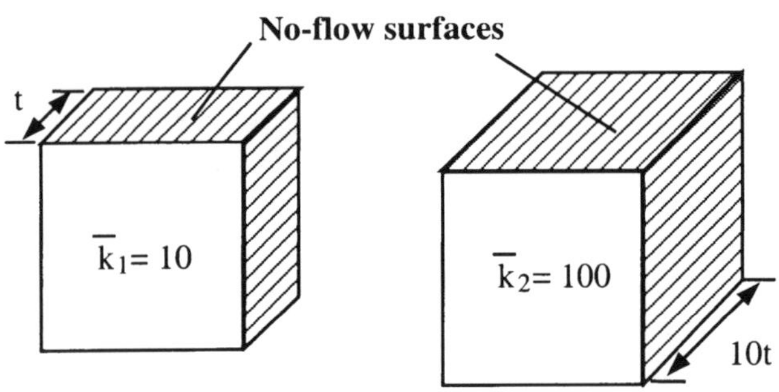

Figure 7-2. Two samples with differing average permeabilities but similar flow properties.

Statisticians have implicitly recognized the need for additivity in hypothesis testing. Their concern, however, usually focuses on knowing the PDF of the difference of the quantities to be compared. For example, problems are often expressed so that two normally distributed variables are compared because the difference of two uncorrelated normal variables is also normal. Besides the normal, the χ^2, Poisson, and other distributions also have this property (see Cramér, 1946; Johnson and Kotz, 1970; and Feller, 1971), but most other PDF's do not. Because the χ^2 PDF is additive and variances of independent variables are additive, hypothesis tests on variabilities compare sample variances rather than standard deviations.

Here, we will take the approach discussed in Chap. 5 and require only local additivity (i.e., small perturbations are additive even though large perturbations may not be additive).

Hypothesis tests are similar in some ways to the judicial process that exists in many countries. When someone is accused of committing a crime, a trial is held to determine the guilt or innocence of the accused. The accused is assumed to be innocent until a jury decides. At the conclusion of the trial, there are usually four possible results to this process:

1. the accused is actually guilty and is found guilty;
2. the accused is actually innocent and is found innocent;
3. the accused is actually innocent but is found guilty; and
4. the accused is actually guilty but is found innocent.

In the first result, the assumption (innocence) was wrong and was found to be wrong. In the second result, the assumption was right and was confirmed. In both cases, the outcomes reflected the true status of the accused and justice was served. In the last two possibilities, however, mistakes were made and the outcomes were wrong. Different countries handle these errors differently. The judicial process may be adjusted to minimize outcome 3 at the expense of increasing the occurrences of outcome 4 by, for example, increasing the burden of proof required for a conviction.

For hypothesis tests, we start with an assumption that is called the *null hypothesis*. This hypothesis is that sampling variability can explain the difference between two estimates. For example, facies f_1 and f_2 have the same mean permeability and $\overline{k_1} \neq \overline{k_2}$ because of sampling variation. The *alternative hypothesis* is that the difference cannot be explained by sampling variability and some other cause (unspecified) must be in effect. We then apply the hypothesis test and obtain a result that either upholds the null hypothesis or shows that the alternative is indicated. Just as in the judicial process, mistakes may result. We may decide the difference in two quantities cannot be explained by sampling variability when, if the truth could be known, it could be so explained (judicial outcome 3). This type of mistaken conclusion is called a *type-1 error*. A *type-2 error* occurs when the null hypothesis is accepted when, in fact, the alternative applies (judicial outcome 4). Before we apply a hypothesis test, we usually stipulate the probability of avoiding a type-1 error, and this probability is called the *confidence level* of the test.

Depending on the hypothesis test used, the number of errors will vary. Different tests may give different outcomes. The choice of test depends upon the information available, the estimates to be tested, and the consequences of making errors. We will only consider the properties of a few hypothesis tests that are commonly used to analyze geoscientific and geoengineering data. We will assume that, when comparisons are made between different sample statistics, the statistics are based on independent samples. Extensive treatments are given by Rohatgi (1984), Snedecor and Cochrane (1980), and Rock (1988). Cox and Hinkley (1974) discuss the theoretical underpinnings of hypothesis tests.

7-1 PARAMETRIC HYPOTHESIS TESTS

These tests depend on knowing the form of the PDF from which the data come. Common tests involve the normal, binomial, Poisson, and uniform distributions. The t and F tests are particularly common; they assume the data come from normal PDF's.

7-1.1 The *t* Test

Populations A and B are assumed normal with means and variances (μ_A, σ_A^2) and (μ_B, σ_B^2). We begin with the case $\sigma_A^2 = \sigma_B^2 = \sigma^2$ and deal later with the case $\sigma_A^2 \neq \sigma_B^2$. From the data sets having I_A and I_B data, respectively, we compute ($\bar{X}_A$, $\hat{s}_A^2$) and ($\bar{X}_B$, $\hat{s}_B^2$) and make the null hypothesis H_0 that the two population means are equal,

null hypothesis, H_0: $\mu_A = \mu_B$

with the alternative hypothesis H_A that the two means are not equal,

alternative hypothesis, H_A: $\mu_A \neq \mu_B$

If we can demonstrate statistically that, at some predetermined confidence level α, the null hypothesis is false, then we have found a "significant difference" in the sample means ($\bar{X}_A$ - $\bar{X}_B$). That is, it is a difference that, at the given level of probability, cannot be explained by sampling variability alone. The roles of the hypotheses and confidence level are quite important because they govern what question the test will answer and with what probability type-1 errors will occur. ($\bar{X}_A$ - $\bar{X}_B$) has a t distribution with (I_A + I_B - 2) degrees of freedom and a variance of $\sigma^2(I_A^{-1} + I_B^{-1})$. Since H_A is satisfied if either $\mu_A > \mu_B$ or $\mu_A < \mu_B$, we must consider both possibilities when choosing the desired confidence level. That is, there is a probability of $\alpha/2$ that $\mu_A > \mu_B$ and a probability of $\alpha/2$ that $\mu_A < \mu_B$. Such a situation requires a "two-tailed" t test.

The appropriate statistic is

$$t = \frac{\bar{X}_A - \bar{X}_B}{\sqrt{\hat{s}^2\left(\dfrac{1}{I_A} + \dfrac{1}{I_B}\right)}}$$

where

$$\hat{s}^2 = \frac{(I_A - 1)\hat{s}_A^2 + (I_B - 1)\hat{s}_B^2}{I_A + I_B - 2}$$

We reject H_0 if

$$\text{Prob}[\,|t| > t(\alpha/2, I_A + I_B - 2)\,] > \alpha$$

and we accept H_0 if

$$\text{Prob}[\,|t| > t(\alpha/2, I_A + I_B - 2)\,] \le \alpha$$

In terms of confidence intervals, H_0 is accepted if

$$-t_0 \sqrt{\hat{s}^2\left(\frac{1}{I_A} + \frac{1}{I_B}\right)} < (\bar{X}_A - \bar{X}_B) < +t_0 \sqrt{\hat{s}^2\left(\frac{1}{I_A} + \frac{1}{I_B}\right)}$$

where $t_0 = |t(\alpha/2, I_A + I_B - 2)|$. If the alternative H_A had admitted only one inequality (e.g., $\mu_A > \mu_B$), then the critical t value would be $t_0 = t(\alpha, I_A + I_B - 2)$.

Example 1a - Comparing Two Averages (Equal Variances Assumed). Consider the following values:

$\bar{X}_A = 50$	$\hat{s}_A^2 = 5$	$I_A = 25$
$\bar{X}_B = 48$	$\hat{s}_B^2 = 3$	$I_B = 30$

For $\alpha = 0.05$ (95% confidence level), $t(\alpha/2, df) = 2.006$ from a table of t values (Table 5-2).

$$\hat{s}^2 = \frac{(24)5 + (29)3}{25 + 30 - 2} = 3.91$$

and

$$t = \frac{50 - 48}{\sqrt{3.91\left(\frac{1}{25} + \frac{1}{30}\right)}} = 3.74$$

Since the absolute value of t is greater than 2.006, we reject H_0 in favor of H_A and conclude with 95% confidence that the means are not equal. Recall from Chap. 5, $\mu = \bar{X} \pm t(\alpha/2, df)s/\sqrt{I}$, and this situation can be represented as in Fig. 7-3.

Unfortunately, errors are the inevitable byproduct of hypothesis testing. No matter how careful we are in selecting the form of the null hypothesis and associated confidence level, we always run the risk of rejecting H_0 when it is true or accepting H_0 when H_A is true. Once an inference is made, there is no way of knowing whether an error was committed; we can only.

compute the probability of committing such an error using the t distribution. In fact, since probabilities and frequencies are equivalent, we can say that if we repeat the above test many times, we will find the sample means to be different 95 times out of 100 trials.

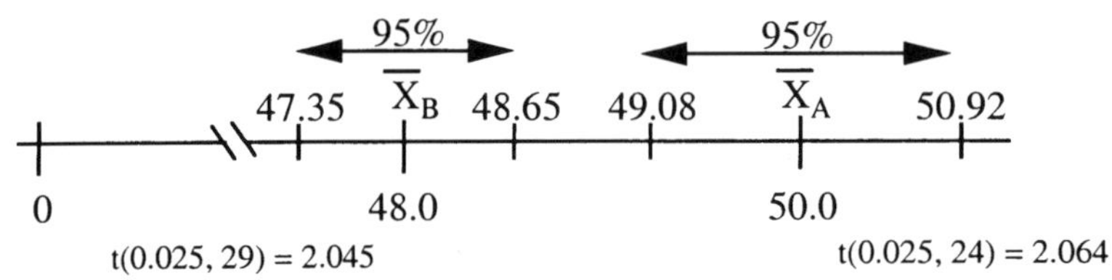

Figure 7-3. Relationships of averages and their 95% confidence limits. Note that $2.045\sqrt{3} \div \sqrt{30} = 0.65$ and $2.064\sqrt{5} \div \sqrt{25} = 0.92$.

The above analysis is largely unchanged when $\sigma_A^2 \neq \sigma_B^2$ except that we can no longer combine the variance estimates $\hat{s}_A^2$ and $\hat{s}_B^2$ to give $\hat{s}^2$ and we have to adjust the degrees of freedom. The appropriate statistic is

$$t = \frac{\overline{X}_A - \overline{X}_B}{\sqrt{\frac{\hat{s}_A^2}{I_A} + \frac{\hat{s}_B^2}{I_B}}}$$

We reject H_0 if $\text{Prob}[\,|t| > t(\alpha/2, df)\,] > \alpha$ and accept H_0 if $\text{Prob}[\,|t| > t(\alpha/2, df)\,] \leq \alpha$ with

$$df = \frac{\left(\frac{\hat{s}_A^2}{I_A} + \frac{\hat{s}_B^2}{I_B}\right)^2}{\frac{\hat{s}_A^4}{I_A^2(I_A - 1)} + \frac{\hat{s}_B^4}{I_B^2(I_B - 1)}}$$

as an "adjusted" degrees of freedom that has been devised to allow us to continue using the conventional t tables (Satterthwaite, 1946).

Example 1b - Comparing Two Averages (Unequal Variances Assumed). Suppose that we have the same data as before but the populations are no longer presumed to have the same variance:

$\overline{X}_A = 50$ $\qquad \hat{s}_A^2 = 5$ $\qquad I_A = 25$

$\overline{X}_B = 48$ $\qquad \hat{s}_B^2 = 3$ $\qquad I_B = 30$

$$df = \frac{\left(\frac{5}{25} + \frac{3}{30}\right)^2}{\frac{5^2}{25^2\,(25 - 1)} + \frac{3^2}{30^2\,(30 - 1)}} = \frac{0.090}{0.0020} = 44.7 \approx 44$$

For $\alpha = 0.05$ (95% confidence level), $t(\alpha/2, df) = 2.015$ (from Table 5-2) and

$$t = \frac{50 - 48}{\sqrt{\frac{5}{25} + \frac{3}{30}}} = 3.65$$

Since the absolute value of t is greater than 2.015, we reject H_0 and accept H_A and conclude with 95% confidence that the means are not equal. Or 95 times out of 100 we will be correct when we conclude the means are not equal.

In the t test, additional information has direct value in terms of the number of data required. For example, in Example 1b, the number of degrees of freedom decreases from 53 to 44 when we remove the information that $\sigma_A^2 = \sigma_B^2$, increasing slightly the possibility of a type-1 error (i.e., a computed $|t|$ that is closer to $t(\alpha/2, df)$). This is in keeping with the intuitive notion that information has value. In this case, the knowledge that $\sigma_A^2 = \sigma_B^2$ is worth 9 data points. Similarly, if we knew that $\mu_A > \mu_B$ was impossible from some physical argument (e.g., a law of physics, thermodynamics, or geology), we could use a one-tailed t test instead of a two-tailed test. In Example 1a, the critical t value, $t(0.05, 53)$, would then be 1.67 instead of 2.006. Alternatively, to have a critical t-value of 2.006, the df could be reduced to 6, so we would need only 8 data ($df = I_A + I_B - 2$). The information that $\mu_A > \mu_B$ can be excluded is therefore worth 45 data points in this case.

The t-test theory is developed on the assumption of normality. It has been found, however, that the test is rather insensitive to deviations from normality (Miller, 1986). For the arithmetic average, the Central Limit Theorem (Chap. 4) suggests that $\overline{X}_A \left(= \sum_{i=1}^{I} X_i / I \right)$ will be much closer to normal than the population from which the samples X_I are taken. This argument, however, may not apply when other quantities, such as the geometric ($\overline{X}_G$) or harmonic ($\overline{X}_H$) averages, are being compared. In that case, transformations might be useful. For example, if the X_i are log-normally distributed, $\ln(\overline{X}_G)$ has a normal PDF, so comparing $\ln(\overline{X}_G)$'s is appropriate. Similarly, $\overline{X}_H^{-1}$ is likely to be more nearly normal than $\overline{X}_H$ because of the central limit effect, so the inverse of harmonic averages may be compared with the t test.

If transformed quantities are being compared, however, the standard deviations also must pertain to the transformed quantity. Box et al. (1978, pp. 122-123) give an example of applying a t test to a log-transformed data set. Transformation before the t test, however, will not work if one average is being compared with another type of average, e.g., $\overline{X_A}$ from one facies (approximately normal) compared with $\overline{X_H}$ of another facies (non-normal) for some property X. If severe nonnormality of the two quantities is suspected and they really are comparable, a nonparametric method, such as the Mann-Whitney-Wilcoxon test (see below) may be used.

When three or more quantities are to be compared, it may be tempting to apply the t test pairwise to the estimates. In general, this is not advisable because of the increasing chance of a type-1 error. If we compare two estimates at the 95% confidence level, there is a 1 - 0.95 = 0.05 probability that H_0 is satisfied but H_A is indicated. If we compare three estimates pairwise at the 95% level, there is a $1 - (0.95)^3 = 0.14$ probability or a 14% chance that H_0 is satisfied but H_A will be indicated by at least one pair of estimates. The preferred approach to testing several estimates is to use the F test. However, the elementary application of the F test is to compare population variances, a subject we deal with next.

7-1.2 The *F* Test

We begin again with the now familiar assumption that populations A and B are normally distributed with means and variances (μ_A, σ_A^2) and (μ_B, σ_B^2). From the data sets, we compute $(\overline{X}_A, \hat{s}_A^2)$ and $(\overline{X}_B, \hat{s}_B^2)$ and we make the null hypothesis H_0 that the two population variances are equal

null hypothesis, H_0: $\sigma_A^2 = \sigma_B^2$

with the alternative hypothesis H_A that the two variances are not equal

alternative hypothesis, H_A: $\sigma_A^2 \neq \sigma_B^2$

The appropriate statistic is

$$F = \hat{s}_A^2 / \hat{s}_B^2$$

where $\hat{s}_A^2 > \hat{s}_B^2$. We reject H_0 if $\text{Prob}[F > F(I_A - 1, I_B - 1, \alpha)] > \alpha$ and accept H_0 if $\text{Prob}[F > F(I_A - 1, I_B - 1, \alpha)] \leq \alpha$. We assume that $\hat{s}_A^2 > \hat{s}_B^2$ because this makes $F \geq 1$, and that simplifies the standard tables. There is one table for each value of α. A limited portion of a standard table is given in Table 7-1. More extensive tables are given in Abramowitz and Stegun (1965) and many other statistics books. Values for F at given degrees of α are also part of many spreadsheet libraries.

Example 1c - Comparing Two Sample Variances. Let us reconsider the previous data to compare the sample variances.

$\overline{X}_A = 50$ $\qquad$ $\hat{s}_A^2 = 5$ $\qquad$ $I_A = 25$

$\overline{X}_B = 48$ $\hat{s}_B^2 = 3$ $I_B = 30$

Let H_0 be $\sigma_A^2 = \sigma_B^2$ and $F(I_A - 1 = 24,\ I_B - 1 = 29,\ \alpha = 0.05) = 1.91$

$$F = \hat{s}_A^2 / \hat{s}_B^2 = 5/3 = 1.67$$

Since 1.67 < 1.91, the critical value, H_0, is accepted. Therefore, the values $\hat{s}_A^2 = 5$ and $\hat{s}_B^2 = 3$ do not indicate that the underlying population variances are unequal. They may, in fact, be unequal, but the variability in the sample variances is sufficiently large to mask the difference. Further samples or other information (e.g., geological provenance of the samples) are needed to make this assessment.

Table 7-1. Critical values for the F test with $\alpha = 0.05$.

I_A-1 → ---------- **I_B-1 ↓**	**1**	**2**	**3**	**4**	**5**	**6**	**8**	**10**	**20**	**30**	**100**	**500**	∞
1	161	200	216	225	230	234	239	242	248	250	253	254	254
2	18.5	19.0	19.2	19.3	19.3	19.3	19.4	19.4	19.4	19.5	19.5	19.5	19.5
3	10.1	9.55	9.28	9.12	9.01	8.94	8.85	8.79	8.66	8.62	8.56	8.54	8.53
4	7.71	6.94	6.59	6.39	6.26	6.16	6.04	5.96	5.80	5.75	5.66	5.64	5.63
5	6.61	5.79	5.41	5.19	5.05	4.95	4.82	4.74	4.56	4.50	4.40	4.37	4.36
6	5.99	5.14	4.76	4.53	4.39	4.28	4.15	4.06	3.87	3.81	3.71	3.68	3.67
8	5.32	4.46	4.07	3.84	3.69	3.58	3.44	3.35	3.15	3.08	2.98	2.94	2.93
10	4.96	4.10	3.71	3.48	3.33	3.22	3.07	2.98	2.77	2.70	2.59	2.55	2.54
20	4.35	3.49	3.10	2.87	2.71	2.60	2.45	2.35	2.12	2.04	1.90	1.85	1.84
30	4.17	3.32	2.92	2.69	2.53	2.42	2.27	2.16	1.93	1.84	1.69	1.64	1.62
100	3.94	3.09	2.70	2.46	2.30	2.19	2.03	1.92	1.68	1.57	1.39	1.30	1.28
400	3.86	3.02	2.62	2.39	2.23	2.12	1.96	1.85	1.60	1.49	1.28	1.16	1.13
∞	3.84	3.00	2.60	2.37	2.21	2.10	1.94	1.83	1.57	1.46	1.24	1.11	1.00

Example 2 - Comparing Dykstra-Parsons Coefficients. Consider the case where two Dykstra-Parsons coefficients (V_{DP}'s) have been estimated from two wells in a field: $\hat{V}_{DP1} = 0.70$ (from $I_1 = 31$ core-plug samples) and $\hat{V}_{DP2} = 0.80$ (from $I_2 = 41$ core-plug samples). We want to determine whether the difference in V_{DP} can be explained by sampling variability at the 5% level.

Since V_{DP} is being used to assess variability, we assume that permeability is log-normally distributed (see Chap. 6). The standard error of V_{DP} estimates is given by

$$\hat{s}^2 = -1.49[\ln(1 - V_{DP})](1 - V_{DP})]/\sqrt{I}$$

where V_{DP} is the "true" population value (Jensen and Lake, 1988). Here we will assume that $V_{DP} = 0.75$, a value between the two estimates and in keeping with our null hypothesis that $\hat{V}_{DP_1}$ and $\hat{V}_{DP_2}$ are not statistically different. The standard errors for the two estimates are

$$\hat{s}^2_{V_1} = -1.49[\ln(1-0.75)](1-0.75)/\sqrt{31} = 0.093$$

and

$$\hat{s}^2_{V_2} = -1.49[\ln(1-0.75)](1-0.75)/\sqrt{41} = 0.081$$

A simple sketch shows that the 95% confidence intervals (of size ± 2 standard errors) for these estimates overlap considerably (Fig. 7-4). Hence, no statistically significant difference appears to exist.

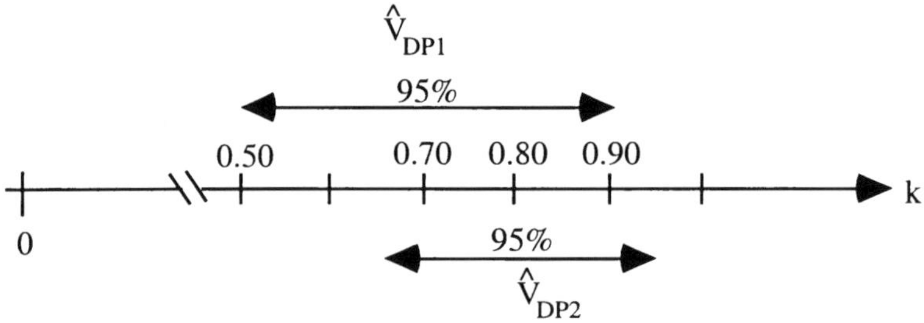

Figure 7-4. Confidence intervals for two Dykstra-Parsons coefficient estimates.

Quantifying this assessment, we have $\Delta\hat{V}_{DP} = \hat{V}_{DP_2} - \hat{V}_{DP_1} = 0.10$ and $Var(\Delta\hat{V}_{DP}) = \hat{s}^2_{V_1} + \hat{s}^2_{V_2} = (0.093)^2 + (0.081)^2 = 0.015$. Hence, the standard error of $\Delta\hat{V}_{DP}$ is about $\sqrt{0.015} = 0.12$, giving $\Delta\hat{V}_{DP} = 0.10 \pm 0.12$, which is not statistically different from 0. Such large standard errors are typical and probably account for the lack of differentiation of formations in the work of Lambert (1981), inasmuch as it is rare that $\hat{V}_{DP}$ is calculated from more than 100 samples.

There is, however, a problem with the above analysis. V_{DP}'s are not additive and, therefore, the implicit comparison shown in Fig. 7-4 may not be meaningful. We see, for example, that $\hat{s}_{V_1}$ is more than 10% of $\hat{V}_{DP_1}$, and we may no longer be in the region of local additivity for this level of V_{DP}. The variances, however, should be additive and can be compared using an F test.

As shown in Chap. 6, $V_{DP} = 1 - \exp(-\sigma)$ where σ^2 is the variance of the log-permeability PDF. We can estimate σ from the $\hat{V}_{DP}$'s and compare the variance estimates.

$$\hat{s}^2_{V_1} = [\ln(1 - \hat{V}_{DP_1})]^2 = 1.45 \qquad \hat{s}^2_{V_2} = [\ln(1 - \hat{V}_{DP_2})]^2 = 2.59$$

$$F = \frac{\hat{s}^2_{V_2}}{\hat{s}^2_{V_1}} = \frac{2.59}{1.45} = 1.80$$

The F statistic for this case is $F(40, 30, 0.05) = 1.79$

Consequently, comparing the variances shows that the difference in the two variabilities is statistically significant at the 95% level. This contradicts the assessment based on the V_{DP} standard errors.

The F test for equality of variances has two weaknesses. First, it is more complicated than the t test because it requires two degrees of freedom and F values from tables. There are no values readily related to the normal-distribution probabilities as has the t distribution. Second, it is very sensitive to nonnormality of the populations (Miller, 1986). The F test is much less sensitive to nonnormality, however, when used for testing equality of means.

When means are being tested, the procedure is termed *analysis of variance* or ANOVA. Once the basis for ANOVA is understood, you can appreciate why it is called an analysis of variance and not an analysis of means. The ANOVA method uses the F distribution in the following manner.

Suppose we have samples from three populations $N(\mu_A, \sigma^2_A)$, $N(\mu_B, \sigma^2_B)$, and $N(\mu_C, \sigma^2_C)$ with $\sigma^2_A = \sigma^2_B = \sigma^2_C = \sigma^2$. The null hypothesis ($H_0$) is that $\mu_A = \mu_B = \mu_C$ with the alternative that either or both equalities are untrue. From the data sets, we compute $(\overline{X}_A, \hat{s}^2_A)$, $(\overline{X}_B, \hat{s}^2_B)$, and $(\overline{X}_C, \hat{s}^2_C)$. The estimated variability of the mean is

$$\hat{s}^2_{\overline{X}} = \frac{I_A(\overline{X}_A - \overline{X})^2 + I_B(\overline{X}_B - \overline{X})^2 + I_C(\overline{X}_C - \overline{X})^2}{2}$$

where the denominator is 2 because we have three averages with (3 - 1) degrees of freedom. The total variability of all the data in the sets is

$$\hat{s}^2_T = \frac{(I_A - 1)\hat{s}^2_A + (I_B - 1)\hat{s}^2_B + (I_C - 1)\hat{s}^2_C}{(I_A - 1) + (I_B - 1) + (I_C - 1)}$$

This gives us all we need to apply the F test. The appropriate quantity is

$F = \hat{s}^2_{\overline{X}} / \hat{s}^2_T$

where we reject H_0 if $\text{Prob}\{\text{F} > \text{F}[2, (I_A - 1) + (I_B - 1) + (I_C - 1), \alpha]\} > \alpha$ and accept H_0 if $\text{Prob}\{\text{F} > \text{F}[2, (I_A - 1) + (I_B - 1) + (I_C - 1), \alpha]\} \leq \alpha$

In this case, F is a ratio comparing two variabilities (hence the name analysis of variance), both of which are estimates of σ^2. The numerator is the variability based on the three averages, $\overline{X}_A$, $\overline{X}_B$, and $\overline{X}_C$, while the denominator is based on pooling all the data together. We will use this approach again in regression, for the coefficient of determination. The above expressions can be extended to tests on many means.

Example 3 - Means Testing Using the F test. Let us retest the means using ANOVA for the data sets from Example 1. H_0: $\mu_A = \mu_B$; H_A: $\mu_A \neq \mu_B$.

$\overline{X}_A = 50 \qquad \hat{s}^2_A = 5 \qquad I_A = 25$

$\overline{X}_B = 48 \qquad \hat{s}^2_B = 3 \qquad I_B = 30$

$$\overline{X} = \frac{I_A \overline{X}_A + I_B \overline{X}_B}{I_A + I_B} = \frac{(25)(50) + (30)(48)}{25 + 30} = 48.9$$

$$\hat{s}^2_{\overline{X}} = \frac{I_A(\overline{X}_A - \overline{X})^2 + I_B(\overline{X}_B - \overline{X})^2}{1} =$$

$$\frac{25\ (50\text{-}48.9)^2 + 30\ (48\text{-}48.9)^2}{1} = 54.5$$

$$\hat{s}^2_T = \frac{(I_A - 1)\hat{s}^2_A + (I_B - 1)\hat{s}^2_B}{(I_A - 1) + (I_B - 1)} = \frac{(24)5 + (29)3}{24 + 29} = 3.91$$

$$F = \frac{\hat{s}^2_{\overline{X}}}{\hat{s}^2_T} = \frac{54.5}{3.91} = 14$$

Since $F(1, 53, 0.05) = 4.03$, H_A is clearly indicated. This is the same result as using the t test. Indeed, it can be shown that $F(1, I, \alpha) = [t(I, \alpha)]^2$.

The F test is moderately insensitive to unequal variances (e.g., $\sigma^2_A \neq \sigma^2_B \neq \sigma^2_C$) among the populations. The ANOVA method can be extended well beyond the limited application shown here. Box et al. (1978) do an excellent job of presenting the various applications and interpretations of ANOVA.

When we want to know more than just that the means are different, a multiple comparison test is needed. It will tell which mean is greater for a stipulated confidence level. Box et al.

(1978) and Rock (1988) discuss several such tests. We reiterate here, however, the earlier caution that hypothesis testing addresses only the question of statistical variability. If the means are found to be significantly different, further analysis should be undertaken. This analysis would include any geological information available for the data sets and whether the result of different means has an engineering significance. Recall our earlier comment (relating to Fig. 7-2) pointing out that statistical significance may not translate to significant differences in flow performance. Facies geometries and the flow process may either mitigate large contrasts or magnify statistically insignificant differences in reservoir properties. See, for example, the Rannoch flood results in Chap. 13.

7-2 NONPARAMETRIC TESTS

These tests do not assume anything about the underlying distributions of the populations under investigation. There is a price to be paid, however, for not providing PDF information in hypothesis tests. Nonparametric tests are not so powerful as parametric tests *if* the data actually come from the stipulated distribution. As remarked with the t and F tests, however, the superiority of parametric tests may quickly vanish (e.g., the F test for equality of variances) if the true PDF's do not accord with the assumed PDF's.

7-2.1 Mann-Whitney-Wilcoxon Rank Sum Test

This hypothesis test can go by several other names, including the *Mann-Whitney* and the *Wilcoxon Rank Sum Test.* The Mann-Whitney test is slightly different from the Wilcoxon test, but they can be shown to be equivalent. It is a test that is sensitive to changes in the median, but it is actually a comparison of the empirical CDF's of the two data sets. Hence, the test will also respond to changes in other features of the distribution, including skewness. It is only slightly less powerful than the t-test when the data are samples from normal populations. It may also be used for ordinal- or interval-type variables, an important tool for comparing geological (indicator) variables.

Suppose we have two data sets, $X_1, X_2, X_3, \ldots, X_I$ and $Y_1, Y_2, Y_3, \ldots, Y_J$, with $I \leq J$. We assume for the null hypothesis that they come from populations with the same CDF. We mix the two sets and order the data from smallest to largest. The smallest datum has rank 1 and the largest has rank $I+J$. Let $R(X_I)$ be the rank of X_I in the combined set and T_X be the sum of ranks of all I X's:

$$T_X = \sum_{i=1}^{I} R(X_i)$$

If $I \geq 10$ and H_0 applies, T_X is approximately normally distributed:

$$T_X \sim N\left(\frac{I(I + J + 1)}{2}, \frac{IJ(I + J + 1)}{12}\right)$$

Therefore, a single- or double-sided test result can be applied. If $10 \geq J > I$, special tables must be consulted (e.g., Rohatgi, 1984). If ties occur, the rank assigned to each value is the average of the ranks. Rice (1988, Chap. 11) gives a good account of the theory behind this test.

There are two points that drive the Mann-Whitney test. First,

$$T_X + T_Y = \frac{I(I+J+1)}{2} + \frac{J(I+J+1)}{2} = \frac{(I+J)(I+J+1)}{2}$$

so the ranks must sum to a number that depends only on the data-set sizes, and each data set *should* contribute to the sum according to the size of the data set. It also suggests that T_X and T_Y are equivalent statistics, and a test on one will yield the same result as a test on the other. Second, if T_X is particularly large or small compared to the size of the data set (i.e., the X data are dominating the high or low ranks, respectively), the suggestion is that the X and Y populations are not the same.

Example 4 - Testing Directional Permeability Data. Suppose that we have the following core-plug permeability measurements from a massive (i.e., structureless) sandstone interval. There is no information about sample depths. Is there a statistically significant difference in the distributions of the horizontal and vertical permeabilities? Our H_0 assumes that there is no statistically significant difference between the distributions against the alternative (H_A), and that there is no preferred direction to the difference.

k_h: 411; 857; 996; 1010; 1510; 1840; 1870; 2050; 2200; 2640; 2660; 3180 mD
k_v: 700; 858; 967; 1060; 1091; 1220; 1637; 1910; 2160; 2320; 2800; 2870 mD

We are testing the medians here because of the structureless nature of the sediment. Testing the arithmetic or harmonic averages would be inappropriate because of the layered nature of the medium that they imply (see Chap. 5). By inspection, the medians are k_{Mh} = 1855 and k_{Mv} = 1428 mD. Comparing these values with the arithmetic averages, $\bar{k}_h = 1768$ and $\bar{k}_v = 1633$ mD, we see that the differences are less than 15%. This suggests that, despite the flow implications for certain averages, the heterogeneity is sufficiently low that any measure of central tendency would suffice. The ranks of the combined set are shown in Table 7-2.

For these data, $T_h = 154$ and $T_v = 146$. Note $T_h + T_v = (12+12)(12+12+1)/2 = 300$, as a quick check of our mathematics. In this case, $I = J$, so we can analyze T_h or T_v, and both samples are larger than 10, so we can use the normal approximation.

$E(T_h) = 12(12+12+1)/2 = 150$ and $Var(T_h) = (12)(12)(12+12+1)/12 = 300$

Hence, $Z = \dfrac{T_h - E(T_h)}{\sqrt{Var(T_h)}} = \dfrac{154 - 150}{17.3} = 0.23$. A table of Z values suggests that there is a 40% chance of observing $T_h \geq 154$ under H_0. Hence, H_0 is upheld.

If we had more information in Example 4, a more sophisticated analysis could be performed to look for systematic differences between k_v and k_h. We might also apply a one-sided t test, allowing only the alternative $k_v < k_h$, if we knew the sediments were not reworked (e.g., dewatered or bioturbated). If we had depth information and the samples were paired (i.e., one k_v and one k_h measurement at each depth), we might have checked for systematic variations. Core-plug depths, however, can be inaccurate and, without reference to the core or core photographs to check sampling locations, so-called "paired" samples may be taken in quite different locations.

Table 7-2. Permeability data and ranks for Example 4.

Data value	Rank	Data type
411	1	*h*
700	2	*v*
857	3	*h*
858	4	*v*
967	5	*v*
996	6	*h*
1010	7	*h*
1060	8	*v*
1091	9	*v*
1220	10	*v*
1510	11	*h*
1637	12	*v*
1840	13	*h*
1870	14	*h*
1910	15	*v*
2050	16	*h*
2160	17	*v*
2200	18	*h*
2320	19	*v*
2640	20	*h*
2660	21	*h*
2800	22	*v*
2870	23	*v*
3180	24	*h*

7-2.2 Kolmogorov-Smirnov Test

This hypothesis test may be used to compare the empirical CDF against either a known CDF or another empirical CDF. The test may be used for nominal, ordinal, or ratio data. It is based on measuring the maximum vertical separation between the two distributions (Fig. 7-5). If d_{max} is small, the CDF's F_1 and F_2 are probably not statistically different. It is a particularly helpful test because, while it may be easy to compare averages or medians of distributions, comparing entire CDF's is more involved.

Each empirical CDF F_i is based on a data set of size I_i so that each F_i will increase in steps of size I_i^{-1} or integer multiples thereof. The product $I_1 I_2 d_{max}$ will be an integer, and that is the way the critical tables are listed. For small samples ($I_I \leq 16$), no approximation is available and tables are needed to interpret the result. Rock (1988) points out that caution is required with tables because some give the value $I_1 I_2 d_{max}$ that must be exceeded while others give the critical value, i.e., the latter give the value $(I_1 I_2 d_{max} + 1)$. If $I_i > 16$, Cheeney (1983) gives the 5% two-sided critical value that must be exceeded as either 1.36 $I_1^{-1/2}$, where the comparison is with a theoretical CDF, or $1.36(I_1 I_2)^{1/2}(I_1 + I_2)^{-1/2}$ for comparing two empirical CDF's. The one-sided 5% test values use 1.22 instead of 1.36 so that approximately 20% fewer data are needed for a one-sided test than for a two-sided test.

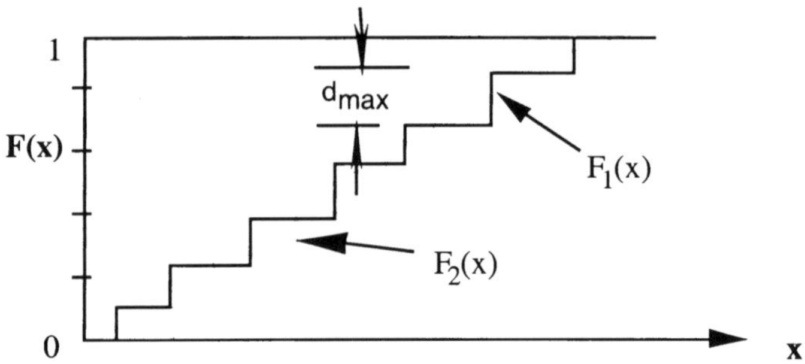

Figure 7-5. Comparison of two CDF's in the Kolmogorov-Smirnov test.

Example 5 - Comparing Wireline and Core-Plug Data. We have core-plug (ϕ_p) and wireline (ϕ_w) porosity measurements over the same interval in a well (Table 7-3). Visual inspection of the core suggests all the data came from similiar facies. At a 5% level, is there a statistically significant difference in the distributional forms of the two porosities? We assume H_0 such that there is no statistically significant difference between the forms against the alternative H_A that they are different.

Because of the different volumes of investigation and measurement conditions, the average porosities and variabilities may be different for ϕ_w and ϕ_p. We will adjust the data so that they are more nearly comparable. Each data set has been normalized by subtracting the average from each datum and dividing by the standard deviation to yield the Z values in Table 7-3. Subtracting the average adjusts for the difference of measurement conditions: *in-situ* for wireline compared to laboratory conditions for plugs. Dividing by the variability adjusts for the different sampling volumes. This is because we assume that the formation can be divided into units of constant porosity (porosity-representative elementary volumes using Bear's (1972) terminology). Both the plug and wireline measurement sample one or more of these units. We further assume that either measurement is the arithmetic average of the formation porosities within its volume of investigation. Each measurement's variability will decrease as n^{-1}, where n is the number of units of constant porosity. Hence, dividing each measurement by the apparent standard deviation of the porosity in the interval will compensate for the difference in volumes.

Table 7-3. Ordered wireline and core-plug data for Example 5.

ϕ_w	ϕ_p	Z_w	Z_p	p_i
32.3	28.4	1.61	1.78	97.8
32.2	28.3	1.56	1.69	93.5
31.7	27.8	1.14	1.21	89.1
31.7	27.5	1.14	0.93	84.8
31.6	27.3	1.04	0.74	80.4
31.3	27.3	0.83	0.74	76.1
31.2	26.7	0.73	0.17	71.7
30.9	26.7	0.41	0.17	67.4
30.7	26.6	0.31	0.07	63.0
30.6	26.6	0.20	0.07	58.7
30.6	26.5	0.20	-0.02	54.3
30.2	26.5	-0.11	-0.02	50.0
30.2	26.4	-0.11	-0.12	45.7
30.1	26.4	-0.22	-0.12	41.3
30.1	26.3	-0.22	-0.22	37.0
30.1	26.3	-0.22	-0.22	32.6
30.0	26.2	-0.32	-0.31	28.3
29.6	26.2	-0.69	-0.31	23.9
29.6	26.0	-0.69	-0.50	19.6
29.1	26.0	-1.10	-0.50	15.2
28.5	25.8	-1.63	-0.69	10.9
28.3	25.2	-1.73	-1.26	6.5
27.9	23.1	-2.15	-3.26	2.2

The CDF's of the normalized data are plotted in Fig. 7-6 using the probabilities shown in Table 7-3. The maximum distance between the CDF's is $d_{max} = 67.39 - 54.35 = 13.0\%$, or $I_1 I_2 d_{max} = (23)(23)(0.130) = 69$. The 5% critical value is $1.36((23)(23))^{1/2}(23+23)^{-1/2} = 5$, which is exceeded considerably by $I_1 I_2 d_{max}$. Hence, the two CDF's are statistically different at the 5% level.

What is surprising about the result of Example 5 is not that the CDF's are different but *where* they are different. Inspection of Fig. 7-6 shows that the CDF's are different at several places; d_{max} occurs near the arithmetic average of the porosities, not at regions in the tails, as might be expected. Clearly, the plug and wireline measurements are not responding similarly and simple shifts or multiplying factors will not reconcile them.

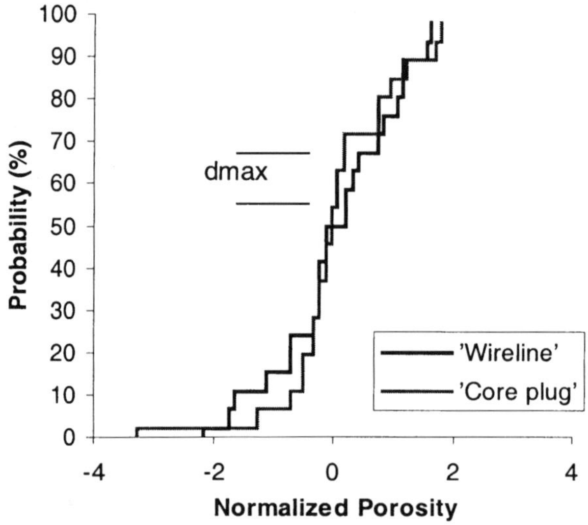

Figure 7-6. Empirical CDF's for Example 5.

7-3 SUMMARY REMARKS

Hypothesis testing has several uses in the analysis of data. You must take care, however, to ensure that the correct problem is being posed that hypothesis testing is equipped to answer. The additivity of variables under test can be a major problem and remedies to this are not always apparent. Many statistical-analysis software packages contain large amounts of hypothesis-testing routines, and it is tempting to apply these to our data. The results, however, may mislead unless we have a clear physical basis for interpretation.

Perhaps one important role that hypothesis testing can fulfill is to raise awareness of the value of auxiliary information. For example, the data requirements for one-sided tests are often considerably smaller than for two-sided tests. The geological characterization of the sediments under analysis may help restrict the analysis so that more may be learned from the data.

CHAPTER 8

BIVARIATE ANALYSIS: CORRELATION

Correlation is the expression of how two or more random variables are related. It is a much less powerful technique than regression, discussed in Chaps. 9 and 10, because correlation does not develop a predictive relationship. It can be useful, however, for establishing the strength and nature of the association, or covariation, between random variables.

Since correlation is about random variables, it is unsuitable for relationships involving deterministic (or "controllable") variables. For example, in many bivariate problems, one of the variables can be considered as a quantity under our control with little or no error (such as a core flood where the amount of fluid injected is a controllable variable). In other situations, however, we simply measure both the X and Y values on a number of samples with no control over either X or Y, e.g., porosity and permeability of core plugs. In these latter situations, correlation analysis is appropriate.

8-1 JOINT DISTRIBUTIONS

We have seen in Chap. 3 that all random variables have a distribution function that characterizes their probabilities of occurrence. If we have two random variables, X and Y, their probabilities of occurrence will be expressed through their *joint probability density function*, $f(x, y)$:

$$\text{Prob}(x < X \le x + \delta x \text{ and } y < Y \le y + \delta y) = f(x, y)\ \delta x\ \delta y$$

If X and Y are independent, a knowledge of X or Y has no effect upon the value of the other variable and the joint PDF is simply the product of the two separate PDF's:

$$f(x, y) = f_X(x)\, f_Y(y)$$

since $\text{Prob}(x < X \le x + \delta x \text{ and } y < Y \le y + \delta y) = \text{Prob}(x < X \le x + \delta x)\ \text{Prob}(y < Y \le y + \delta y)$. In this case, as shown in Chap. 4, $E(XY) = E(X)E(Y)$.

The joint PDF of X and Y is an extension of the univariate situation where X and Y each have their own PDF, f_X and f_Y. X and Y may both be continuous, discrete, or mixed with X of one type and Y of another type of variable (see Chap. 3 for a description of these variable types). Plotting $f(x, y)$ requires three axes (Fig. 8-1). The projections of $f(x, y)$ onto the f—x and f—y planes produce the *marginal* distributions of X and Y. These are the PDF's f_X and f_Y

that we observe when we ignore the other variable. The areas under each of these distributions must be one to ensure that they represent probabilities.

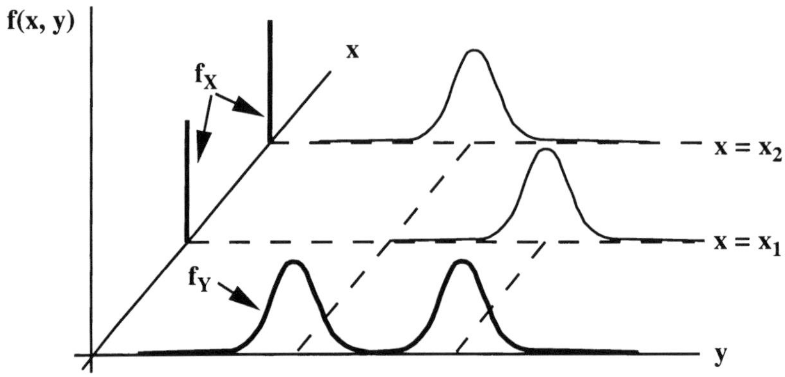

Figure 8-1. A mixed joint PDF for X discrete and Y continuous, with their bimodal marginal PDF's.

Several features of the X-Y relationship can be obtained from the joint PDF. If one of the two variables is fixed within a small range, say $x_0 < X \le x_0 + \delta x$, while Y is allowed to vary, we can obtain the *conditional PDF* of Y, $f_{Y|X}(y \mid X = x_0)$. Geometrically, $f_{Y|X}(y \mid X = x_0)$ is the plane $X = x_0$ that "slices through" $f(x, y)$. The conditional PDF can be used to determine how the mean value of Y varies with x_0:

$$E(Y \mid X = x_0) = \int_{-\infty}^{+\infty} y f_{Y|X}(y \mid X = x_0) dy$$

$E(Y \mid X = x_0)$ is called the *conditional expectation* of Y. It is a function of x_0. There is a similar conditional PDF and expectation of X. The joint PDF is also used to measure the X-Y relationship strength, which we discuss further below.

Just as with the univariate case, there are also joint CDF's. Conditional CDF's are obtained from the joint CDF by holding one of the variables constant. Recall that in Chap. 3 (Example 3), we used conditional CDF's to relate the sizes of shales to their depositional environments. In that case, the depositional environment and shale size are the two random variables. The depositional environment variable has a strong effect on the shape of the shale-size CDF.

A plane can be drawn parallel to the x-y axes such that $f(x, y)$ is constant (the "tombstone" sketch of Fig. 8-2). The plane represents the probability p, where

$$p = \text{Prob}(\, y_1 < Y \le y_2, X = x_2) + \text{Prob}(\, y_3 < Y \le y_4, X = x_1)$$

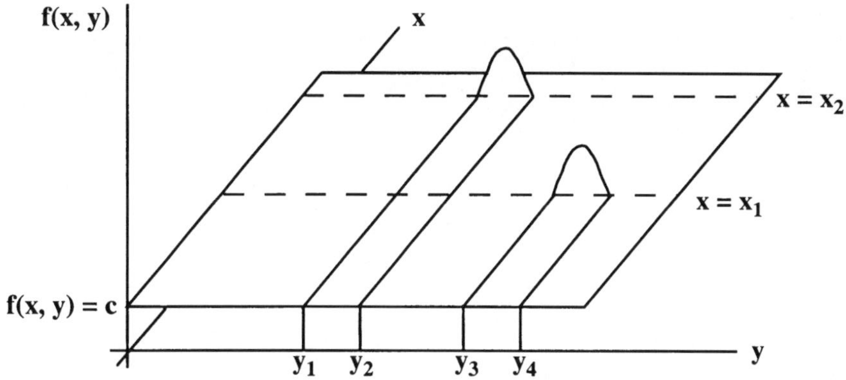

Figure 8-2. A plane with constant $f(x, y)$ represents (X, Y) pairs with a specific probability of occurrence.

Since probabilities are areas under the PDF, p can be interpreted as the following sum of areas:

$$p = \int_{y=y_1}^{y_2} f(x_2, y)dy + \int_{y=y_3}^{y_4} f(x_1, y)dy$$

In a common case where both X and Y are continuous variables and the joint PDF is unimodal, $f(x, y) = c$ will be a single closed curve, often elliptical in shape. We will see examples of this below.

8-2 THE COVARIANCE OF *X* AND *Y*

How do we measure the strength of covariation between the random variables X and Y? One such measure of this relationship is the *covariance* of X and Y, defined as

$$Cov(X,Y) = E[(X - \mu_X)(Y - \mu_Y)] = \int_{-\infty}^{+\infty} (x - \mu_X)(y - \mu_Y) f(x, y)dxdy \tag{8-1}$$

where $E(X) = \mu_X$ and $E(Y) = \mu_Y$. The covariance can also be written as

$$Cov(X,Y) = E(XY) - \mu_X \mu_Y$$

from the properties of the expectation operator (Chap. 4). The covariance is one of the most important devices in all of multivariable statistics; it will be used extensively in the remainder of this text, so it is fitting to devote some space to its properties.

The covariance is a generalized variance. That is,

$$Cov\,(X, X) = Var(X)$$

from its definition and that of the variance. Thus, we would expect *Cov* and *Var* to share many of the same properties. As we shall see, this is only partially true, since the added generality makes *Cov* substantially different from *Var* except in the degenerate case mentioned above. For example, whereas we must have $Var(X) \geq 0$, *Cov* can be either negative or positive. Additional properties are

covariance is commutative: $Cov(X, Y) = Cov(Y, X)$

from the definition (Eq. (8-1)), and

covariance with a constant c is zero: $Cov(X, c) = 0$.

This latter property is reasonable once it is understood that *Cov* measures the extent to which the first argument varies (covaries) with the second. Any fluctuations in the random variable X will not have any like variations with a constant.

The reader might notice that, although the covariance was introduced and continues to be based on the notion of the joint PDF, we begin to use it without showing the PDF explicitly. This practice will continue to the end of the text. It is a good idea, however, to keep the joint PDF and the associated probability definitions in mind, since these form the basic link between probability and statistics.

One manifestation of this is the covariance with multiple random variables X_i and Y_j:

$$Cov\left(\sum_{i=1}^{I} a_i X_i, \sum_{j=1}^{I} b_j Y_j\right) = \sum_{i=1}^{I}\sum_{j=1}^{I} a_i b_j Cov(X_i, Y_j)$$

which is a relationship that occurs in several applications.

The most important property of the covariance, however, lies in its ability to measure the association between two random variables. If X and Y are independent, then $Cov(X, Y) = 0$, a property that follows from the independence property of the joint PDF given above. The converse, however, is not necessarily true; if $Cov(X, Y) = 0$, then X and Y might not be independent. Meyer (1966, pp. 144-145) gives an example.

The covariance has units of X times Y. To form a useful basis of comparison, the measure of relationship strength should be dimensionless. The covariance of X and Y is often normalized by dividing by the variabilities of X and Y. The resulting quantity is the *correlation coefficient* of X and Y, given by

$$\rho(X, Y) = \frac{Cov(X,Y)}{\sqrt{Var(X)Var(Y)}}$$

From the properties of the expectation (Chap. 4), this equation can be rewritten as

$$\rho(X, Y) = E\left[\left(\frac{X - \mu_X}{\sqrt{Var(X)}}\right)\left(\frac{Y - \mu_Y}{\sqrt{Var(Y)}}\right)\right]$$

or $\rho(X, Y) = E(X^*Y^*)$

where

$X^* = (X - \mu_X)/\sqrt{Var(X)}$

and

$Y^* = (Y - \mu_Y)/\sqrt{Var(Y)}$

Since X^* and Y^* are standard random variables with zero mean and unit variance, $\rho(X, Y)$ can be viewed as the expectation of the product of two standardized variables. Standardization shifts the center and scale of a cloud of (X, Y) values to the origin with equal scales on both axes (Fig. 8-3).

Each point (X^*, Y^*) will lie in one of the four quadrants formed by the axes (Fig. 8-3). In quadrants 1 and 3, the product X^*Y^* is positive, while X^*Y^* is negative in quadrants 2 and 4. Thus the sign of $\rho(X, Y)$ suggests, taking all possible pairs (X, Y), whether the relationship is mostly in quadrants 1 and 3 or quadrants 2 and 4. The narrower the cloud is around either the 45° or 135° lines, the larger ρ or $-\rho$, respectively, will be.

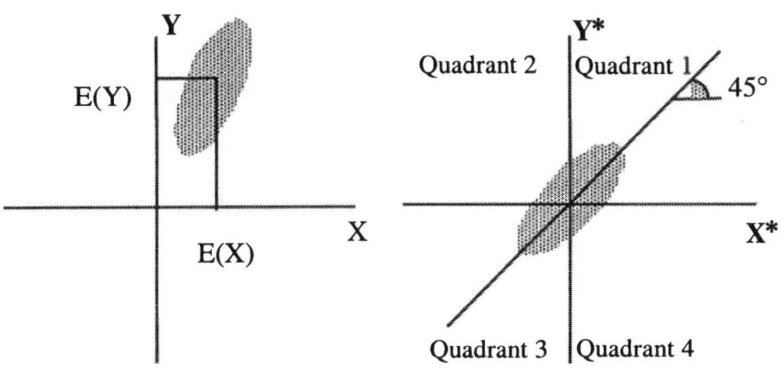

Figure 8-3. Standardized variables X^* and Y^* shift the original X-Y relationship (left) to the origin and produce a change of scales. The shaded region represents all (X, Y) pairs within a given probability of occurrence.

Some properties of the correlation coefficient include

1. $|\rho(X, Y)| \leq 1$ and
2. $|\rho(X, Y)| = 1$ if and only if $Y = aX + b$ for some constants a and b, $a \neq 0$.

Although ρ is not in general a good indicator of relationship strength, property 2 indicates that it is useful if the relationship being assessed is linear. Even when this is true, however, the correlation coefficient should be used with caution. The behavior of $\rho(X, Y)$ depends on the joint distribution of X and Y, so what may work well with one set of data may not work well in another case, as shown below.

Example 1 - Comparing Correlation Coefficients. Consider the case of three variables, X, Y, and Z, where we have $0 < |\rho(X, Y)| < |\rho(X, Z)| < 1$ (Fig. 8-4). Do X and Z have a stronger linear element to their relationship than do X and Y?

The X-Y figure (left) has the same X_i coordinates as the X-Z figure for $-1 < X < 1$. Thus $Y = Z$ for $-1 < X < 1$. $Z = \rho X + (1 - \rho^2)^{1/2}\varepsilon$, where X and $\varepsilon \sim N(0, 1)$, making X and Z joint normally distributed (Morgan, 1984, p. 89). For this distribution, the Z-X covariation is strictly linear (see below). Thus, both data sets have identical linear elements to their relationships, while the estimated ρ's are different.

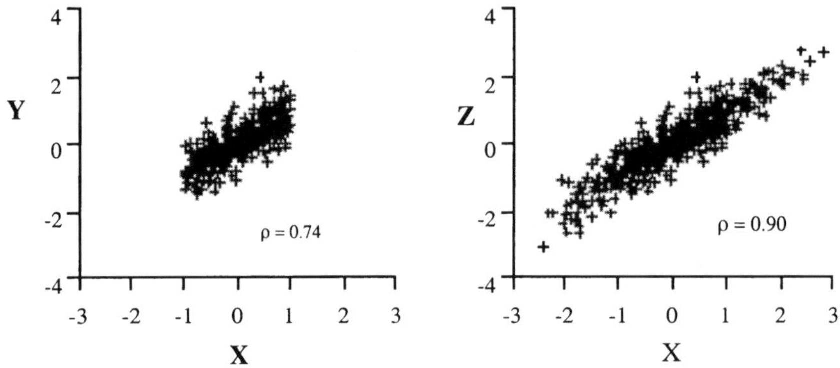

Figure 8-4. Bivariate data sets for Example 1.

8-3 THE JOINT NORMAL DISTRIBUTION

One case for which the correlation coefficient ρ is important is if X and Y are *joint normally distributed* (or *bivariate normally distributed*). If X and Y are joint normally distributed (JND), then

$$f(x, y) = \frac{1}{2\pi\sigma_X\sigma_Y\sqrt{1-\rho^2}} \exp\left[-\frac{x^{*2} - 2\rho x^* y^* + y^{*2}}{2(1-\rho^2)}\right] \tag{8-2}$$

where $x^* = \dfrac{x - \mu_X}{\sigma_X}$ with $\mu_X = E(X)$ and $\sigma_X^2 = Var(X)$

$y^* = \dfrac{y - \mu_Y}{\sigma_Y}$ with $\mu_Y = E(Y)$ and $\sigma_Y^2 = Var(Y)$

and $\rho = \frac{\sigma_{XY}}{\sigma_X \sigma_Y}$ with $\sigma_{XY} = Cov(X,Y)$

ρ is the correlation coefficient between Y and X. It is a population parameter of the distribution, just as μ_X and μ_Y are.

The joint normal PDF looks like Fig. 8-5 when $\rho = 0.7$. When $\rho = 0$, a plane $f(x^*, y^*) =$ constant makes a circle; when $\rho \neq 0$, the form is an ellipse. The closer $|\rho|$ is to 1, the more elongated the ellipse becomes. The major axis of the ellipse is at 45° or 135° to the X-Y axes. The interior of any ellipse represents those (X, Y) pairs having a given probability of occurrence. Hence, each ellipse has a probability level associated with it. This probability is given by the volume of the PDF within the boundary of the ellipse (recall Fig. 8-2 and associated discussion). Formulas for the ellipses and their associated probabilities are given in Abramowitz and Stegun (1965, p. 940).

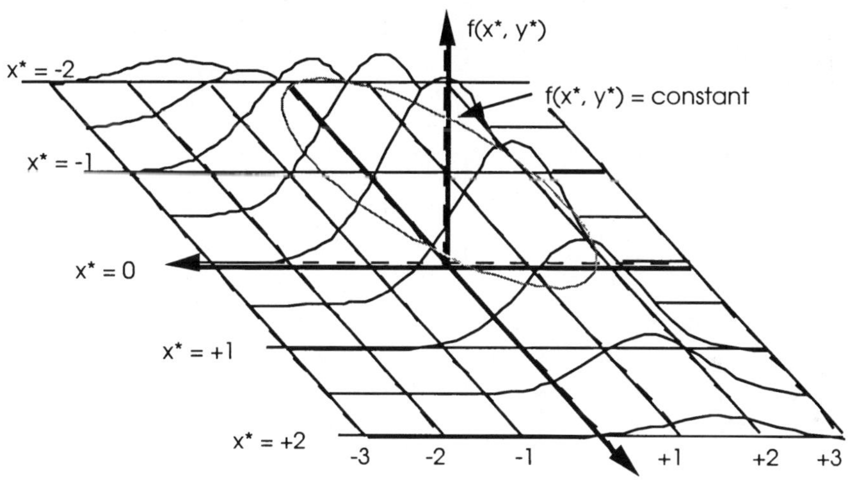

Figure 8-5. A joint normal PDF with $\rho = 0.7$. The marginal PDF's of X^* and Y^* are both $N(0, 1)$.

The JND ellipses are often plotted with data on scatter plots. The ellipses are a diagnostic tool to help detect significant deviations from joint normality. Drawing such an ellipse does not imply that the data come from a JND population, however. For example, if 5 out of 40 points fall outside the 95% ellipse, it is rather unlikely that the data are JND. A test for X and Y JND is to examine the marginal PDF's; if either X or Y is not normal, they cannot be JND. Marginal normality is necessary but not sufficient, however. Hald (1952, pp. 602-606) discusses a test (the χ^2 test) for joint normality.

Example 2 - Assessing Joint Normality with Probability Ellipses. Figure 8-6 shows two 100-point data sets with their 90% probability ellipses. As in Example 1, we have constructed the Y data from $Y = \rho X + (1 - \rho^2)^{1/2}\varepsilon$, where X and $\varepsilon \sim N(0, 1)$ and $\rho = 0.9$. For Z, however, we have $Z = \rho X + 3.46\,(1 - \rho^2)^{1/2}\delta$, where δ is a uniformly distributed variable with upper and lower endpoints of 0.5 and -0.5, respectively.

The factor 3.46 is to keep $Var(\delta) = Var(\varepsilon) = 1$. X and Y are JND, but X and Z are not. Both Y and Z have marginal distributions very close to normal, as judged by probability plots.

When comparing distributions, differences are usually most pronounced at the extremes, rather than near the middle of PDF's. Therefore, we choose a large probability ellipse (e.g., 90, 95, or 99%) and examine the behavior of the points falling outside this ellipse. The more data there are, the larger probability level we can choose. In this case with 100 data, a 90 or 95% level is appropriate, and we have chosen the 90% level (Fig. 8-6).

In line with the probability level, about ten points lie outside the ellipse for the JND data (Fig. 8-6 left). These points are not clustered at one area and are near the ellipse. In contrast, the non-JND data set (Fig. 8-6 right) has fewer data outside the ellipse, these deviations are concentrated at the extreme upper and lower parts of the ellipse, and there is one point (X = -2.5, Y = -2.9) well away from it. In this case, the non-JND data fail the test. But differences are such that more formal tests (such as the χ^2 test) may be needed to determine the joint normality of X and Z.

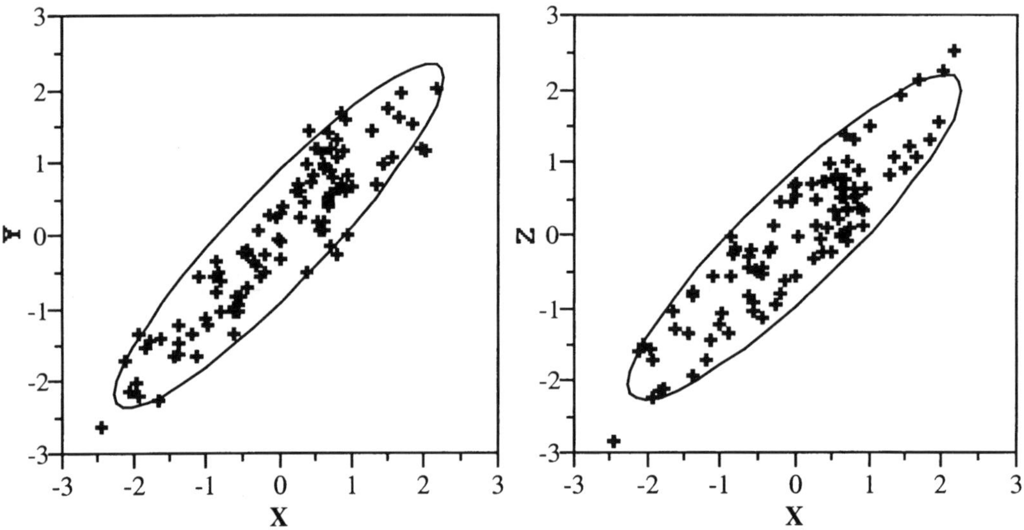

Figure 8-6. Two data-set scatter plots with normal probability ellipses.

When X and Y are JND, several nice properties apply to X and Y (see, for example, Hald, 1952, pp. 590-601, for a detailed discussion with figures):

1. The conditional PDF's of X and Y have a normal (Gaussian) shape, but these shapes do not have unit area. (The volume under the full PDF is equal to one.) This is plausible from inspection of Eq. (8-2) and Fig. 8-5.

2. The marginal PDF's of X and Y are both normal. To obtain this result, we integrate Eq. (8-2) over all values of one variable.

3. The standardized variable X^*-Y^* covariation is linear, along a 45° or 135° line.

4. The conditional means are linear functions of y_0 and x_0:

$$E(X \mid Y = y_0) = \mu_X + \rho\frac{\sigma_X}{\sigma_Y}(y_0 - \mu_Y) \text{ for variable } x \text{ and fixed } y \tag{8-3a}$$

and

$$E(Y \mid X = x_0) = \mu_Y + \rho\frac{\sigma_Y}{\sigma_X}(x_0 - \mu_X) \text{ for variable } y \text{ and fixed } x \tag{8-3b}$$

These are the equations for *regression* lines in the case X and Y JND. Thus, a linear regression model is automatically appropriate. See Chap. 9 where we will derive these equations.

An estimator for ρ is the sample correlation coefficient, given by

$$r = \frac{S_{XY}}{\sqrt{S_{XX}S_{YY}}} = \frac{\Sigma[(X_i - \overline{X})(Y_i - \overline{Y})]}{\sqrt{\Sigma(X_i - \overline{X})^2\Sigma(Y_i - \overline{Y})^2}}$$

where S is the sum of squares of the products shown in its subscripts and the summations are over the I data. $I^{-1}S_{AB}$ is an estimate for σ_{AB}, where A and B represent X or Y. r^2 is equal in value to the *coefficient of determination* (R^2) when X and Y are JND. R^2, however, has a different interpretation from that of r, which is an estimate of a population parameter when X and Y are JND. Otherwise, r can be used as a measure of the linear interdependence of X and Y. R^2 has an interpretation no matter what the joint PDF of X and Y is or even if X is deterministic, not random. R^2, however, must have a model (which might not be linear) to indicate what the anticipated form of the covariation is. R^2 will be discussed further in Section 9-9. The estimator given above is a standard function in most spreadsheets, as are estimators for the covariance and coefficient of correlation.

There are significance tests available for r. The PDF of r is rather complicated, but the transformed quantity $(1/2)[\ln(1 + r) - \ln(1 - r)]$ is approximately normal with mean $(1/2)[\ln(1+\rho) - \ln(1 - \rho)] + \rho/[2(I - 1)]$ and variance $1/(I - 3)$ for I data (Hald, 1952, p. 609). Tables are widely available (e.g., Snedecor and Cochran, 1980, pp. 477-478). A common test is to determine whether $|r|$ is sufficiently large for the sample to indicate that ρ is significantly different from zero. These tests, however, are only applicable if X and Y are JND (or nearly so). Variabilities and significance tests for $\overline{X}$, $\overline{Y}$, S_{XY}, S_{XX}, and S_{YY} are fairly complicated and are given in Anderson (1984, Chaps. 4 and 7).

The estimated slopes of the regression lines are given by

$$r\sqrt{S_{YY}/S_{XX}} \quad \text{and} \quad r\sqrt{S_{XX}/S_{YY}}$$

for Y upon X and X upon Y, respectively. We can calculate r without calculating any regression lines, since it requires neither the slope nor the intercept of either line. This points out a fundamental distinction between correlation and regression. Correlation indicates only the presence of an association between variables and has no predictive power. Regression permits the quantification of relationships among variables (the appropriate model being known) and their exploitation for prediction.

8-4 THE REDUCED MAJOR AXIS (RMA) AND REGRESSION LINES

The line at 45° or 135° in the X^*-Y^* plane (Fig. 8-3) represents the form of the X-Y covariation if X and Y are approximately JND. This line, the major axis of all ellipses formed from $f(x^*, y^*)$ = constant (Fig. 8-5), is called the *line of organic correlation* or the *reduced major axis line* (see Sokal and Rohlf, 1981, p. 550 for other names). It is related to the major axis line but has the advantage of being scale invariant (Agterberg, 1974, pp. 119-123). The RMA line equation is

$$Y = \mu_Y + \frac{\sigma_Y}{\sigma_X}(X - \mu_X) \tag{8-4}$$

and is similar in form, but with a different slope, to the regression lines of Eqs. (8-3). The RMA line lies between the two regression lines (see Fig. 8-7). All the estimated lines pass through the point ($\overline{X}$, $\overline{Y}$) in the X-Y plane. The RMA line represents the relationship such that one standard deviation change in X (i.e., σ_X) is associated with a unit standard deviation change in Y (i.e., σ_Y). When perfect linear correlation exists, all three lines are coincident. The RMA line is insensitive to the strength of correlation between X and Y since its slope does not change as σ_{XY}, the covariance of X and Y, changes.

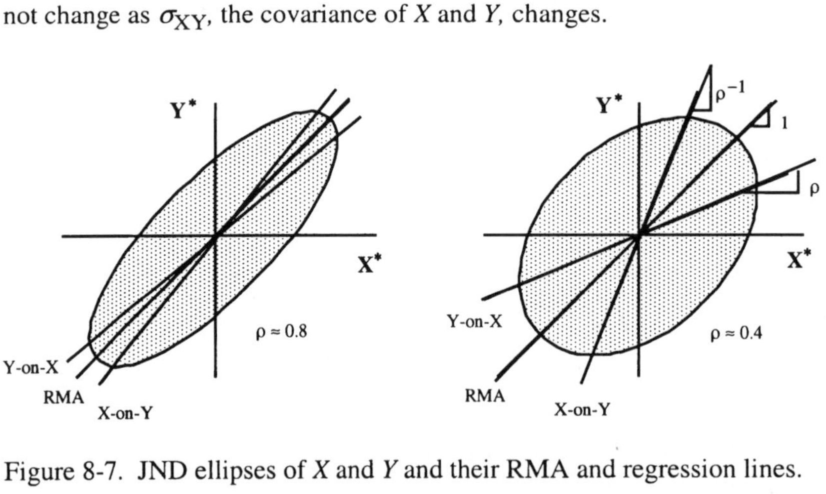

Figure 8-7. JND ellipses of X and Y and their RMA and regression lines.

The RMA line represents how X and Y covary; as X changes by σ_X, Y tends to change by σ_Y. This gives the slope S_{YY}/S_{XX} for the estimated line. The "tightness" of the X-Y

covariation is estimated by r. There is no suggestion in this relationship that one variable is a predictor while the other is a predicted variable. The RMA line is usually more appropriate for applications where prediction is *not* involved. Such applications include estimating an underlying relationship or comparing a fitted line to a theoretical line. For example, suppose we had porosity and permeability data from core plugs coming from two different lithofacies. To compare the porosity-permeability relationships of the two lithofacies, the RMA line would be an appropriate choice. See Example 3 below.

The two regression lines also define linear relationships between X and Y. The predictor-predicted roles of the variables, however, are inherent in the lines. The Y-on-X line, representing $E(Y \mid X = x_0)$, shows how the mean value of Y varies when X is fixed at the value x_0. Thus X must be known in order to establish the mean of Y. As the correlation diminishes to zero, the regression lines become coincident with the axes, suggesting that the best estimate of X or Y is $E(X)$ or $E(Y)$, respectively. In this case, a knowledge of one variable gives little help in predicting the value of the other. The RMA line, however, remains the same and is always coincident with the major axis of the ellipse.

A geometrical interpretation exists for calculating the RMA and regression lines for a standardized (X^*, Y^*) data set (Agterberg, 1974, Chap. 4; Troutman and Williams, 1987). The Y-on-X regression line is obtained by minimizing the sum of the squared vertical distances (line ad in Fig. 8-8) from the line to the data points. The X-on-Y regression line is given by minimizing the sum of the squared horizontal distances (line ab in Fig. 8-8). The RMA line obtains from minimizing the sum of the squared distances from the data points to the line ac in Fig. 8-8. Because of these geometrical interpretations, it is sometimes argued that the RMA line is the more appropriate line for prediction when both variables are subject to substantial error. The relative merits of these lines for prediction will be discussed in Chap. 10.

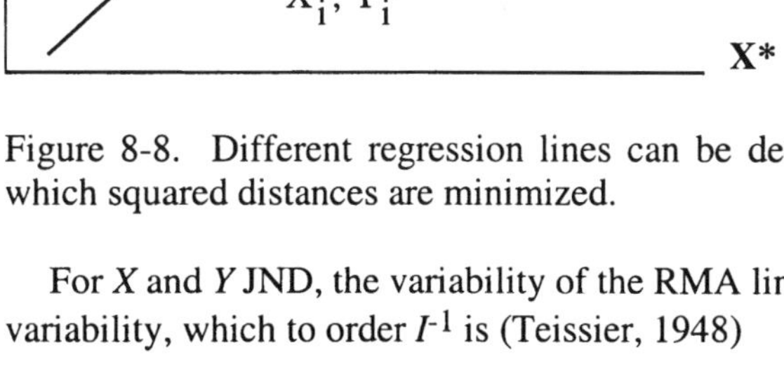

Figure 8-8. Different regression lines can be determined from a data set, depending upon which squared distances are minimized.

For X and Y JND, the variability of the RMA line slope is the same as the Y-on-X line slope variability, which to order I^{-1} is (Teissier, 1948)

$$Var\left(\frac{S_{YY}}{S_{XX}}\right) \approx \frac{\sigma_Y^2}{\sigma_X^2} \frac{1-\rho^2}{I} \tag{8-5}$$

The RMA *Y*-intercept variability is (Kermack and Haldane, 1950)

$$Var\left(\bar{Y} - \frac{\sigma_Y}{\sigma_X}\bar{X}\right) \approx \frac{\sigma_Y^2}{I}(1-\rho)\left[2 + \frac{\mu_X^2(1+\rho)}{\sigma_X^2}\right]$$

Kermack and Haldane (1950) also give results for *X* and *Y* not JND.

Example 3 - Regression and RMA Lines for Core Data. Figure 8-9 shows scatter plots for two sets of core-plug porosity and permeability ($\log_{10}$) data. Both sets are from the Rannoch Formation (see Example 6, Chap. 3 for more details of the Rannoch) and come from similar facies within one well from each field. Field *A*, however, has a slightly coarser sediment with less mica than Field *B*. As expected, the average porosity and permeability of Field *B* are less than those of Field *A*.

All marginal distributions appear approximately normal on probability plots, and slightly fewer than the expected number of data fall outside the 90% ellipses: 22 data for *A* and 5 data for *B*. The cluster of *A* data near $X = 21$ and $Y = 1.7$ suggests that these data may not be from a JND population but are reasonably close for our purposes.

Figure 8-9 also shows the RMA (dashed) and *Y*-on-*X* regression (solid) lines for both data sets. The RMA and *Y*-on-*X* lines are close for Field *B*, reflecting the large correlation coefficient, $r_B = 0.97$. Field *A* has a smaller *r*, $r_A = 0.84$, so the lines are more distinct. Both *r*'s are significantly different from zero (no correlation). For example, for Field *A*, $0.5[\ln(1 + r_A) - \ln(1 - r_A)] = 2.4$ with standard error $(I - 3)^{-1/2} = 0.06$.

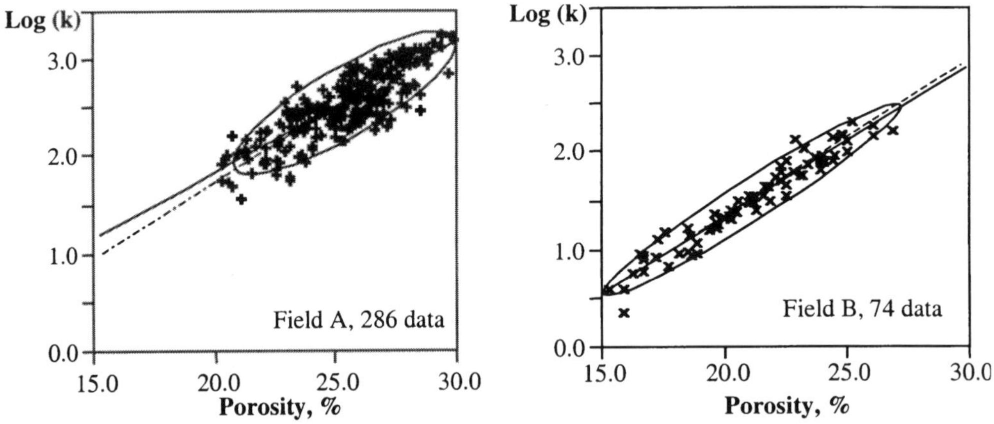

Figure 8-9. Porosity (*X*)-permeability (*Y*) scatter plots and lines for Example 3.

The RMA lines have different intercepts because of the textural differences of the two fields. The RMA line slopes are statistically similar, however: 0.162 ± 0.005

for Field *A* and 0.161 ± 0.005 for Field *B*. (The standard errors are from Eq. (8-5).) This suggests that the same mechanism may be controlling the permeability-porosity covariation in both wells. A different conclusion might be reached, however, using the *Y*-on-*X* lines: Field *A* has slope 0.135±0.005 while Field *B* has slope 0.156±0.005. The regression-line slopes are related to the strength of correlation, which reflects the amount of variability within the relationship as well as the nature of the relationship. For the purposes of comparing relationships, the RMA lines are the easier to interpret.

The similarity of the permeability-porosity relationships of Fields *A* and *B* in Example 3 suggests that Field *B* does not have lower porosity and permeability from causes other than the grain size and mica. We would, therefore, expect that porosity and permeability estimates from wireline or other measurements could be similarly interpreted in both fields.

8-5 SUMMARY REMARKS

Correlation analysis provides several procedures for assessing joint variation (correlation). Most of these methods are best applied when the data are from a joint normally distributed sample space because the statistics have a particularly helpful interpretation and the sampling errors can be evaluated. We will see correlation measures developed and extended further in Chap. 11, where sampling locations are also involved.

CHAPTER 9

BIVARIATE ANALYSIS: LINEAR REGRESSION

Probably no other statistical procedure is used and abused as much as regression. The popularity of regression stems from its purpose, to produce a model that will predict some property from the measurements of other properties, from its generality, and from its wide availability on calculators and computer software. Regression is especially useful for reservoir description.

Because reservoirs are below the Earth's surface, measurements made *in situ* are often restricted in type and number. Therefore, it is desirable to exploit any relationship between a property that can be measured and another that is needed but cannot be measured. It is, however, open to abuse because of numerous aspects of which the user need not be aware in order to apply the results. It may be only after the unsatisfactory predictions are applied that a problem becomes apparent.

For two reasons, we will concentrate here on linear regression with one predictor variable. The first reason is clarity. The conclusions for the bivariate and multivariate cases are similar, but the matrix expressions can obscure the underlying concepts as more variables are introduced. The second reason is *parsimony,* a term borrowed from time-series analysis, which means that the simplest model that explains a response should be used. Often, the bivariate model will meet this requirement. More elaborate models may fit the data better, but complications may arise when, unwittingly, the model is used outside the range of the data on which it was developed.

We begin by covering the basic features of least-squares bivariate regression. Mathematically, the procedure is straightforward and can be covered in a few lines. Statistically, however, regression has a number of aspects that we consider in detail to get the most benefit from the data collected. Because of the amount of material involved, more advanced considerations are deferred to Chap. 10.

9-1 THE ELEMENTS OF REGRESSION

The regression procedure consists of the following elements:

1. A *model* (derived or assumed), an equation that relates one or more observed quantities to a quantity to be predicted. The model contains unknown parameters that must be evaluated.

2. Measurements of both observed and predicted quantities.

3. A method to reverse the role of the variables and the parameters in the model to determine the latter, based on the measurements.

4. Application of the model to predict the desired quantity.

Regression is based on using data to determine some unknown parameters in a model. Hence, a model must exist or be developed (element 1) for the regression to proceed. The model represents the *a priori* information we bring to the data analysis, but it is incomplete by itself. Unknown parameters in the model must be determined from measurements (element 2) of all the quantities involved. The regression centers on element 3, in which the model parameters can be determined from statistical arguments, provided that the data satisfy certain conditions. The procedure has great practical use because it provides measures to indicate the appropriateness of the model and it provides parameter values. It does not, however, of itself determine the model.

Element 4 in the above list is also very important. If we are developing a model for a purpose other than prediction, another procedure may be more appropriate than regression (e.g., the RMA line discussed in Chap. 8). Recall from Chap. 8 that regression implies that we want to model how the mean value (i.e., the expected value) of the predicted quantity varies with the observed quantities.

Element 1 might be unnecessary in some situations. Consider the plots in Fig. 9-1, which have a large number of measurements for *X* and *Y*, two variables of interest. It is apparent from the left diagram that there is some relationship between the two variables. Hence, a knowledge of one could be profitably used to predict the value of the other. Assume that we wish to predict *Y* from measurements of *X*. There are enough data that we could divide up the "cloud" into sections or bins (right figure) and calculate the arithmetic average of the *Y*'s in each section. This procedure would give us a set of $\overline{Y}_i$'s for the $i = 1, 2, \ldots, I$ sections. We could then use this set to predict *Y* from *X*: when $X_{i-1} \leq X \leq X_i$ for some *i*, then $\hat{Y} = \overline{Y}_i$. The coarseness of the approximations thus depends on how many data we have; if we have many data, the sectioning could be quite fine and we would have a good approximation to how $\overline{Y}$ varies with *X*, the *regression* relationship, without recourse to a model.

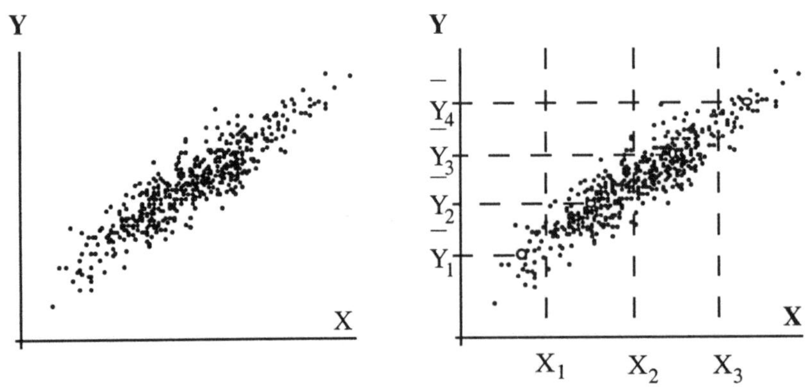

Figure 9-1. Many data in an *X-Y* scatter plot (left) provide directly for an *X-Y* relationship.

Of course, because this procedure involves step changes in $\hat{Y}$, it requires a considerable amount of data to develop. It also should not be used outside the measured data range; the procedure would require that $\hat{Y} = \bar{Y}_I$ for any $X > X_I$, no matter how large X is, and it ignores any information we have about the X-Y relationship. Instead, we may wish to introduce an equation (element 1) that will tell us, in a continuous manner, how $\bar{Y}$ varies with X. Therefore, we want to use prior knowledge (the form of the equation) to replace the need for so many data and obtain a continuous regression relation.

9-2 THE LINEAR MODEL

The most common type of regression is linear regression. Furthermore, of all possible models, the linear model is the most frequently used:

$$Y = \beta_0 + \beta_1 X + \varepsilon \tag{9-1}$$

where X is the *predictor* or *explanatory variable*, Y is the *response*, ε is a *random error*, and β_0 and β_1 are the *regression model parameters*. X is called an explanatory variable because a knowledge of how X varies will be used to explain variations in Y. β_1 is also called the regression coefficient.

Note that "linear" can be used in two ways when considering regression. The first way describes how the unknowns (β_0 and β_1 in the above equation) appear in the problem. The second way describes how the predictor variable appears in the problem. For example, the model $Y = \beta_0 + \beta_1 X^2 + \varepsilon$ is still a linear regression problem, since the unknown β's appear linearly in the model, but the model is nonlinear in X, the predictor. Since ε is a random variable in the model, Y is also a random variable. X might or might not be a random variable. See Graybill (1961, Chap. 5) for an excellent discussion of the various situations Eq. (9-1) might represent.

If the model, Eq. (9-1), is describing the data behavior well, the errors are random. This means that they are independent: individual deviations do not depend on other deviations or on the data values. It is also common to assume that errors are normally distributed with zero mean ($E(\varepsilon) = 0$). The term "error" here covers a multitude of sins. It includes measurement errors in the observed values of Y and inadequacies in the model. For example, ε includes the effects of variables that influence the response but that were not measured.

Model inadequacy is often a significant contributor to the "error." Typically, what we set out to measure is measured with reasonably good accuracy. The properties that we cannot measure, coupled with the use of simplistic models, usually account for most of the prediction error. This aspect of the error is often overlooked in discussions of its role and magnitude. However, even if we cannot measure the predictor accurately, regression is usually the appropriate method when we want to develop a model for prediction.

We can take the conditional expectation of both sides of Eq. (9-1) to give

$$E(Y \mid X = X_0) = \beta_0 + \beta_1 X_0$$

since we assume that $E(\varepsilon) = 0$. Likewise, the variance is

$$Var(Y \mid X = X_0) = Var(\beta_0 + \beta_1 X + \varepsilon \mid X = X_0) = \sigma_\varepsilon^2$$

We are using conditional operators here because, by assumption, Y depends on X. Recall that, from Chap. 8, $E(Y \mid X = X_0)$ is the mean value for Y when X takes the value X_0 and involves the conditional PDF of Y, $f_{Y|X}(Y \mid X = X_0)$. The last equation yields the variance of errors because the errors are assumed independent of X.

9-3 THE LEAST-SQUARES METHOD

We use the method of *least squares* to estimate the model parameters β_0 and β_1 in Eq. (9-1) from a set of paired data $\{(X_1, Y_1), (X_2, Y_2), \ldots, (X_I, Y_I)\}$. The least-squares criterion is a minimization of the sum of the squared differences between the observed responses, Y_i , and the predicted responses, $\hat{Y}_i$, for each fixed value of X_i (Fig. 9-2). These differences, $Y_i - \hat{Y}_i$, are called *residuals*.

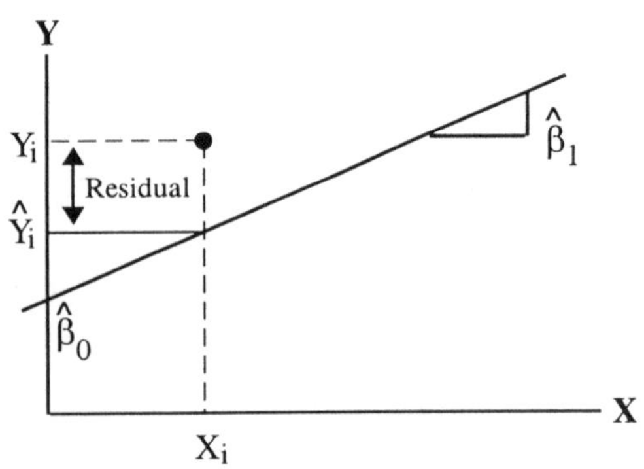

Figure 9-2. The least-squares procedure determines a line that minimizes the sum of all the squared differences, $(Y_i - \hat{Y}_i)^2$, between the data points and the line.

In mathematical terms, the preceding description can be expressed as

$$S(\beta_0, \beta_1) = \sum_{i=1}^{I} \left(Y_i - \hat{Y}_i\right)^2 = \sum_{i=1}^{I} (Y_i - \beta_0 - \beta_1 X_i)^2 \tag{9-2}$$

Equation (9-2) is I times the estimate of $Var(Y_i - \hat{Y}_i)$, since $Var(Y_i - \hat{Y}_i) = E[(Y_i - \hat{Y}_i)^2]$ and $E(Y_i - \hat{Y}_i) = 0$. Since S is proportional to a variance, it is always nonnegative.

We would like to find the values of the slope (β_1) and intercept (β_0) that make S as small as possible (Fig. 9-3).

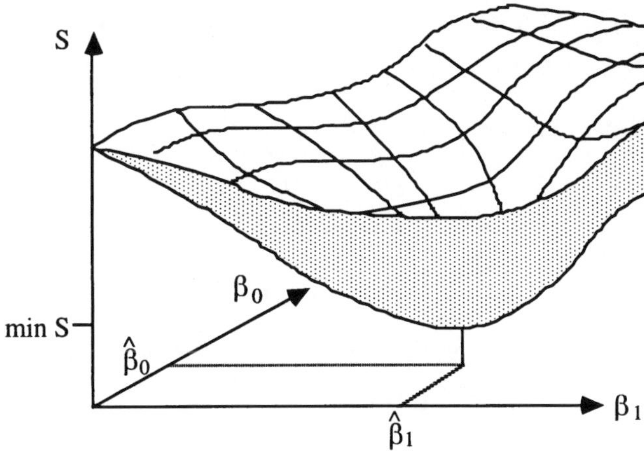

Figure 9-3. The least-squares procedure finds the minimum value of S.

A differential change in S, dS, can be written in terms of differential changes in β_0 and β_1 as

$$dS = \frac{\partial S}{\partial \beta_0} d\beta_0 + \frac{\partial S}{\partial \beta_1} d\beta_1$$

This is a local condition since it applies only to small changes in the variables. At the minimal S, we must have $dS = 0$. Since the slope and intercept are independent (i.e., $\lambda_0 d\beta_0 + \lambda_1 d\beta_1 = 0$ only for $\lambda_0 = \lambda_1 = 0$), this can only be true if each partial derivative is zero.

$$\text{From } \frac{\partial S}{\partial \beta_0} = 0: \qquad I\hat{\beta}_0 + \hat{\beta}_1 \sum X_i = \sum Y_i \quad \text{or} \quad \Sigma (Y_i - \hat{Y}_i) = 0 \tag{9-3a}$$

$$\text{From } \frac{\partial S}{\partial \beta_1} = 0: \qquad \hat{\beta}_0 \sum X_i + \hat{\beta}_1 \sum X_i^2 = \sum Y_i\, X_i \quad \text{or} \quad \Sigma (Y_i - \hat{Y}_i)\, X_i = 0 \tag{9-3b}$$

where the summations are from $i = 1$ to $i = I$. These equations are the least-squares *normal equations*. $\hat{\beta}_0$ and $\hat{\beta}_1$ are the values of β_0 and β_1 that minimize S, but they also have a statistical interpretation. $\hat{\beta}_0$ and $\hat{\beta}_1$ are the slope and intercept estimates derived from a sample of I data; they may differ in value from the population slope and intercept values. Because the Y_i are samples of a random variable, $\hat{\beta}_0$ and $\hat{\beta}_1$ are also samples of random variables. Similarly, S is a random variable, related to the variability of the residuals.

The version of Eq. (9-3b) on the right makes it apparent that the residuals $Y_i - \hat{Y}_i$ and X_i are orthogonal (i.e., have zero estimated covariance). This means that the least-squares procedure is giving residuals that have no further relationship with X. Thus, the model and least-squares estimator have "squeezed out" all the explanatory power of X (see Box et al., 1978, Chap. 14, for further details).

Both of the normal equations are linear; hence, they can be solved as

$$\hat{\beta}_0 = \frac{1}{I} \sum Y_i - \hat{\beta}_1 \frac{1}{I} \sum X_i \qquad \text{(9-4a)}$$

and

$$\hat{\beta}_1 = \frac{\Sigma Y_i X_i - \frac{1}{I}(\Sigma X_i)(\Sigma Y_i)}{\Sigma X_i^2 - \frac{1}{I}(\Sigma X_i)^2} \qquad \text{(9-4b)}$$

to give

$$\hat{Y} = \hat{\beta}_0 + \hat{\beta}_1 X$$

the *sample regression of Y on X*. Equation (9-4b) can also be written as $\hat{\beta}_1 = S_{XY} / S_{XX}$, where S_{XY} represents the sum $\Sigma(X_i - \bar{X})Y_i$ and S_{XX} is the sum of the squares $\Sigma(X_i - \bar{X})^2$.

The form of Eq. (9-2) ensures that, within the constraints of the model, the values $\hat{\beta}_0$ and $\hat{\beta}_1$ will approximate the mean value of Y at any given $X = x$. Recall that, from Chap. 4, the mean value minimizes the variance. Since S is a measure of the variance of the data about the line, $\hat{\beta}_0 + \hat{\beta}_1 x$ thus approximates $E(Y \mid X = x)$. Equation (9-4a) reflects this property in that $\hat{\beta}_0$ is computed to ensure that the line includes the point $(\bar{X}, \bar{Y})$.

Example 1a - Regression Using a Small Data Set. Compute the least-squares estimates $\hat{\beta}_0$ and $\hat{\beta}_1$ for the model $Y = \beta_0 + \beta_1 X + \varepsilon$ with the data set

$\{(X, Y)\} = \{(0, 0), (1, 1), (1, 3), (2, 4)\}$.

We have $\Sigma X_i = 4$, $\Sigma Y_i = 8$, $\Sigma X_i Y_i = 12$, and $\Sigma X_i^2 = 6$. From Eqs. (9-4),

$$\hat{\beta}_1 = \frac{12 - \frac{1}{4}(4)(8)}{6 - \frac{1}{4}4^2} = \frac{4}{2} = 2 \quad \text{and} \quad \hat{\beta}_0 = \frac{1}{4}(8) - 2\left(\frac{1}{4}\right)(4) = 0$$

A similar approach can be taken with models having more than two unknowns, although the expressions are best handled using matrix notation. Suppose the model is

$$Y = \beta_0 + \beta_1 X^{(1)} + \beta_2 X^{(2)} + \cdots + \beta_J X^{(J)} + \varepsilon$$

where we have J explanatory variables and a set of I data. Let $\hat{Y}$ and Y be the I x 1 vectors of predicted and actual responses, X is the I x (J + 1) matrix of explanatory values, and $\hat{\beta}$ is a (J + 1) x 1 vector of the estimated model parameters. $\hat{Y} = X\hat{\beta}$ thus represents the following equation:

$$\begin{bmatrix} \hat{Y}_1 \\ \hat{Y}_2 \\ \cdot \\ \cdot \\ \cdot \\ \hat{Y}_I \end{bmatrix} = \begin{bmatrix} 1 & X_1^{(1)} & X_1^{(2)} & \cdot & \cdot & \cdot & X_1^{(J)} \\ 1 & X_2^{(1)} & X_2^{(2)} & \cdot & \cdot & \cdot & X_2^{(J)} \\ \cdot & \cdot & \cdot & \cdot & \cdot & \cdot & \cdot \\ \cdot & \cdot & \cdot & \cdot & \cdot & \cdot & \cdot \\ \cdot & \cdot & \cdot & \cdot & \cdot & \cdot & \cdot \\ 1 & X_I^{(1)} & X_I^{(2)} & & & & X_I^{(J)} \end{bmatrix} x \begin{bmatrix} \hat{\beta}_0 \\ \hat{\beta}_1 \\ \cdot \\ \cdot \\ \cdot \\ \hat{\beta}_J \end{bmatrix}$$

Using this shorthand notation, the normal equations, Eqs. (9-3), can be written as $X^T(Y - \hat{Y}) = 0$ or, since $\hat{Y} = X\hat{\beta}$, the normal equations become $X^T(Y - X\hat{\beta}) = 0$, where X^T is the transpose of X. If we solve the normal equations for $\hat{\beta}$, we obtain

$$\hat{\beta} = [X^T X]^{-1} X^T Y$$

Example 1b - Regression Using a Small Data Set. We re-solve the previous problem using matrix notation. In this case, $J = 1$,

$$Y = \begin{bmatrix} 0 \\ 1 \\ 3 \\ 4 \end{bmatrix}, \quad X = \begin{bmatrix} 1 & 0 \\ 1 & 1 \\ 1 & 1 \\ 1 & 2 \end{bmatrix}, \quad \text{and} \quad \hat{\beta} = \begin{bmatrix} \hat{\beta}_0 \\ \hat{\beta}_1 \end{bmatrix}$$

Hence,

$$[X^T X]^{-1} = \left\{ \begin{bmatrix} 1 & 1 & 1 & 1 \\ 0 & 1 & 1 & 2 \end{bmatrix} \begin{bmatrix} 1 & 0 \\ 1 & 1 \\ 1 & 1 \\ 1 & 2 \end{bmatrix} \right\}^{-1} = \begin{bmatrix} 4 & 4 \\ 4 & 6 \end{bmatrix}^{-1} = \begin{bmatrix} 3/4 & -1/2 \\ -1/2 & 1/2 \end{bmatrix}$$

and

$$\hat{\beta} = [X^T X]^{-1} X^T Y = \begin{bmatrix} 3/4 & -1/2 \\ -1/2 & 1/2 \end{bmatrix} \begin{bmatrix} 1 & 1 & 1 & 1 \\ 0 & 1 & 1 & 2 \end{bmatrix} \begin{bmatrix} 0 \\ 1 \\ 3 \\ 4 \end{bmatrix} = \begin{bmatrix} 0 \\ 2 \end{bmatrix}$$

This is the same result as obtained in Example 1a.

The approach embodied by the above equations, which assume that all the data (X_i, Y_i) are equally reliable, is called *ordinary least squares* (OLS). The equations for the more general case where different points have differing reliabilities (*weighted least squares,* or WLS) is treated in Chap. 10. For the linear OLS model, many hand calculators have these equations already programmed as function keys.

9-4 PROPERTIES OF SLOPE AND INTERCEPT ESTIMATES

Understanding the origin of variances in the slope and intercept estimates takes some doing, because we have been treating them as single-valued parameters used to describe a set of data. Recalling the probabilistic approach described in Chap. 2, however, the observations Y_i and X_i are really only the results of experiments from two sample spaces that are correlated. If we were to pick another set of (X_i, Y_i), we would get another slope and intercept. If we pick many sets of (X_i, Y_i), we get a set of slopes and intercepts that are distributed with mean and variance as given below.

When $E(\varepsilon) = 0$ and $Var(\varepsilon) = \sigma_\varepsilon^2$, we can estimate the bias and precision of the slope and intercept obtained by OLS. See Montgomery and Peck (1982, Chap. 2), Box et al. (1978, Chap. 14), or Rice (1988, Chap. 14) for derivations of these results. Both $\hat{\beta}_0$ and $\hat{\beta}_1$ are unbiased:

$$E(\hat{\beta}_0) = \beta_0 \quad \text{and} \quad E(\hat{\beta}_1) = \beta_1$$

More importantly, the precision of the estimates is given by

$$Var(\hat{\beta}_0) = \sigma_\varepsilon^2 \left[\frac{1}{I} + \frac{\overline{X}^2}{\sum_{i=1}^{I} (X_i - \overline{X})^2} \right] \tag{9-5a}$$

and

$$Var(\hat{\beta}_1) = \frac{\sigma_\varepsilon^2}{\sum_{i=1}^{I} (X_i - \overline{X})^2} \tag{9-5b}$$

These assume that the ε_i values are independent and we know the value of σ_ε^2. Usually, we have to depend on the data to estimate σ_ε^2. The σ_ε^2 estimate is based on the residual sum of squares:

$$S_\varepsilon = \Sigma\, \varepsilon_i^2 = \Sigma\, (Y_i - \hat{Y}_i)^2$$

S_ε is the value of S in Eq. (9-2) when the values $\hat{\beta}_0$ and $\hat{\beta}_1$ are chosen for the intercept and slope, respectively. The expected value of the residual sum of squares is

$$E(S_\varepsilon) = (I - 2)\sigma_\varepsilon^2$$

so that σ_ε^2 is estimated as

$$\hat{s}_\varepsilon^2 = \frac{S_e}{I-2} = MS_\varepsilon \qquad (9\text{-}6)$$

S_ε is the estimate of the error variance and MS_ε is called the *mean square error.* The square root of MS_ε is the *standard error of regression.* S_ε has $(I - 2)$ degrees of freedom because two degrees of freedom are associated with the regression estimates of the slope and intercept.

Returning now to the sample slope and intercept variabilities, our estimates are based on Eqs. (9-5) and (9-6).

$$s_{\beta_0}^2 = MS_\varepsilon \left[\frac{1}{I} + \frac{\overline{X}^2}{\sum_{i=1}^{I}(X_i - \overline{X})^2} \right] \qquad (9\text{-}7a)$$

and

$$s_{\beta_1}^2 = \frac{MS_\varepsilon}{\sum_{i=1}^{I}(X_i - \overline{X})^2} \qquad (9\text{-}7b)$$

The variabilities of both sample parameters are directly proportional to the variability of ε and inversely related to the variability in X. Thus, a greater spread in X's will give more reliable parameter estimates. This means that a well-designed experimental program will interrogate the explanatory variables (the X's in this case) over as wide a range as possible.

Example 1c - Estimating Parameter Variability. We now estimate the variabilities of the sample intercept and slope for the small data set of Example 1a. The residuals are shown in the table below.

i	X_i	Y_i	$\hat{Y}_i$	$Y_i - \hat{Y}_i$	$(Y_i - \hat{Y}_i)^2$	$X_i - \overline{X}$	$(X_i - \overline{X})^2$
1	0	0	0	0	0	-1	1
2	1	1	2	-1	1	0	0
3	1	3	2	1	1	0	0
4	2	4	4	0	0	1	1

Hence, $S_\varepsilon = \Sigma(Y_i - \hat{Y}_i)^2 = 2$, $MS_\varepsilon = SS_\varepsilon/(I - 2) = 1$, $\bar{X} = 1$ and $\Sigma(X_i - \bar{X})^2 = 2$. This gives

$$s^2_{\beta_0} = 1(\frac{1}{4} + \frac{1^2}{2}) = 3/4 \quad \text{and} \quad s^2_{\beta_1} = 1/2$$

If we assume that the errors are normally distributed, we can produce confidence intervals for $\hat{\beta}_0$ and $\hat{\beta}_1$ from the above results (Chap. 5). The 95% interval for $\hat{\beta}_0$ is $\pm\, t(0.025, 2)\, s_{\beta_0} = \pm 4.3\ (0.75)^{1/2} = \pm 3.7$. The t value chosen, $t = 4.3$, reflects the two-sided nature of the interval and that, with 4 data and 2 parameters estimated, there are only two degrees of freedom remaining. A similar treatment for $\hat{\beta}_1$ yields the 95% interval $\pm t(0.025, 2)\, s_{\beta_1} = \pm 4.3(0.5)^{1/2} = \pm 3.0$. From these data and the assumptions mentioned, $\beta_0 = 0 \pm 3.7$ and $\beta_1 = 2 \pm 3$ with a 95% level of confidence. Such large errors on the slope and intercept reflect the small number of data in this example.

If the errors remaining from the regression are independent and normally distributed, then the estimates of the slope and intercept are called *BLUE*. This stands for *b*est (minimum variance), *l*inear, *u*nbiased *e*stimators. We will encounter BLUE estimators again in Chap.12.

9-5 SEPARATELY TESTING THE PRECISION OF THE SLOPE AND INTERCEPT

Suppose we wish to test whether the slope $\hat{\beta}_1$ is significantly different from some *a priori* constant, β_1'. If the errors are $N(0, \sigma^2_\varepsilon)$ and the observations Y_i are uncorrelated, $\hat{\beta}_1 \sim N(\beta_1, s^2_{\beta_1})$. To test whether the estimate and the constant are different, we form the statistic

$$t_{\beta_1} = \frac{|\hat{\beta}_1 - \beta_1'|}{\sqrt{s^2_{\beta_1}}}$$

The t_{β_1} statistic has a t distribution with $(I - 2)$ degrees of freedom (Chap. 7). The t distribution is used here to account for the added variability in the test resulting from the MS_ε approximation of s^2 for small data sets. As I becomes large, t_{β_1} approaches an $N(0,1)$ distribution. The regression estimates account for two degrees of freedom; therefore, the t ratio has $df = I - 2$. We complete the test by comparing the computed value of t_{β_1} with $t(\alpha/2, df)$. The proposed value and the estimated value are different to the α probability if t_{β_1} is greater than the tabulated value.

A similar approach can be taken for the confidence limits of $\hat{\beta}_0$. To compare $\hat{\beta}_0$ with any reference value β_0', we use

$$t_{\beta_0} = \frac{|\hat{\beta}_0 - \beta_0'|}{\sqrt{s_{\beta_0}^2}}$$

Example 2a - Separately Testing Intercept and Slope Estimates. A set of core-plug (ϕ_p) and wireline (ϕ_w) porosity data (I = 41) were depth-matched and plotted to assess their relationship. The aim of the analysis is to see if ϕ_w could be used to predict ϕ_p. If both measurements are responding to the same features of the formation, we would expect a regression line to have both unit slope and zero intercept. The result (Fig. 9-4), however, suggests the measurements may not be similar; the regression-line intercept is not zero and the slope is 16% greater than one. The slope and intercept estimate variabilities should be evaluated to help decide what these data are indicating. We will test the intercept and slope to see if either one is statistically different from 0 or 1, respectively.

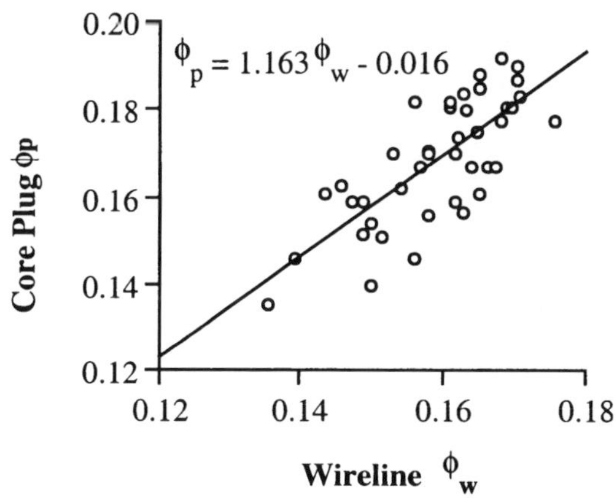

Figure 9-4. Wireline and core-plug porosities with regression line.

We first check the residuals, $\phi_{pi} - \hat{\phi}_{pi}$. A probability plot (Fig. 9-5 top left) suggests that they appear normally distributed. A plot of residuals against ϕ_w (Fig. 9-5 top right) shows no particular pattern, suggesting the linear model is adequate. Similarly, there appears to be no pattern in a scatter plot of the ith residual versus the $(i+1)^{th}$ residual (Fig. 9-5 bottom), suggesting the errors are uncorrelated to each other. These checks help to confirm that the assumptions of $\varepsilon \sim N(0,\sigma_\varepsilon^2)$ and the appropriateness of the linear model apply.

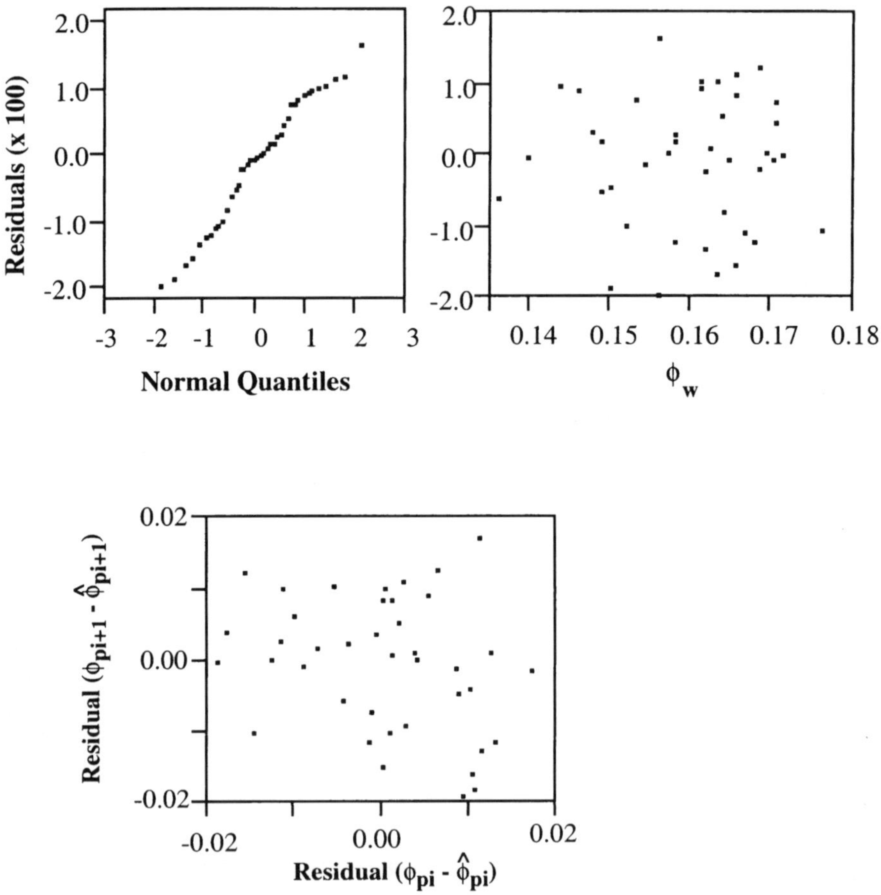

Figure 9-5. Diagnostic probability (top left) and residuals plots (top right and bottom) suggest the errors are normally distributed, show no systematic variation over the range of X, and are uncorrelated, respectively.

For these 41 data, $MS_\varepsilon = 0.0000892$, $\bar{\phi}_w = 0.159$, and $\Sigma(\phi_{w_i} - \bar{\phi}_w)^2 = 0.00342$. From Eqs. (9-7),

$$s^2_{\beta_0} = 0.0000892\ \left(\frac{1}{41} + \frac{0.159^2}{0.00342}\right) = 0.00066$$

and

$$s^2_{\beta_1} = \frac{0.0000892}{0.00342} = 0.0261$$

The critical t value for the 95% level is $t(0.025, 39) = 2.02$, so that

$$-0.016 - 2.02\sqrt{0.00066} \le \beta_0 \le -0.016 + 2.02\sqrt{0.00066} \quad \text{or} \quad -0.068 \le \beta_0 \le 0.036$$

and

$$1.163 - 2.02\sqrt{0.0261} \le \beta_1 \le 1.163 + 2.02\sqrt{0.0261} \quad \text{or} \quad 0.83 \le \beta_1 \le 1.5$$

Hence, the data do not contradict either of the hypotheses that $\beta_0 = 0$ or $\beta_1 = 1$. The t tests confirm this:

$$t_{\beta_0} = \frac{|-0.016 - 0|}{\sqrt{0.00066}} = 0.62 \quad \text{and} \quad t_{\beta_1} = \frac{|1.163 - 1|}{\sqrt{0.0261}} = 0.99$$

Referring to a table of t values (e.g., Abramowitz and Stegun, 1965), we find that t_{β_0} is only significant at approximately the $\alpha = 50\%$ level, and t_{β_1} is significant at the 70% level. This analysis suggests that the wireline measurements can be used interchangeably with core-plug porosities provided the formation character does not change from that represented by these data (e.g., similar levels of porosity heterogeneity).

As an alternative to plotting the residuals as in Example 2a, we could have estimated the covariances between the residuals and ϕ_w and between the ith and $(i+1)^{th}$ residuals to perform these functions. Plotting is actually a more powerful way to check for independence because it will show all error types.

9-6 JOINTLY TESTING THE PRECISION OF THE SLOPE AND INTERCEPT

In Example 2a, we separately tested the slope and intercept estimates and concluded that, at a 95% level (i.e., $\alpha = 5\%$), the sample line does not have a slope or an intercept appreciably different from what we would expect. $\hat{\beta}_0$ was tested with the null hypothesis (H_0) that $\beta_0 = 0$ and the alternative hypothesis (H_A) that $\beta_0 \ne 0$. For $\hat{\beta}_1$, H_0 is $\beta_1 = 1$ and H_A is $\beta_1 \ne 1$. What we did not do, however, was test together (jointly) the null hypothesis that $\beta_0 = 0$ *and* $\beta_1 = 1$.

In Chap. 7, we explained that applying a t test to each part of a joint problem does not give the same confidence level that a joint test does. The *Bonferroni* inequality (Sec. 2-3) provides a useful indicator of what may happen to the confidence level when separate tests are applied instead of using a joint test. For example, applying two tests with each having $\alpha = 5\%$ equates to one test having $\alpha \le 10\%$. The lower limit is very close in some situations, e.g., independent events. In the case of testing the slope and intercept individually at $\alpha = 5\%$, our results might be no better than 10%. Trying to obtain the exact confidence level, however, is tricky because $\hat{\beta}_0$ and $\hat{\beta}_1$ are correlated.

The problem of joint confidence tests can be described in geometrical terms. Separate confidence intervals (C. I.) on $\hat{\beta}_0$ and $\hat{\beta}_1$ provide a rectangular confidence region in the β_0 –

β_1 plane (Fig. 9-6 left). This region represents the values for β_0 and β_1 that the data would support within the specified confidence level. The value for one parameter takes no account of the effect it has on the other parameter. The borders of the region represent different values of S in Eq. (9-2), where S is the sum of squared deviations in Y between the regression line and the data.

Regions of constant S are ellipses (Fig. 9-6 right), and it is values of S that are a better measure for permissible combinations of β_0 *and* β_1. Why is this so? It is clear that S is a measure of the joint suitability of $\hat{\beta}_0$ and $\hat{\beta}_1$ with the data (Eq. (9-2)), but S is also a random variable. Therefore, S has an associated confidence interval so that $S \le S_\alpha$, where S_α is the maximum permitted value of S at the $(1 - \alpha)$ level of confidence. S may not achieve its minimum value when $\hat{\beta}_0 = \beta_0$ and $\hat{\beta}_1 = \beta_1$. The data and the least-squares estimator, however, dictate the values for $\hat{\beta}_0$ and $\hat{\beta}_1$. Therefore, this confidence level for S represents all pairs (β_0, β_1) that make the joint null hypothesis acceptable such that $S \le S_\alpha$. For example, points A and B are both on the border of the independent assessments region (Fig. 9-6 left) but have different values of S. Point A is well outside the permissible variation of S at the specified confidence level whereas Point B just qualifies.

The equation for the elliptical region, derived in Montgomery and Peck (1982, pp. 389-391), is

$$\frac{I(\hat{\beta}_0 - \beta_0)^2 + 2(\hat{\beta}_0 - \beta_0)(\hat{\beta}_1 - \beta_1)\Sigma X_i + (\hat{\beta}_1 - \beta_1)^2 \Sigma X_i^2}{2MS_\varepsilon} \le F(2, I - 2, \alpha) \tag{9-8}$$

where $F(2, I - 2, \alpha)$ is the F statistic (see Chap. 7) for the desired confidence level $(1 - \alpha)$.

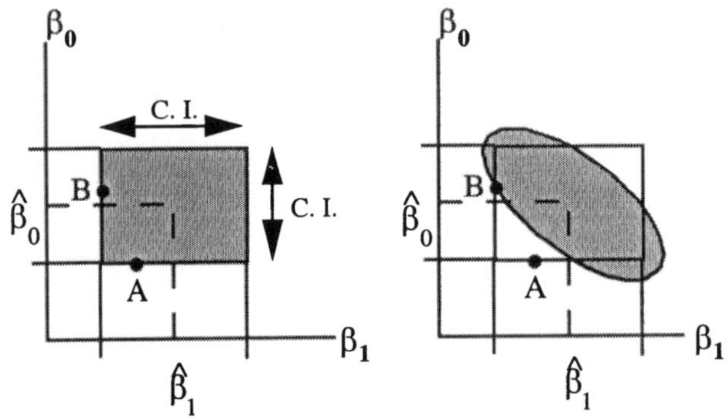

Figure 9-6. Confidence regions for β_0 and β_1, as reflected by the data, for independent assessments (left) and joint assessment (right) are shown by the shaded regions. (C. I. = confidence interval.)

Example 2b - Jointly Testing Intercept and Slope Estimates. We now test the slope and intercept estimates jointly at a 95% level of confidence with H_0: $\beta_0 = 0$ and β_1

= 1. We use Eq. (9-8) with the following values:

$\hat{\beta}_0 - \beta_0 = -0.016 - 0 = -0.016$; $\hat{\beta}_1 - \beta_1 = 1.163 - 1 = 0.163$

$I = 41$ $MS_\varepsilon = 0.0000892$; $\Sigma X_i = I\,\bar{X} = 41(0.159) = 6.5$

$\Sigma X_i^2 = \Sigma(X_i - \bar{X})^2 + I\,\bar{X}^2 = 0.00342 + 41(0.159^2) = 1.04 F(2, 39, 0.05) = 3.23$

The left side of Eq. (9-8) gives

$$\frac{41(-0.016)^2 + 2(-0.016)(0.163)6.5 + (0.163)^2\,1.04}{2(0.0000892)} = 22$$

which exceeds the critical F value of 3.23 by a considerable margin. Hence, the data do not support the model $Y = X$ (i.e., a model with unit slope *and* zero intercept).

The contours of constant S (confidence regions) are rather long, narrow, and inclined with a negative slope for this data set (Fig. 9-7) because $\hat{\beta}_0$ and $\hat{\beta}_1$ are strongly anticorrelated: if $\hat{\beta}_1$ increases, $\hat{\beta}_0$ will decrease. From Eq. (9-5a), the variability of $\hat{\beta}_0$ is caused by the variabilities in $\bar{Y}$ and $\hat{\beta}_1$. For this data set, where $\bar{X}$ is far from $X = 0$ (about two standard deviations), the variability of $\hat{\beta}_1$ dominates. In the context of Example 2, we can conclude that there is not a one-to-one correspondence between core-plug and wireline porosities, contrary to the independent assessment results, because this test overlooked the dependency between slope and intercept.

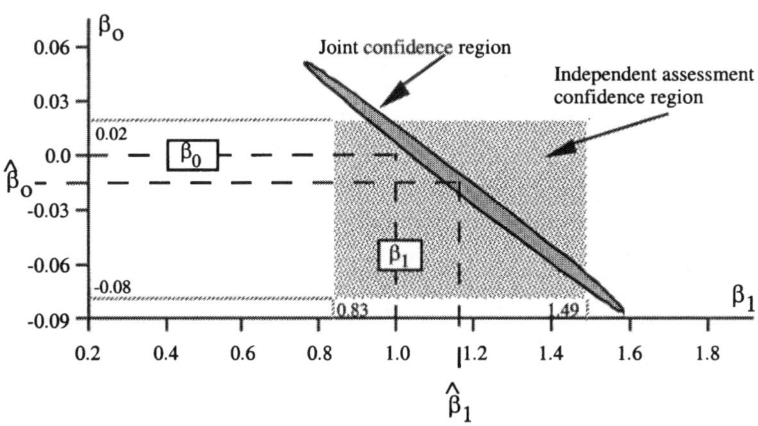

Figure 9-7. The rectangular confidence region obtained from independent assessments of $\hat{\beta}_0$ and $\hat{\beta}_1$ and the joint, elliptical confidence region of $\hat{\beta}_0$ and $\hat{\beta}_1$.

There are other methods for producing joint confidence regions and testing besides the constant-S approach discussed here (e.g., Montgomery and Peck, 1982, pp. 393-396). They are generally simpler but may not give the desired result if a specific confidence level is required; they only assure that the test will be at least at the specified level. Joint confidence regions are also useful for testing the significance of different regression lines (Seber, 1977, Chap. 7).

9-7 MEAN RESPONSE AND NEW OBSERVATION CONFIDENCE INTERVALS

If we wish to estimate $E(Y \mid X = X_0)$, an unbiased point estimator is

$$\hat{Y}_0 = \hat{\beta}_0 + \hat{\beta}_1 X_0$$

$\hat{Y}_0$ has variance

$$Var(\hat{Y}_0) = Var\,(\hat{\beta}_0 + \hat{\beta}_1 X_0)$$

$$= Var[\,\bar{Y} + \hat{\beta}_1(X_0 - \bar{X})]$$

$$= \sigma_\varepsilon^2\left[\frac{1}{I} + \frac{(X_0 - \bar{X})^2}{\Sigma(X_i - \bar{X})^2}\right] \qquad (9\text{-}9)$$

When we replace σ_ε^2 by its estimate, this gives the following sample variance:

$$s_{\hat{Y}_0}^2 = MS_\varepsilon\left[\frac{1}{I} + \frac{(X_0 - \bar{X})^2}{\Sigma(X_i - \bar{X})^2}\right]$$

The estimate for $Var(\hat{Y}_0)$ is then used to define the $(1 - \alpha)$ confidence bands for the estimate of $E(Y \mid X = X_0)$:

$$\hat{Y}_0 \pm t(\alpha/2, I - 2)\, s_{\hat{Y}_0}$$

The confidence interval widens as $(X_0 - \bar{X})^2$ increases and, when $X_0=0$, $Var(\hat{Y}_0)=Var(\hat{\beta}_0)$. Intuitively, this result seems reasonable since we would expect to estimate $E(Y \mid X = X_0)$ better for X values near the center of the data than for X values near the extremes.

To produce a confidence interval about the line to define where a new observation (X_0, Y^*) might lie, we add the error component variability to Eq. (9-9).

$$Var(Y^*) = Var(\hat{Y}_0) + Var(\varepsilon) = \sigma_\varepsilon^2\left[1+\frac{1}{I}+\frac{(X_0-\overline{X})^2}{\Sigma(X_i-\overline{X})^2}\right]$$

Addition of variances is permissible here because of the independence of the error ε. Using sample values, this leads to the confidence interval

$$\hat{Y}_0 \pm t(\alpha/2, I-2)\sqrt{MS_\varepsilon\left[1+\frac{1}{I}+\frac{(X_0-\overline{X})^2}{\Sigma(X_i-\overline{X})^2}\right]}$$

about the line for a new observation, Y^*.

9-8 RESIDUALS ANALYSIS

As suggested in Example 2, plots of the residuals can be very helpful in determining how well a model captures the behavior of the data. Residual plots can also convey whether parameters that were not included in the regression could help to better predict the Y's.

Example 3 - Residual Analysis of a Data Set. From the scatter plot in Fig. 9-8 left, a linear relation looks appropriate and a least-squares line may be calculated. A residuals plot (Fig. 9-8 right), however, shows a problem. For $-1 \leq X \leq 4$, the residuals appear to systematically increase with X, suggesting the calculated line is not increasing as quickly as the Y's are. The point at $X = 5.8$ (a *high leverage* point) has a large influence on the line and has reduced the slope. Consequently, there is a significant relationship between the residuals and X. We should now decide whether to keep the point, which requires further assessment of how representative it is. It would be inappropriate to continue with the line as it is because the assumption of normally distributed, independent errors has been violated.

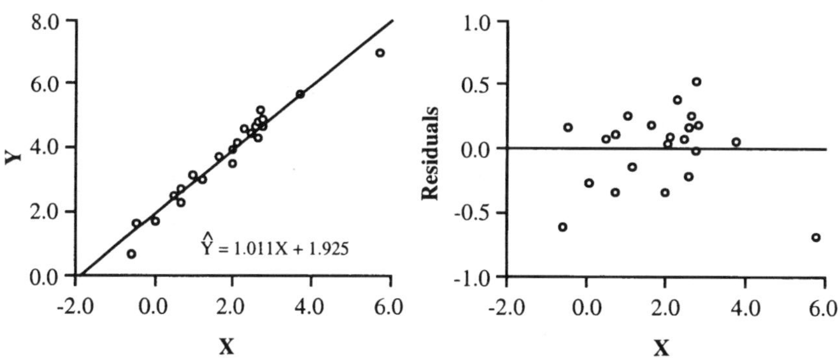

Figure 9-8. While a linear model (left) appears suitable, a residual plot suggests otherwise. Note the change of scale for the vertical axis between the plots.

As already suggested, we can plot residuals against any other variable, including those not considered in the model. By doing so, we will be able to tell whether there is any variation in

Y that can be explained by a missing variable. This approach is an improvement over multiple regression procedures that throw all possible variables in at once without considering whether they have any explanatory power. We should approach multiple regression systematically, variable by variable, choosing first those in which we have the most confidence and those that have the strongest engineering and geological reasons for being included. Further aspects of residual analysis are discussed in many regression texts, including Box et al. (1978), Montgomery and Peck (1982), and Hoaglin et al. (1983).

Residual analysis is particularly important when confidence intervals are desired. Confidence intervals can be used, when the assumptions are fulfilled, for many purposes. They help to decide whether a predictor has any explanatory power, to determine required numbers of samples, and to compare parameter estimates with other values. Confidence intervals, however, may easily mislead if the residuals are not examined first to ensure the model and errors are behaving suitably.

A simple example of misleading results occurs when confidence intervals are used to decide whether X has any explanatory power for Y. Figure 9-9 shows two scatter plots and fitted lines for which the slope confidence intervals include zero. This result for the left figure correctly leads to the conclusion that X has no power to predict Y. On the other hand, the right figure suggests that X has considerable explanatory power, but the model is inappropriate. In both cases, a plot of $(Y_i - \hat{Y}_i)$ versus X would immediately show whether the fitted line properly reflected the X-Y association depicted by the data.

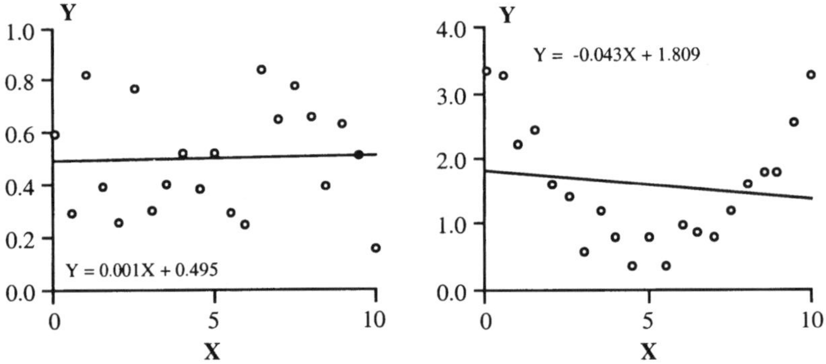

Figure 9-9. Two data sets with near-zero regression line slopes but different explanatory powers of X.

9-9 THE COEFFICIENT OF DETERMINATION

Often we would like to know how much of the observed variability can be explained by a given regression model with fitted parameters. The *coefficient of determination* is that proportion of the variability in Y explained by the model. It is defined as

$$R^2 = 1 - \frac{S_\varepsilon}{S_{YY}}$$

where S_{YY} is the total variability in Y, given by $\sum (Y_i - \bar{Y})^2$, so that

$$R^2 = 1 - \frac{\Sigma(\hat{Y}_i - Y_i)^2}{\Sigma(Y_i - \bar{Y})^2}$$

The total variability in Y, S_{YY}, can be broken down into two components: the part that the model can explain, when X is known and can be used to predict Y, and the remainder (residual) that the model cannot explain. R^2, so-called because of its relationship to r^2 when X and Y are joint normally distributed (Chap. 8), measures the residual part compared to the total variation. If $R^2 = 1$, there is no residual and the model explains all variation in Y. When $R^2 = 0$, the residual and total variability are equal, thus the model has no explanatory power.

With a little reflection, we see that the coefficient of determination

1. increases as the inclination of the cloud of points increases and vice-versa,
2. does not measure the magnitude of the slope of the regression line, and
3. does not measure the appropriateness of a model.

The first point suggests that, given a cloud of points and a regression line with a positive slope, R^2 will increase as the cloud is rotated counter-clockwise about the point $(\bar{X}, \bar{Y})$. Thus, while the appropriateness of the line to the cloud does not change, R^2 changes. The second point follows because the definition of the coefficient does not contain the slope. The third point is a form of the oft-repeated caveat about regression not being able to determine the correct model.

Caution should be used when comparing R^2 values. We have just observed that comparisons of identical models between different data sets may be invalid using R^2. But comparisons of different models using the same data set may also be misleading. For example, suppose we wish to determine whether the model $Y = \beta_0 + \beta_1 X$ is better than the model $\sqrt{Y} = \beta_2 + \beta_3 X$ for a particular data set. If we simply compute the values of R^2 for the two cases and compare them, we will be comparing two different things. In the first case, R^2 gives us the proportion of variability in Y that is accounted for when X is known. In the second case, it is the proportion of variability in $\sqrt{Y}$ that is accounted for by knowing X. Because the square root is a nonlinear function, the variability in Y is quite different from the variability in $\sqrt{Y}$ (e.g., compare Y decreasing from 100 to 0 with what happens to $\sqrt{Y}$). To compare the two models, we should compare R^2 for the model $Y = (\beta_2 + \beta_3 X)^2$, where β_2 and β_3 are estimated by regressing $\sqrt{Y}$ upon X, with the R^2 value obtained using the model $Y = \beta_0 + \beta_1 X$.

Example 4 - R^2 and Porosity-Permeability Relationships. Example 6 of Chap. 3 discussed the lower Brent sequence, showing that geological information can help to separate porosity and permeability data from geologically different elements. This is also true for porosity–permeability relationships. Figure 9-10 left shows a typical plot and regression line that would be obtained if all data from the Rannoch formation were included. A large R^2 value obtains for the regression line because the line connects two clouds of data, well-separated in terms of their porosities. The lower-left cloud represents carbonate-cemented regions (concretions) with very little porosity and permeability. The upper-right cloud represents micaceous, fine-grained sandstones with no permeability below 7 mD. The regression line connects these two groups as if they represented a continuum of changing porosity and permeability, yet they represent very different lithologies with no permeability between 0.03 and 7 mD. The line poorly represents the sandstone porosity–permeability relationship (e.g., it underpredicts permeability for porosities below 25%) and is not needed to predict the concretion properties. The concretions are nonproductive intervals, easily detected with resistivity or acoustic measurements. A more appropriate regression line for the sandstones is obtained ignoring the concretion data (Fig. 9-10 right) but, for this line, the R^2 has decreased considerably.

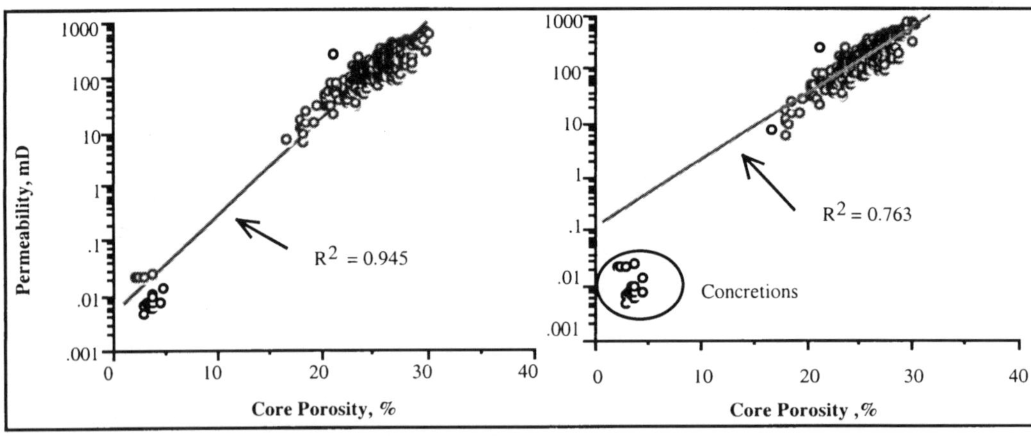

Figure 9-10. Rannoch formation porosity–permeability data and regression lines including (left) and excluding (right) the carbonate concretions.

9-10 SIGNIFICANCE TESTS USING THE ANALYSIS OF VARIANCE

The breakdown of the total variability of Y into two parts can be used to develop a significance test for the model. The total variation in Y can be expressed as $S_{YY} = S_{reg} + S_{\varepsilon}$, where S_{reg} is the regression sum of squares, $\sum (\hat{Y}_i - \bar{Y})^2$. Following the discussion in Chap. 7, the ratio of the regression to the error sums of squares forms a statistic suitable for the F test.

$$F = \frac{S_{reg}}{S_e/(I-2)}$$

can be tested against $F(1, I - 2, \alpha)$ at the desired confidence level, $1 - \alpha$, with the null hypothesis H_0: $\beta_1 = 0$ and the alternative H_A: $\beta_1 \neq 0$. S_{reg} has only one degree of freedom, since the line must pass through ($\bar{X}$, $\bar{Y}$), and S_ε has I-2 degrees of freedom for a linear model (Sec. 9-4).

Example 5 - Interpreting Computer Package Regression Output. In Example 1, a regression line was obtained for the data set $\{(X, Y)\} = \{(0, 0), (1, 1), (1, 3), (2, 4)\}$. A typical computer printout of this process gives an analysis of the slope and intercept estimates, $\hat{\beta}_1$ and $\hat{\beta}_0$, along with their standard errors. The t ratios and probabilities commonly assume the null hypotheses are $\hat{\beta}_0 = \hat{\beta}_1 = 0$, and care is needed to remember that these null values might not suit the user. For example, we may be more interested in whether $\hat{\beta}_1 = 1$.

Y by *X* Linear Fit

Summary of Fit

Rsquare	0.8	Root Mean Square Error	1
Mean of Response	2	Observations	4

Analysis of Variance

Source	DF	Sum of Squares	Mean Square	F Ratio
Model	1	8.000000	8.00000	8.0000
Error	2	2.000000	1.00000	Prob>F
C Total	3	10.000000		0.1056

Parameter Estimate

Term	Estimate	Std Error	t Ratio	Prob>|t|
Intercept	0	0.86603	0.00	1.0000
X	2	0.70711	2.83	0.1056

The analysis of variance F test gives the same result as the t test on the slope. The model sum of squares is S_{reg} and the error sum of squares is S_ε. The mean square values are sums of squares divided by their respective degrees of freedom.

9-11 THE NO-INTERCEPT REGRESSION MODEL

A frequently encountered special case of the linear regression model is where the intercept is zero. The number of degrees of freedom for the model is now (I - 1) since we have fixed the intercept, but all of the other equations are the same. The model is

$$Y = \beta_1 X + \varepsilon$$

The least-squares estimate of the slope is

$$\hat{\beta}_1 = \frac{\Sigma(Y_i X_i)}{\Sigma X_i^2}$$

which is unbiased as before. The estimate for σ_ε^2 is

$$MS_\varepsilon = \frac{1}{I-1}\Sigma(Y_i - \hat{Y}_i)^2$$

The confidence interval for the slope is

$$\hat{\beta}_1 \pm t(\alpha/2, I-1)\sqrt{\frac{MS_\varepsilon}{\Sigma X_i^2}}$$

The confidence interval for the mean response at $X = X_0$ (which becomes wider as X_0 moves away from the origin) is

$$Y_0' \pm t(\alpha/2, I-1)\sqrt{MS_\varepsilon \frac{X_0^2}{\Sigma X_i^2}}$$

For the no-intercept model, the coefficient of determination is given by

$$R^2 = 1 - \frac{\Sigma(\hat{Y}_i - Y_i)^2}{\Sigma Y_i^2}$$

The denominator of this expression is a sum of squares about the origin, not the mean of Y. Hence, comparisons between coefficients of determination from a no-intercept model and a with-intercept model are not meaningful.

All the above variabilities also apply for the more general case where the line is forced through an arbitrary point (X_0, Y_0), not just the origin. In this case,

$$\hat{\beta}_1 = \frac{\Sigma X_i Y_i - X_0 \Sigma Y_i - Y_0 \Sigma X_i + I X_0 Y_0}{\Sigma X_i^2 + I X_0^2 - 2X_0 \Sigma X_i} \quad \text{and} \quad \hat{\beta}_0 = Y_0 - \hat{\beta}_1 X_0$$

9-12 OTHER CONSIDERATIONS

As we noted in the introduction to this chapter, least-squares regression is mathematically quite straightforward but it has numerous statistical ramifications. We will cover some of the more advanced aspects in Chap. 10, but there are basic topics we still have not yet discussed. We now consider some of these issues.

9-12.1 Model Validity

Strictly speaking, regressed models may not be valid outside the range of the variables used to fit the model. Of course, if the model to be regressed is based on physical principles, this is much less of a factor, but regression-only models are often not so based.

The risk of extrapolation is especially large when a regression model is developed from data in one portion of a reservoir and then applied blindly throughout the reservoir. The computer predictions may be based upon values of X well outside the range for which the original relation was developed. The geological properties of the rock can also change, invalidating the X–Y model, as was shown in Example 4. The fundamental problem is that we often employ bivariate models to multivariate problems when we cannot measure the other explanatory variables. Some of these hidden variables are geological factors such as grain size and sorting, minerals, and diagenetic alteration that, in nearby wells, may not change significantly. Over larger distances, however, these variables cause the reservoir petrophysical properties to change considerably, invalidating predictions that do not take these factors into account.

9-12.2 Situations With Errors in the Explanatory Variable(s)

We have only considered those situations where the X values can be perfectly measured; all the error was associated with Y or the model (e.g., unmeasured variables). In practice, this means that the error in X should be much less than the errors in Y and in the model for these techniques to apply. When substantial errors exist in X, more involved methods are available that either require that the ratio of the errors in Y to those in X be known or that X no longer be an arbitrary value. If conventional procedures such as ordinary least squares described above are used, the relation will not appear as strong as it should be, causing $\hat{\beta}_1$ to be too small (negative bias). Seber (1977, pp. 155-160) has further details. In any case, if X cannot be measured with reasonable accuracy, its value as a predictor is diminished and it may not be worthwhile to develop a predictive model including that variable.

9-12.3 Predicting *X* From *Y*

Although we have only considered the case of using least-squares regression with a model to predict $E(Y)$ on the basis of X, the alternative situation (predicting $E(X)$ on the basis of Y) may also occur. The lines are different for these two cases. Although the line will still pass through ($\bar{X}$, $\bar{Y}$), the slope will change from SS_{XY}/SS_X to SS_{XY}/SS_Y. This is because the minimization is now on the X residuals, $X_i - \hat{X}_i$. Compare this to Fig. 9-2 and Eq. (9-2). In this case, Y is assumed to be known with negligible error. Algebraic manipulation of the Y-on-X model to predict X (i.e., $\hat{X}_i = (Y_i - \hat{\beta}_0)/\hat{\beta}_1$) will give biased results except at ($\bar{X}$, $\bar{Y}$) and is not recommended One exception to this rule will be discussed in Chap. 10 concerning prediction of the X-axis intercept.

If the Y's are measured for prespecified values of X, then the least-squares regression procedure to predict X from Y does not apply. The procedure assumes that the response variable is a random variable. For example, suppose we have a core flood for which, for a given amount of injected fluid (X pore volumes), we measure the oil recovery, Y. Developing a regression model to predict Y from a knowledge of X is appropriate, whereas applying regression to develop a predictor of X from Y violates the procedure and gives a meaningless result.

It may seem curious that more than one line is available to express relationships between two variables. In deterministic problems, one and only one solution exists; in statistical problems, each line represents a different aspect of the relationship. Y-on-X regression estimates $E(Y \mid X = X_0)$, X-on-Y regression gives $E(X \mid Y = Y_0$, while other lines express yet other aspects of the X–Y relationship. This variation between different lines is in addition to the variability that exists because we have a limited sample of X–Y values.

9-12.4 Discrete Variables

While we have considered only continuous explanatory variables, it is also possible to use discrete variables. For example, we may wish to develop a drilling model relating the penetration rate (Y) to bit type (X). Clearly, bits come only in distinct types, so we have to consider discrete variables of the form $X = 1$ (roller cone) or $X = 0$ (synthetic diamond). If J bit types are involved, $(J - 1)$ regressors will be required.

9-12.5 Variable Additivity

Once again, during both the model development and parameter estimation phases, the role of additivity of the variables should be considered. Chapter 5 discussed the considerations and these apply in regression as well. The problems are greater with regression, however, because of the presence of cross-product terms ($\Sigma X_i Y_i$) and the fact that estimates $\hat{\beta}_0$ and $\hat{\beta}_1$ have standard errors.

When a physically based model exists for the X-Y relationship, additivity questions are usually less problematic. The model parameters and cross-product terms may have a physical interpretation. For example, porosity ϕ and bulk density ρ_b are related by mass balance through $\rho_b = \rho_{ma} + \phi(\rho_f - \rho_{ma})$ for a given volume of material, V_T, where ρ_{ma} is the matrix density and ρ_f is the pore-fluid density. A regression of ρ_b upon ϕ for samples of equal bulk volume provides estimates of ρ_{ma} and $\rho_f - \rho_{ma}$. Standard errors of these estimates are masses, since multiplication by V_T is implied, and so are additive. The cross-product term $\phi\rho_b$ also represents mass.

Without a physically based model, regression introduces questionable operations. For example, a permeability predictor in the form $\log(k)$ using porosity is often sought. Regressing $\log(k)$ upon ϕ involves calculations with terms $\phi\log(k)$ for which a physical interpretation is difficult.

9-13 SUMMARY REMARKS

Bivariate regression is an elegant method for eliciting a predictive relationship from data and a model. The results, however, are only as good as the care taken to ensure the model is appropriate and the underlying assumptions are tolerably obeyed. Diagnostic procedures are readily available for ensuring the model and data are consistent and for examining for outliers

In this regard, the residuals are extremely valuable and will show model adequacy as well as the value of including further explanatory variables in the model.

Estimates of the model parameters can be assessed for variability, either separately or jointly. These variabilities, based on the residual variability, give a valuable indication of the explanatory power of the predictor variable. R^2 also uses the residual variability to indicate what proportion of the total variability in Y can be explained by a knowledge of X. Inappropriate comparisons of R^2 are easily made, however, because it is dimensionless and the scale of variability is not apparent.

CHAPTER 10

BIVARIATE ANALYSIS: FURTHER LINEAR REGRESSION

Having discussed the mathematical basis and statistical interpretations for the ordinary least-squares (OLS) estimator in Chap. 9, we now explore more advanced topics of regression These topics do not require further mathematical methods or sophisticated statistical analyses. They involve commonly encountered issues that require a more careful consideration of the assumptions involved in OLS and their implications. We will try here to highlight the statistical aspects of the problem for which the OLS method provides estimates.

10-1 ALTERNATIVES TO THE LEAST-SQUARES LINES

The least-squares procedure, described in Chap. 9, is remarkably robust even if there are deviations from the basic assumptions (Miller, 1986, Chap. 5). The most prominent weakness of the procedure is for possibly erroneous data with large $|X - \bar{X}|$. These extreme data points have a large leverage (Chap. 9, Example 3). Robust methods exist to develop regression lines, but variabilities in slope and intercept are more difficult to obtain. Hoaglin et al. (1983, Chap. 5) is a useful introduction.

Other lines, such as the major axis and reduced major axis lines (RMA) also exist but, as shown in Chap. 8, they are not regression lines. One attraction of the RMA line is that it lies between the *Y*-on-*X* and *X*-on-*Y* lines (Fig. 8-6 and Eqs. (8-3) and (8-4)). The RMA line slope (σ_Y/σ_X) is between the *Y*-on-*X* line slope ($\rho\sigma_Y/\sigma_X$) and *X*-on-*Y* line slope ($\rho^{-1}\sigma_Y/\sigma_X$) and they all pass through $(\bar{X}, \bar{Y})$. The RMA line might provide better estimates for *Y* than the OLS line for the case where *X*, the predictor, is measured with significant error. Since the *Y*-on-*X* slope will underpredict $E(Y \mid X = X_0)$ and the *X*-on-*Y* slope will overpredict this value, a line with an intermediate slope (the RMA line) might be better. This logic, however, has two weaknesses.

The first weakness concerns the assessment of the relative errors of measuring *X* and *Y*. If the error term ε in the model $Y = \beta_0 + \beta_1 X + \varepsilon$ represents only measurement error of *Y*, then substantial errors in *X* would affect the estimates $\hat{\beta}_0$ and $\hat{\beta}_1$ obtained from *Y*-on-*X* regression. The error term ε, however, often includes model inadequacy. Hence, the errors of *X* should be compared to the errors arising from model inadequacy as well as *Y*-measurement errors.

The second weakness concerns the wisdom of developing a model that is to be used with unreliable predictor values. When *X* is poorly measured, the *Y*-on-*X* slope will approach zero, indicating that, under the circumstances, the best estimate of *Y* is $\hat{Y} = E(Y)$. This is a useful diagnostic feature indicating that, no matter how good the underlying *X*-*Y* relationship is, the measurement errors of *X* render it nearly useless as a predictor of *Y*.

The relative merits of lines should be assessed in the context of the specific problem. As a general rule, remember that, while the RMA line portrays the X-Y covariation without regard to the roles of the variables, the regression lines X-on-Y and Y-on-X do assign explicit roles to the variables. Thus, use of the RMA line in a prediction role should be carefully justified.

A frequently overlooked limitation of OLS lines is that they may lead to biased results for some applications. Recall from Chap. 9 that the regression line is an estimate of the conditional expectation $E(Y \mid X = X_0)$ and the estimates $\hat{Y}$ are unbiased. Thus, any linear combination of estimates $\hat{Y}$ is also unbiased but nonlinear combinations will be biased. The following simplified example illustrates this.

Example 1a - Applying Regression Line Estimates for Prediction of Effective Permeability. Permeability can be difficult to measure *in situ*. Therefore, it is common in reservoir characterization to predict permeability from an OLS regression line and *in situ* porosity measurements. The regression line is based on permeability and porosity data taken from geologically similar portions of the reservoir. These predictions are then often combined, using the arithmetic and harmonic averages, to predict the larger-scale aggregate horizontal and vertical permeabilities of stratified rocks for subsequent applications. (Recall from Chap. 5 that, in a layered medium, the arithmetic average of permeability is an appropriate estimator for layer-parallel flow. The harmonic average estimates aggregate permeability for layer-transverse flow.)

A portion of a hypothetical porosity–permeability regression line and the "data" are shown in Fig. 10-1 (left). The line is passing midway between the two points shown, as we would expect (on average). Looking more closely, the data indicate k may be 1 or 100 with equal probability when $\phi = 0.1$. At $\phi = 0.1$, however, the line provides an estimate $\hat{k} = 50.5$, which is neither 1 nor 100. Nonetheless, 50.5 is a reasonable value since regression lines are designed to predict $E(k \mid \phi = \phi_0)$ and the average of 1 and 100 is 50.5.

Suppose now we have a rock unit composed of four equally thick and equally porous layers (Fig. 10-1 right) with porosity measurements but the permeabilities are unknown. Therefore, we must use the regression-derived estimated permeability for each layer to predict the aggregate horizontal and vertical permeabilities of the unit. We compare the estimated horizontal and vertical permeabilities with the true values given as the arithmetic mean $E(k \mid \phi = 0.1) = 50.5$ (horizontal) and the harmonic mean $[E(k^{-1} \mid \phi = 0.1)]^{-1} = 1.98$ (vertical).

For the aggregate horizontal permeability, we have

$$k_A = \frac{1}{4}\sum_{i=1}^{4}\hat{k}_i = 50.5$$

This agrees exactly with $E(k \mid \phi = 0.1) = 50.5$. For the vertical permeability of the unit, the harmonic average gives

$$k_A = \left(\frac{1}{4} \sum_{i=1}^{4} \hat{k}_i^{-1} \right)^{-1} = 50.5$$

which is a poor estimate of the harmonic mean. From the regression estimates, the rock unit appears to have no permeability anisotropy (vertical-to-horizontal permeability ratio equals one), while the unit in fact has a vertical-to-horizontal permeability ratio of 0.04.

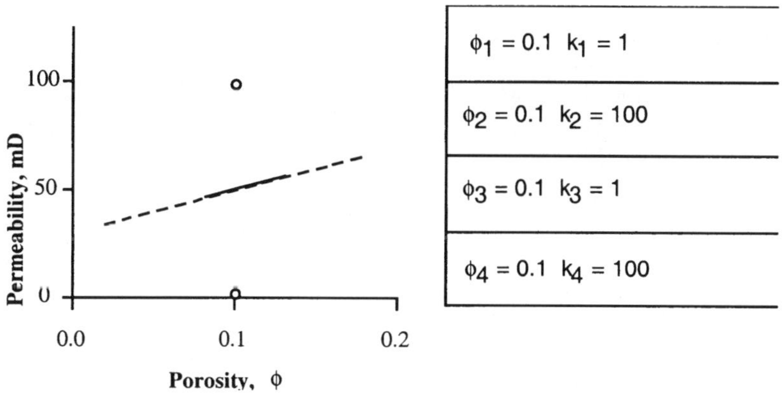

Figure 10-1. Regression line and data (left) and layered formation (right).

In Example 1a, the regression line gave an unbiased estimate for the arithmetic average because this average is a linear combination of the estimates. This is a result of the Gauss-Markov Theorem (Graybill, 1961, pp. 116-117), which assures that all linear combinations of the estimates $\hat{\beta}_0$ and $\hat{\beta}_1$ are unbiased. The harmonic average is a nonlinear combination of the estimates, and use of the *Y*-on-*X* line could give highly misleading results. The following example shows an alternative procedure for use with the harmonic average.

Example 1b - A Regression Estimator for the Harmonic Average. A regression line based on resistivity to flow, $r = k^{-1}$, and using the data of Example 1a would estimate $E(r \mid \phi = 0.1)$ as the average of 1 and 100^{-1}, or 0.505. Thus, for the four-layer model of Fig. 10-1 right, $\hat{r}_i = 0.505$. This gives, for the harmonic average,

$$\bar{k}_H = \left(\frac{1}{4} \sum_{i=1}^{4} \hat{r}_i \right)^{-1}$$

$$= 1.98$$

which agrees with the harmonic mean.

When the harmonic average is needed, better results are obtained using a Y^{-1}-on-X line. The line then provides estimates of Y^{-1} for use in the harmonic average. This is still not entirely satisfactory, because the inverse of the sum of estimates is required. Nonetheless, the sum would be much less variable than any one estimate of Y^{-1}, so it would be less biased. A similar approach, using a log(Y)-on-X line, would be suitable for geometric mean estimation.

In Chap. 9, we remarked that the possibility of two lines (Y-on-X and X-on-Y) is a curious feature of regression. This may be especially so for those more accustomed to deterministic methods. Here, we have another example of regression being able to produce several lines. These lines arise because the application of the estimates influences the choice of estimator. This choice must be made on geological and physical reasons; the statistics do not help.

10-2 VARIABLE TRANSFORMATIONS

The linear model is a compact and useful way of expressing bivariate relationships. In cases where a linear model is inappropriate, some transformation of the variables may permit a linear model to be used if the application for the estimates is compatible with the transforms. For example, while porosity ϕ and permeability k are usually observed to be nonlinearly related, a logarithmic transform (to any base) applied to permeability may linearize the relationship: $\log(k) = \beta_0 + \beta_1\,\phi + \varepsilon$, and predicting log($k$) is appropriate for estimating the geometric mean. A particular case of linear bivariate relationships occurs when random variables are joint normally distributed. If random variables are transformed to become joint normally distributed, the linear model will automatically apply (Chap. 8). There are, however, some precautions that should be observed when selecting a transformation and applying the results of the regression model.

Variable transformations affect the way that the noise, ε, enters into the model. For example, the model $\ln(Y) = \beta_0 + \beta_1 X + \varepsilon$ assumes the noise is additive for ln(Y). This means that the noise is multiplicative for Y because $Y = \exp(\beta_0 + \beta_1 X)\exp(\varepsilon)$. This model is often appropriate when the size of error is proportional to the magnitude of the quantity measured, e.g., $\varepsilon = \pm 0.10Y$. Regressing ln(Y) upon X then gives the parameters $\hat{\beta}_0$ and $\hat{\beta}_1$. If, on the other hand, an additive noise model is more suitable, we can regress Y upon exp(X). This implies the model $Y = \beta_0 + \beta_1\exp(X) + \varepsilon$. The proportional effect of noise in permeability measurements may be why ln(k) versus porosity (ϕ) plots are more common than k versus exp(ϕ) plots. Recall from Chap. 9, however, that the term ε includes all errors, including model deficiencies as well as measurement errors. Thus, while a small component of the k error may be multiplicative from measurement error, the predominant component may still be additive, justifying a regression of k upon exp(ϕ).

What happens if we get the noise model wrong? Usually, the spread of residuals varies with X (Fig. 10-2). In this case, the assumption of OLS that $\varepsilon \sim N(0, \sigma_\varepsilon^2)$ (see Chap. 9) is being violated since the variance is not constant. Alternative transformations of X and Y are needed to produce a more appropriate noise model.

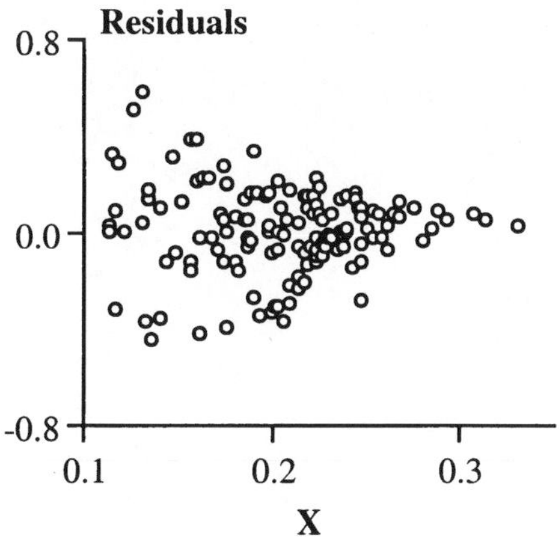

Figure 10-2. Regression residuals showing an inappropriate noise model.

Transformations of the response variable also affect the least-squares criterion and the resulting line. This feature is quite evident, for example, with a logarithmic transformation (Fig. 10-3). Deviations above the regression line ($\log_{10}(Y_i) - \log_{10}(\hat{Y}_i) > 0$) do not represent the same magnitude as deviations below the line ($\log_{10}(Y_i) - \log_{10}(\hat{Y}_i) < 0$) in the Y domain. This behavior is because of the nonlinearity of the logarithmic function. The effect can produce a considerable bias, especially with highly variable data having a weak X -log(Y) relationship, but, in certain circumstances, can be compensated for during detransformation.

Detransformation is the process whereby estimates of $W = f(Y)$ produced by the regression line are converted back to estimates of Y. The algebraic approach $\hat{Y} = f^{-1}(\hat{W})$ does not always work well when f is strongly nonlinear or the relationship is weak. However, when the assumption of $\varepsilon \sim N(0, \sigma_\varepsilon^2)$ applies for the model $W = \beta_0 + \beta_1 X + \varepsilon$, we can use the theory of the normal distribution to derive a bias-corrected Y.

For any regression of W upon X, $E(W \mid X = X_0) = \beta_0 + \beta_1 X_0$. In the common case $W = \ln(Y)$, Y will be log-normally distributed and $W \sim N(\beta_0 + \beta_1 X_0, \sigma_\varepsilon^2)$. Thus we have, from the properties of the log-normal PDF (Chap. 4),

$$E(Y \mid X = X_0) = \exp(\beta_0 + \beta_1 X_0 + 0.5\sigma_\varepsilon^2)$$

This result indicates that there is an additional multiplicative factor of $\exp(0.5\sigma_\varepsilon^2)$ to determining the conditional mean of Y besides the factor $\exp(\beta_0 + \beta_1 X_0)$ we might naively assume. σ_ε^2 is approximated from the data by the mean squared error, MS_ε (Eq. (9-6)).

The preceding equations apply equally well using base-10 logarithms as natural logarithms. We suggest that, to avoid making mistakes by overlooking factors of 2.3 to convert ln to $\log_{10}$, the natural (base-e) form be used. The following example illustrates the procedure for data transformed using base-10 logarithms.

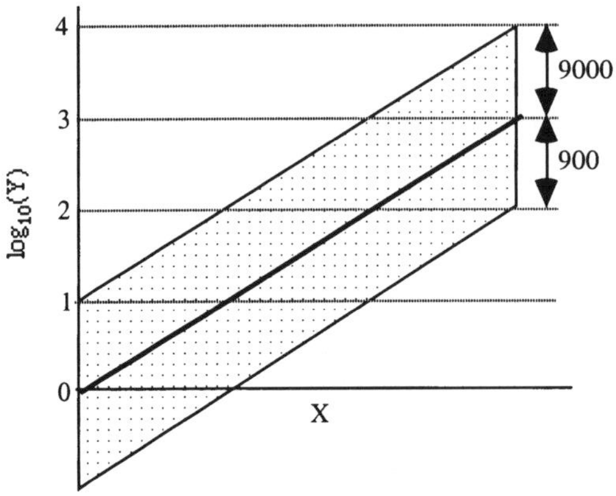

Figure 10-3. The regression-line procedure gives equal weight to deviations of 9,000 and 900.

> *Example 2 - Detransformation of a Log-Transformed Data Set.* The data of Fig. 10-4 left indicate that the $\log_{10}(Y)$ - X relationship is linear, with the line explaining about 79% of the variability in $\log_{10}(Y)$. The sample variance of $\log_{10}(Y)$ is 0.733 and, with the line, this is reduced to $0.733(1\text{-}R^2) = 0.153$. This gives a standard deviation of about 0.4 decades $(= (0.153)^{1/2})$ about the regression line. A naive detransformation suggests the relation $\hat{Y} = 10^{15.4X-1.41}$ (Fig. 10-4 right). A probability plot indicates that the residuals of log(Y) are approximately normally distributed so that the log-normal correction applies. This gives an additional factor of $\exp(0.5\sigma_\varepsilon^2) = \exp((0.5)(2.3^2)(0.153)) = 1.50$, where the factor 2.3^2 converts the $\log_{10}$ variability to natural units (ln). The bias-corrected line relation, $\hat{Y} = (1.5)10^{15.4X-1.41}$, is also shown in Fig. 10-4.

At first glance, it may be difficult to discern whether the bias-corrected line is better than the naive predictor in Fig. 10-4 (right). More careful inspection reveals that the latter line is consistently near the bottom of the scatter for large X; the bias-corrected line passes more nearly through the center of the large-X data. Nonetheless, the linear plot shows a considerable variability of the Y values that is not readily apparent on the *log*(Y) plot.

$W = \ln(Y)$ is one example of the more general case of power transformations, $W = Y^p$, discussed in Chap. 4. If we assume $\varepsilon \sim N(0,\sigma_\varepsilon^2)$, Jensen and Lake (1985) give a general expression for any value of p. If $p > 0$, the correction is additive rather than multiplicative. For example, when $p = 1/2$,

$$E(Y \mid X = x_0) = E(W \mid X = x_0)^2 + \sigma_\varepsilon^2/4 \qquad (10\text{-}1)$$

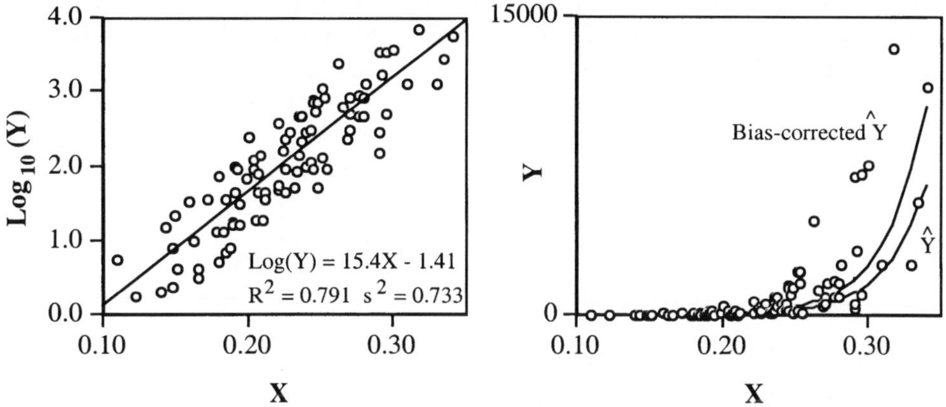

Figure 10-4. Data set for Example 2 with log- and untransformed plots.

Jensen and Lake (1985) suggest transforming both predictor and response variables prior to regression if the predictor is a random variable. The objective is to find functions $f(Y)$ and $g(X)$ that make the joint PDF approximately normal. As described in Chap. 8, joint normal variables automatically have linear regression relationships, which helps to keep the X-Y relationship as parsimonious as possible. The noise model for $f(Y)$ upon $g(X)$ will be appropriate and detransformation will be simplified since the residuals are normally distributed. It is usually sufficient to select f and g so that the marginal histograms are approximately symmetrical. This is not strictly adequate, however, because symmetry of the marginal distributions is necessary but not sufficient for joint normality. The following example demonstrates this procedure. Further examples are contained in Jensen and Lake (1985).

Example 3 - Transformations for Joint Normality. Scatter plots of a data set (Fig. 10-5) suggest that neither untransformed nor log-transformed versions of the response are entirely suitable, despite the impressive values of R^2. The Y plot shows a concave-upward form while the log(Y) plot is concave downwards.

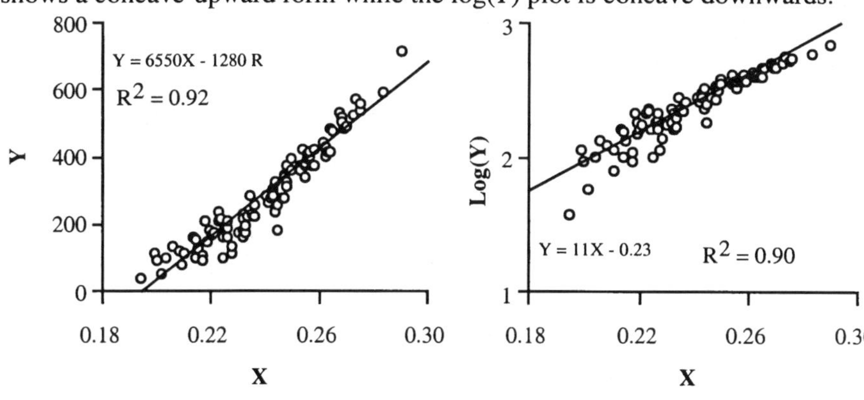

Figure 10-5. Scatter plots and regression lines for Y and log(Y).

Histograms of *Y* and *log*(*Y*) (Fig. 10-6 left and center) are significantly skewed while the histogram of *X* (Fig. 10-6 right) is reasonably symmetrical. Consequently, only *Y* need be transformed. Since the sample PDF of *Y* is right-skewed while *log*(*Y*) is left-skewed, an exponent between 0 and 1 is indicated. A square-root transform (Fig. 10-7 left) for *Y* gives a nearly symmetrical histogram.

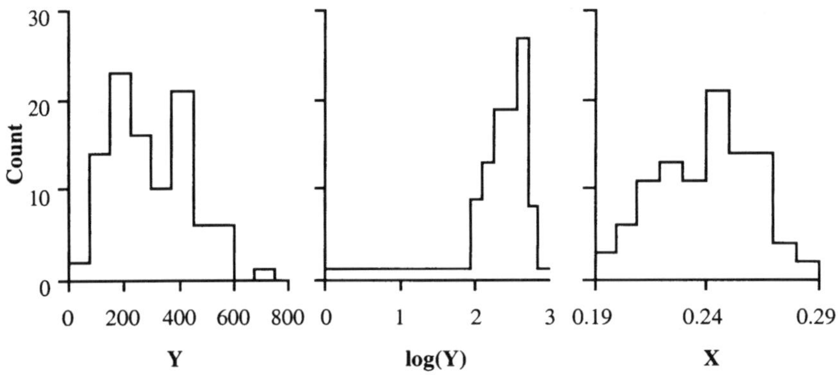

Figure 10-6. Histograms for the data in Fig. 10-5.

The scatter plot and regression line for $\sqrt{Y}$ (Fig. 10-7 right) suggest a better linear relationship with *X*. There is not an even variability about the line; lower values of *X* appear to have about twice the variability that larger *X* values have. This might be corrected with more careful selection of transformations, but it is unlikely to produce much benefit. If estimates of *Y*, instead of $Y^{1/2}$, are needed, the final predictor should be bias-corrected for the detransformation. The large R^2 value and moderate variability of $Y^{1/2}$, however, show that the correction is quite small (Eq. (10-1)):

$$\hat{Y} = (200X - 31)^2 + \frac{1}{4}(1.16)^2 = (200X - 31)^2 + 0.34$$

giving a correction of less than 1%.

In summary, there are three important aspects to using transformed variables in developing models through OLS regression. The first is that using transformed variables may help to keep the model simple because a linear model is sufficient for variables with symmetrical histograms. Second, transformations help us to honor the noise behavior assumed for OLS regression. Third, detransforming predicted values requires some care, because the algebraically correct detransformation of predictions (i.e., naive detransformation) may give significantly biased estimates.

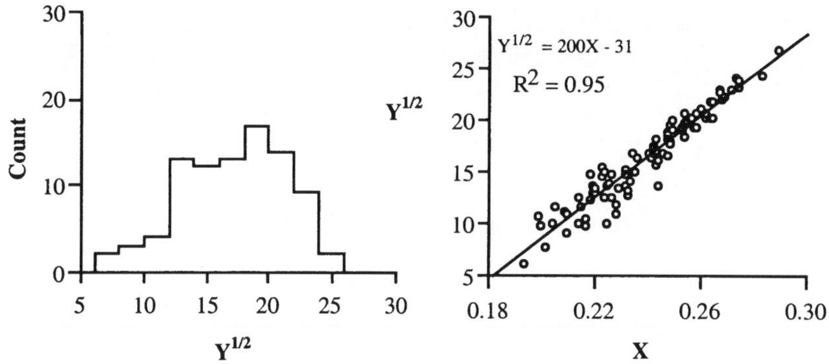

Figure 10-7. Histogram, scatter plot, and regression line for $\sqrt{Y}$.

10-3 WEIGHTED LEAST SQUARES

For ordinary least squares, we assume that all the data used in the regression method are equally reliable. This assumption appears in the sum of the squared differences (Eq. (9-2)):

$$S(\beta_0, \beta_1) = \sum_{i=1}^{I} (Y_i - \beta_0 - \beta_1 X_i)^2$$

because all the differences (Y_i - β_0 - $\beta_1 X_i$) are equally influential on S. We can assign a weight, w_i, to each term in the sum

$$S^*(\beta_0, \beta_1) = \sum_{i=1}^{I} w_i (Y_i - \beta_0 - \beta_1 X_i)^2$$

If we calculate estimates $\hat{\beta}_0$ and $\hat{\beta}_1$ that minimize S^*, the equations for the $\hat{\beta}$'s become

$$\hat{\beta}_0 = \frac{\Sigma w_i Y_i - \hat{\beta}_1 \Sigma w_i X_i}{\Sigma w_i}$$

and

$$\hat{\beta}_1 = \frac{\Sigma w_i \Sigma w_i Y_i X_i - \Sigma w_i X_i \Sigma w_i Y_i}{\Sigma w_i \Sigma w_i X_i^2 - (\Sigma w_i X_i)^2}$$

where all summations are from $i = 1$ to $i = I$.

The w_i's can be obtained in several ways. One way is to use experience. If some data are thought to be twice as reliable as others, then the w's for the more reliable data should be twice the w's for the less reliable points. It does not matter what values are chosen for the w's (as long as they are not all zero), just make some twice the value of the others. Another way is if quantitative assessments of the relative reliabilities of the Y_i are available, based on theory, for example, then more careful estimates can be made for the w's. If Y_i has variability σ_i^2, then $w_i = \sigma_i^{-2}$.

Small differences among the w's usually do not influence the estimates $\hat{\beta}_0$ and $\hat{\beta}_1$ by much, compared to the OLS values. The changes will depend, however, on X_i and the residual $(\hat{Y}_i - Y_i)$. Residual analysis from weighted least squares requires that variables be weighted by $\sqrt{w_i}$. Hence, plot $\sqrt{w_i}\ (\hat{Y} - Y_i)$ versus $\sqrt{w_i}\ X_i$ or any other desired variable to assess the model fit.

10-4 REDUCING THE NUMBER OF VARIABLES

In a multivariate model such as $Y = \beta_0 U + \beta_1 V$, it is tempting to reduce the number of predictors by dividing by one of the predictor variables. Hence, $Y = \beta_1 V + \beta_0 U$ could become

$$\frac{Y}{U} = \beta_1\left(\frac{V}{U}\right) + \beta_0$$

By setting $Y^* = Y/U$ and $X^* = V/U$, we obtain $Y^* = \beta_1^* X^* + \beta_0^*$. Thus, we have changed a three-dimensional problem into a two-dimensional problem and can use simpler equations to estimate the β's. While this approach is algebraically correct, it may not be statistically sound. That is, will $\hat{\beta}_1 = \hat{\beta}_1^*$ and will $\hat{\beta}_0 = \hat{\beta}_0^*$?

Unfortunately, the answer is probably not. If U is constant, then the above procedure is correct. If, however, U varies significantly compared to V, then the estimated β's will differ. The reason is because of a change in the role of the errors in the models. When we use OLS regression to estimate β_0 and β_1 in $Y = \beta_0 U + \beta_1 V$, we are implicitly assuming that the model is $Y = \beta_0 U + \beta_1 V + \varepsilon$. Furthermore, we're also assuming that ε does not change with U, V, or Y. Dividing the original model through by U gives us the model

$$Y^* = \beta_1 X^* + \beta_0 + \varepsilon^*$$

where $\varepsilon^* = \varepsilon/U$, which is not the same model as

$$Y^* = \beta_1^* X^* + \beta_0^* + \varepsilon$$

When U is constant, the noise ε^* is independent of V and Y but, if U is variable, then ε^* is variable and is related to X^* and Y^*. Hence, because we're solving two different equations, we cannot expect $\hat{\beta}_1 = \hat{\beta}_1^*$ and $\hat{\beta}_0 = \hat{\beta}_0^*$.

This situation often arises in reservoir-engineering material-balance calculations. Tehrani (1985) shows that the volumetric-balance equation can take the form $Y = \beta_0 U + \beta_1 V$. He proceeds to show that the revised model $(Y/U) = \beta_0 + \beta_1(V/U)$ does not give the same $\hat{\beta}_0$ and $\hat{\beta}_1$ as the original model and large errors can occur. The following example illustrates the problem for a gascap-drive reservoir and shows that, in addition, confidence intervals are also affected.

Example 4 - Material-Balance Errors in Gascap-Drive Reservoirs. The material-balance equation for gascap-drive reservoirs is (Dake, 1978, pp. 78-79)

$$F = NE_o + mNE_g \qquad (10\text{-}2)$$

where F is the underground withdrawal (reservoir barrels, rb), N is the initial oil in place (stock tank barrels, STB), E_o is the oil and dissolved gas expansion factor (RB/STB), E_g is the gascap expansion factor (rb/STB), and m is the ratio of the initial gascap-hydrocarbon to the initial oil-hydrocarbon volumes prior to production. F, E_o, and E_g are known quantities based on measurements of the produced oil and gas and their properties. m may be estimated from seismic or other geological data but, depending upon circumstances, the estimate may be rather poor.

If m is not known or poorly known, we have a two-unknown, nonlinear problem because Eq. (10-2) has the unknowns, m and N, appearing as a product in the second term. Thus, the best solution procedure requires nonlinear regression, a procedure outside the scope of this book. See Bard (1974) for a good discussion of nonlinear regression methods and analysis. We can solve linearized forms of Eq. (10-2) that, as we will see, have some limitations. The linear approaches require treating the product mN as one regression unknown. If we do this, Eq. (10-2) can be solved directly as a multiple-linear-regression problem with two explanatory variables, E_g and E_o, or by reducing the number of variables using the form

$$F/E_o = N + mN(E_g/E_o) \qquad (10\text{-}3)$$

Thus, an OLS F/E_o-on-E_g/E_o line would also appear to give an intercept N and slope mN.

For the following data, taken from Dake (1978, p. 92), we calculate lines for both Eqs. (10-2) and (10-3) to examine the differences.

$10^{-6}F$	E_o	E_g	$10^6 F/E_o$	E_g/E_o
5.807	0.01456	0.07190	398.8	4.938
10.671	0.02870	0.12942	371.8	4.509
17.302	0.04695	0.20133	368.5	4.288
24.094	0.06773	0.28761	355.7	4.246
31.898	0.09365	0.37389	340.6	3.992
41.130	0.12070	0.47456	340.8	3.932

Because the units of F are 10^6 or million rb, the corresponding units of N will be million STB. Using multiple-regression (discussed briefly later in this chapter), the zero-intercept, two-variable model (Eq. (10-2)) gives (with standard errors in brackets)

$N = 105$ [±34] and $mN = 59.6$ [±8.4]

With bivariate regression, the free-intercept, single variable model (Eq. (10-3)) gives

$N = 109$ [±23] and $mN = 58.8$ [±5.4]

The initial oil figures differ by about 3%, which is well within the statistical variability of either estimate. The precision of the second model is substantially less than that of the first model. For either model, it is clear that material balance can only estimate the oil in place to within, say, 30 or 40% based on such few data.

m can be estimated by taking the ratio mN/N (i.e., 59.6/105 = 0.57). Standard errors of m are more difficult to calculate because the regression coefficients are correlated. A simplistic approach is to ignore the correlation and use the sensitivity analysis described in Chap. 5. Let $Z = mN$ so that

$$\frac{\Delta Z}{Z} = \frac{\Delta m}{m} + \frac{\Delta N}{N}$$

where the Δ's are perturbations in the appropriate quantity. We have

$$\frac{8.4}{59.6} = \frac{\Delta m}{m} + \frac{34}{105}$$

giving $\Delta m/m = -0.18$. Since $m = mN/N = 59.6/105 = 0.57$, $m = 0.57$ [$\mp$ 0.1]. Note the errors in m are in antiphase with the errors of N.

A nonlinear regression with Eq. (10-2) gives the results

$N = 105$ [±34] and $m = 0.57$ [$\mp$ 0.26]

The multivariate linear (the first model, Eq. (10-2)) and the nonlinear regression results are identical for N. The estimates of m are identical but the simplified perturbation analysis has considerably understated the standard error. The simpler the solution methods to solve this problem, the less reliable have been the results.

Why are the standard errors so different for the two linear regression models, Eqs. (10-2) and (10-3)? This is because, by dividing by a common factor (E_o), the correlation between the response and explanatory variables has been improved. A simple example discussed by Hald (1952, p. 614) helps to explain the problem. Consider the case where a response (Y) and an explanatory (X) variables are unrelated so that the true regression-line slope is zero. If new response and explanatory variables are created by dividing by a common factor, $Y^* = Y/Z$ and $X^* = X/Z$, where Z is not constant, then Y^* and X^* will appear to be correlated and the regression-line slope will no longer be zero. See Kenney (1982) for further examples of this effect.

10-5 THE X-AXIS INTERCEPT AND ERROR BOUNDS

Chapter 9 presented an expression, Eq. (9-5a), for the variability of the Y-axis intercept, β_0. A fairly common but more complicated situation is to obtain a confidence interval for the X-axis intercept estimate $-\hat{\beta}_0/\hat{\beta}_1$. This estimate is biased but there are few attractive alternatives. In the statistical literature, this problem is often known as the calibration or discrimination problem. There are several expressions for confidence intervals for $-\hat{\beta}_0/\hat{\beta}_1$. Miller (1981, pp. 117-119) shows that, when they exist, limits of the $(1 - \alpha)$ confidence region are given by

$$x' = \overline{X} + \frac{\overline{Y}}{a\hat{\beta}_1} + \frac{\sqrt{\overline{Y} - a[t^2 MS_\varepsilon(1 + 1/I) - \overline{Y}^2]}}{a\hat{\beta}_1} \qquad (10\text{-}4a)$$

and

$$x'' = \overline{X} + \frac{\overline{Y}}{a\hat{\beta}_1} - \frac{\sqrt{\overline{Y} - a[t^2 MS_\varepsilon(1 + 1/I) - \overline{Y}^2]}}{a\hat{\beta}_1} \qquad (10\text{-}4b)$$

where

$$a = \frac{t^2 MS_\varepsilon}{\hat{\beta}_1^2 \sum_{i=1}^{I} (X_i - \overline{X})^2} - 1$$

and $t = t(\alpha/2, I - 2)$, the t- statistic.

x_{lo}, x_{hi}, or both may not exist. This is because, if the line slope could be zero at the required level of confidence, the X intercept may be at infinity. That is, if the OLS regression line could be parallel to the X axis, the intercept will be unbounded. Consequently, only meaningful confidence intervals can be obtained from Eqs. (10-4) if $\hat{\beta}_1$ is significantly nonzero.

Montgomery and Peck (1982, p. 402) and Seber (1977, p. 189) discuss another approach for estimating the X intercept and its confidence interval. If the explanatory variable is also a random variable (i.e., not a controllable variable), an X-on-Y regression could be performed to estimate the X intercept and its confidence interval. They suggest that the method of Eqs. (10-4) is generally preferred but there is little difference when R^2 is large.

Example 5 - Estimating Grain-Density Variability from Core and Wireline Data. The bulk-density log is often used to estimate formation porosity if certain parameters of the lithology are known. Conversion of the bulk density (ρ_b) to porosity (ϕ_D) requires the densities of the pore fluid (ρ_{fl}) and matrix (ρ_{ma}) be known or estimated:

$$\phi_D = (\rho_{ma} - \rho_b) / (\rho_{ma} - \rho_{fl})$$

For this model, $\rho_b = \rho_{ma}$ when $\phi_D = 0$. If ρ_b is plotted against porosity measured by another method (ϕ_{plug}), the linear model may apply and ρ_{ma} estimated from the ρ_b-axis intercept.

Figure 10-8 shows a scatter plot of core-plug porosities and the corresponding wireline density measurements for a well. A strong linear relation appears, despite the large difference in volumes of investigation. If the porosity heterogeneity were large at the density-log scale, this relation might have been much weaker. The data here suggest that ρ_b and the model could form a useful porosity predictor. As a check on the quality of the line fit, however, the ρ_b intercept ($\hat{\rho}_{ma}$) should match the grain density measured from core samples ($\hat{\rho}_{maC}$). In this case, an average of laboratory measurements of core samples gives $\hat{\rho}_{maC}$ = 2.67 g/cm^3, while $\hat{\rho}_{ma}$ = 2.69 g/cm^3. Is this difference significant at the 95% level?

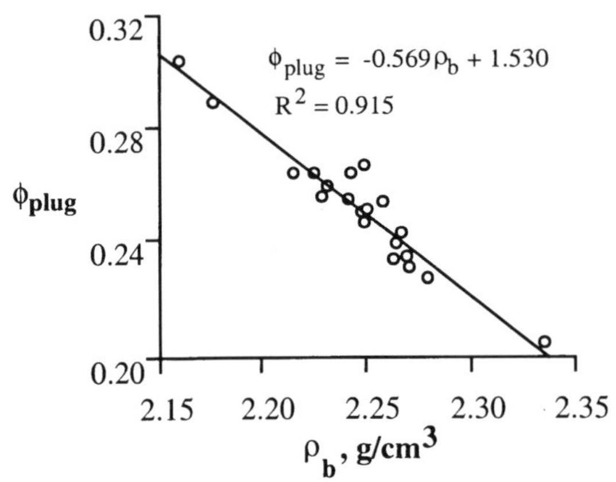

Figure 10-8. Wireline density versus core-plug porosity scatter plot and regression line.

For this data set, we have I = 20, $t(0.025, 18)$ = 2.101, $\bar{X}$ = 2.246, $\Sigma(X_i - \bar{X})^2$ = 0.0256, $\bar{Y}$ = 0.252, MS_ε = 0.0000425, and $\hat{\beta}_1$ = -0.569. Using Eqs. (10-4), these values give $a = -0.977$, $x_{lo} = 2.63$, and $x_{hi} = 2.77$. Thus, the results $\hat{\rho}_{maC} = 2.67$ and $\hat{\rho}_{ma} = 2.69$ are not significantly different at the 95% level, since $\hat{\rho}_{maC}$ falls within the 95% confidence interval for $\hat{\rho}_{ma}$.

Since neither ϕ_{plug} nor ρ_b are controlled variables, we can regress ρ_b on ϕ_{plug} to obtain another estimate of ρ_{ma}. These data give $\hat{\rho}_b = -1.61\phi_{plug} + 2.65$, implying that $\hat{\rho}_{ma} = 2.65$ g/cc. The 95% confidence interval for this $\hat{\rho}_{ma}$ is 2.65 ± 2.101(0.029) or 2.59 to 2.71, calculated using Eq. (9-7a). This interval is slightly smaller than the one obtained using Eqs. (10-4), but both methods suggest the core and log matrix densities are similar.

In the above example, the 95% confidence region does not lie symetrically about the X intercept: $|2.63 - 2.69| \neq |2.69 - 2.77|$. This is reasonable since $\hat{\beta}_1 - \beta_1 = \delta\beta_1$ is equally likely to be positive or negative, but the commensurate change in the X-axis intercept, δX, is nonlinearly related to $\delta\beta_1$. Because the line must pass through the point ($\bar{X}$, $\bar{Y}$), by geometry we have that $\delta X = -\bar{Y}\delta\beta_1/[\beta_1(\beta_1 + \delta\beta_1)]$. Thus, except for very small $\delta\beta_1$, δX and $\delta\beta_1$ are nonlinearly related and, consequently, an asymmetrical confidence interval occurs.

Another situation involving discrimination is the definition of nonproductive intervals. Frequently, nonproductive portions of the formation are defined on the basis of permeability (see Sec 2-4). For example, 1 mD is a common cut-off value with only intervals having permeabilities above that value being declared as "net pay" for liquid flow. As described in Example 1, however, permeability is difficult to measure *in situ*, so porosity is often used to predict the permeability. In the following brief example, we obtain the best estimate of porosity for the cutoff value and its range of uncertainty.

Example 6 - Defining the Net Pay Cutoff Value Using a Porosity-Permeability Correlation. We use the data set presented in Example 2 and interpret Y as the permeability and X as porosity. We assume that a cutoff value of approximately 1 mD is required. We use the term "approximately" here because the bias of the line may also be a consideration. In this case, log(1) = 0 actually represents 1.5 mD permeability. However, as we see below, the bias of the predictor is an unimportant issue here.

In this case, the X axis ($\log(Y) = 0$) is the desired cutoff value, so the porosity corresponding to this permeability is $-\hat{\beta}_0/\hat{\beta}_1 = 1.41/15.4 = 0.092$. Eqs. (10-4) provide the 95% confidence interval endpoints as $x'' = 0.144$ and $x' = 0.037$. In other words, an interval with permeability 1 mD has a 95% probability that its porosity will fall somewhere in the interval [0.037, 0.144].

It is surprising that even with nearly 100 data points and a strong porosity-permeability correlation, this example gives a wide porosity interval. The variation in possible X intercept values represents approximately two orders of magnitude in permeability variation. This large variability is why the estimator bias is of relatively little importance.

The primary source for this X intercept variability can be explained by thinking of the porosity variability about the line for a fixed permeability. The standard deviation of the porosity is about 0.05 and the line, having $R^2 = 0.79$, reduces this variation to about $0.05\sqrt{1-0.79} = 0.023$ at a fixed permeability. Multiplying this variation by $t = 2$ for the 95% interval suggests the precise porosity for a given permeability is not going to be pinned down to better than ±0.05 or so unless we can include more information into the prediction.

Examples 5 and 6 differ in an important aspect. The confidence interval of Example 5 is genuinely used in only one test to compare the X intercept with a value determined by another method. The confidence interval determined in Example 6, however, might not be used just once; it could be applied on a foot-by-foot basis across the entire interval of interest. In this case, if a type 1 error at the 95% level is to be avoided for all intervals simultaneously, a modified form of Eqs. (10-4) needs to be used. Miller (1981, pp. 117-119) has the details.

10-6 IMPROVING CORRELATIONS FOR BETTER PREDICTION

In several examples of this and other chapters, we allude to a common problem in reservoir characterization. Frequently, we attempt to relate measurements having very different volumes of investigation. In homogeneous materials, this would not be a problem. In heterogeneous media such as sedimentary rocks, however, differences in volumes of investigation tend to weaken relationships among measurements.

Reconciling and comparing different measurements on different volumes of rock is a broad and difficult area. Much research remains to be done. We can, however, give a few tips and illustrate the ideas with an example.

Relationships between measurements are stronger when

1. there is a physically based reason for the two quantities to be related (e.g., porosity and bulk density, resistivity and permeability, radioactivity and clay content);

2. the measurement volumes are similar (e.g., core-plug porosity and permeability, probe permeability and microscale resistivity, grain size and probe permeability); and

3. the measurements have sample volumes consistent with one of the geological scales (e.g., probe permeability and laminations, resistivity and beds, transient tests and bed sets).

The first item avoids exploiting "happenstance" relationships that may disappear without warning. The second item ensures that, in heterogeneous rocks, one measurement is not responding to material to which the other sensor cannot respond, or responds differently. The third item exploits the semirepetitive nature of geologic events by making measurements on similar rock units.

Example 7 - Assessing Probe Permeameter Calibration from Core-Plug Data. While plug and probe permeameters both measure permeability, their volumes of investigation and the boundary conditions of the measurements differ considerably. This would suggest that there is not necessarily a one-to-one correspondence between the measurements.

Figure 10-9 (left) is a plot of probe and plug data taken from the Rannoch formation (discussed in Example 6 of Chap. 3). Each core plug had four probe measurements on each face, giving eight probe permeabilities per plug. Some plugs showed considerable variability; Fig. 10-9 (right) is a histogram of the probe-based C_V's from each plug, in which we see that 29% of the plugs have C_V • 0.5. This variability arises because each core plug is sufficiently large to contain several laminations, which the probe is detecting individually.

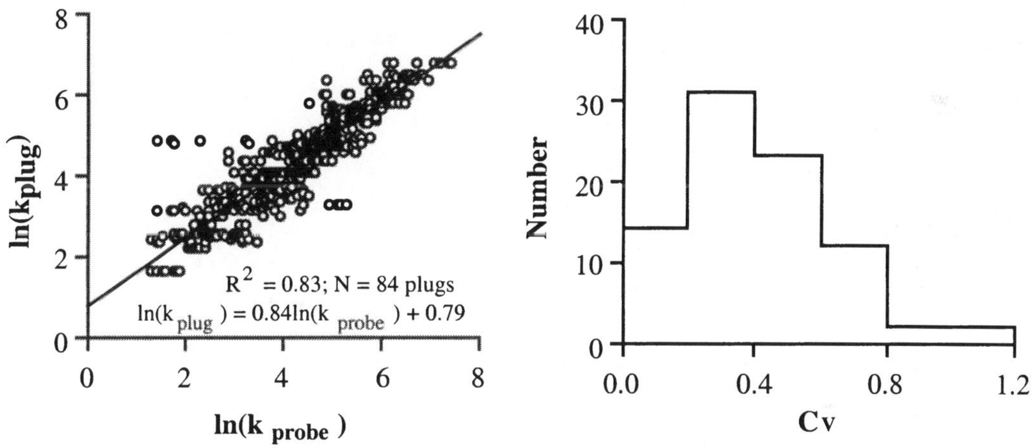

Figure 10-9. Scatter plot and regression line (left) and histogram of probe-based variability (right) of Rannoch formation core plugs.

A plot of probe-measured variability (Fig. 10-10 left) supports this observation. The average and ±1 standard deviation points are shown for the C_V's of three rock types common to the Rannoch formation. Rocks with stronger visual heterogeneity tend to exhibit greater permeability variation.

A regression line using all the data (Fig. 10-9 left) suggests that the probe and plug data are systematically different (i.e., $0.84 \neq 1$ and $0.79 \neq 0$). Some of this difference, however, may arise because of the highly heterogeneous nature of some samples. A regression line using only the plugs with $C_V \leq 0.4$ shows the probe and plug measurements are similar and, therefore, no further calibration is required; the slope and intercept are nearly 1 and 0, respectively. Thus, there is good cause for some probe and plug data to differ, and the cause here lies with the scales of geological variation to which each measurement is responding.

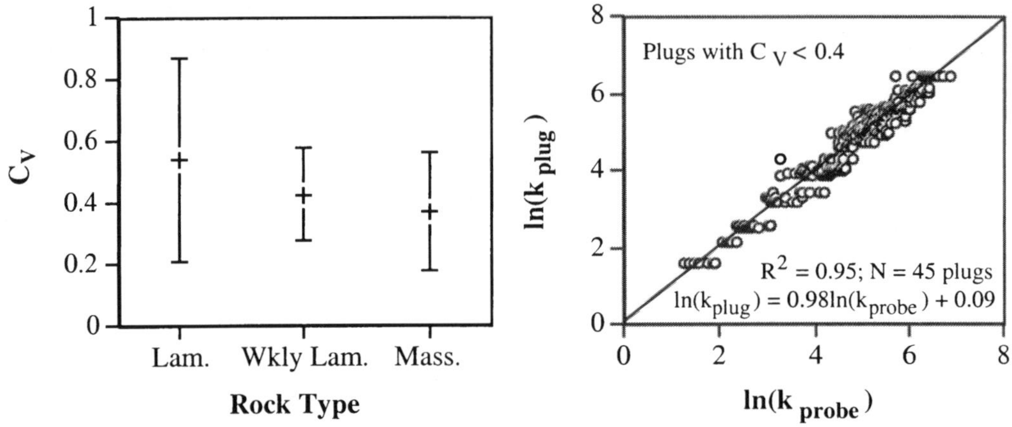

Figure 10-10. Probe permeability variation observed as a function of visually assessed rock type (left) and a probe–plug permeability scatter plot (right) for plugs with limited variability. Lam. = laminated, Wkly lam. = weakly laminated, and Mass. = massive.

10-7 MULTIVARIATE REGRESSION

As explained in the introduction to Chap. 9, multivariate linear regression is not covered in detail in this book. Nonetheless, we will say a few words about it because it is very common and we will use many of its concepts when we discuss Kriging in Chap. 12.

The model now is

$$\hat{Y} = \beta_0 + \sum_{m=1}^{M} \beta_m X^{(m)} + \varepsilon \tag{10-5}$$

where $X^{(m)}$ is a set of the observed data $\{X_1^{(m)}, X_2^{(m)}, \ldots, X_I^{(m)}\}$, and $\hat{Y}$ is a set containing estimates of the variable to be predicted. ε is the white noise whose elements have zero mean and are uncorrelated with anything except themselves. The procedure is the same as before; we take the expectation of Eq. (10-5) to solve for β_0.

$$\beta_0 = E(\hat{Y}) - \sum_{m=1}^{M} \beta_m E(X^{(m)}) \tag{10-6}$$

After inserting the model and eliminating β_0 with Eq. (10-6), we have

$$Var(Y - \hat{Y}) = Var(Y) - 2\sum_{m=1}^{M} \beta_m Cov(Y, X^{(m)}) + \sum_{m=1}^{M} \sum_{m=1}^{M} \beta_n \beta_m Cov(X^{(n)}, X^{(m)})$$

This expression is the same as the sum-of-the-squares operator S given in Chap. 9 if we replace the expectations with their estimates. To find the regression coefficients, set the derivative with respect to each regression coefficient to zero:

$$\frac{\partial S}{\partial \beta_1} = \frac{\partial S}{\partial \beta_2} = \cdots = \frac{\partial S}{\partial \beta_M} = 0$$

These operations lead to the following normal form of the multivariate linear model:

$$\begin{bmatrix} C_{X^{(1)}X^{(1)}} & \cdots & C_{X^{(1)}X^{(M)}} \\ \cdot & \cdots & \cdot \\ \cdot & \cdots & \cdot \\ \cdot & \cdots & \cdot \\ C_{X^{(M)}X^{(1)}} & \cdots & C_{X^{(M)}X^{(M)}} \end{bmatrix} \begin{bmatrix} \beta_1 \\ \cdot \\ \cdot \\ \cdot \\ \beta_M \end{bmatrix} = \begin{bmatrix} C_{X^{(1)}Y} \\ \cdot \\ \cdot \\ \cdot \\ C_{X^{(M)}Y} \end{bmatrix} \tag{10-7}$$

where we have substituted the covariance definitions for the expectation operators. $C_{X^{(1)}X^{(M)}} = Cov(X^{(1)}, X^{(M)})$, etc. Equation (10-7) represents a set of M linear equations that can be solved by standard means. The coefficient β_0 follows from Eq. (10-6) after determining β_1 through β_M.

The matrix on the left of Eq. (10-7) is the *covariance matrix*; its elements represent the degree of association between the "independent" variables $X^{(1)}$ through $X^{(M)}$. If the variables are independent (of each other), the matrix is diagonal and the resulting solution much easier. In any event, the diagonal elements are the variances of each variable.

Example 8a - Multivariate Regression to Examine Relationship of Oil Recovery to Slug Size and Pressure. We can illustrate the above procedure and introduce two other statistical insights with a fairly simple example.

The following table represents the incremental oil recovery (Y) from seven immiscible-gas-injection projects (extracted from Quigley, 1984) and their corresponding slug size ($X^{(1)}$) and average pressure ($X^{(2)}$).

Incremental oil recovery, fraction	Slug Size, fraction	Average Pressure, psig
0.394	0.3742	3671
0.06	0.00157	1695
0.4	1.515	4046
0.189	0.04707	4200
0.091	0.00002	2285
0.0997	0.0031	2049
0.128	0.5764	1050

For this problem, $M = 2$ and $I = 7$. We seek the model coefficients β_0, β_1, and β_2.

For numerical reasons, it is often advantageous to work in standardized normal variates, particularly when the range of data is large as it is here. To do this we require for Y

$$Y = (0.394 + 0.06 + \cdots + 0.128)/7 = 0.195$$

and

$$\sigma_Y^2 = [(0.394)^2 + (0.06)^2 + \cdots + (0.128)^2]/7 - (0.195)^2 = 0.0207$$

from which we have, for example,

$$Y_1 = (0.394 - 0.195)/(0.0207)^{1/2} = 1.386$$

and similarly for the other values and for $X^{*(1)}$ and $X^{*(2)}$. The following table shows the standardized variates.

Y^*	$X^{*(1)}$	$X^{*(2)}$
1.386	0.02615	0.7676
-0.9347	-0.6423	-0.8168
1.428	2.073	1.068
-0.0384	-0.5607	1.191
-0.7193	-0.6451	-0.3438
-0.6589	-0.6396	-0.533
-0.4622	0.3889	-1.334

Such a transformation also renders the model equation dimensionless. $\beta_0 = 0$ in the standardized variates, since the expectations of the new variables are 0. Also, their variances are 1.

The normal equations require the covariances; an example calculation for one element is

$$Cov(X^{*(1)}, X^{*(2)}) = [(0.02615)(0.7676) + (-0.6423)(-0.8168) + \cdots]/7 = 0.3049$$

which leads to the following normal equations:

$$\begin{bmatrix} 1 & 0.3049 \\ 0.3049 & 1 \end{bmatrix} \begin{bmatrix} \beta_1 \\ \beta_2 \end{bmatrix} = \begin{bmatrix} 0.6176 \\ 0.6460 \end{bmatrix}$$

Owing to the standardization of the variables, the elements of the matrix on the right side are now correlation coefficients. The solution to these leads to the following regression equation:

$$\hat{Y}^* = 0.4637\, X^{*(1)} + 0.5046\, X^{*(2)}$$

Apparently, $\hat{Y}$ is about equally sensitive to both $X^{*(1)}$ and $X^{*(2)}$. However, $X^{*(1)}$ and $X^{*(2)}$ are somewhat correlated.

Physical arguments on this type of process suggest that incremental oil recovery (IOR) should increase roughly linearly with slug size as the model suggests (Lake, 1989, Chap. 7). The dependency of IOR on pressure is more complex, but the two should be positively correlated also. Moreover, there is no obvious reason for slug size and pressure to be correlated; hence, the 0.3049 correlation coefficient is likely to be not significant. Of course, from a statistical point of view, we should interrogate ε for independence and normality (Chap. 9).

Analogous to the analysis for bivariate models, we can obtain confidence intervals for the parameter estimates of multivariate regression. This topic is covered in Montgomery and Peck (1982) and other regression texts.

If the covariance (correlation) matrix is nonsingular, it is possible to again redefine the variables to make them independent. We need a little development to see this. Write the original model as

$$\hat{Y} = \boldsymbol{X}^{\mathrm{T}} \boldsymbol{\beta}$$

and the normal equations as

$$\boldsymbol{C}_{\mathbf{XX}}\, \boldsymbol{\beta} = \boldsymbol{C}_{\mathbf{XY}}$$

in matrix form. For any nonsingular matrix $\boldsymbol{P}$ we can write

$$\boldsymbol{C}_{\mathbf{XX}}\, \boldsymbol{P}^{-1}\boldsymbol{P}\, \boldsymbol{\beta} = \boldsymbol{C}_{\mathbf{XY}}$$

It is also possible to multiply both sides by $\boldsymbol{P}$ to arrive at

$$\tilde{\boldsymbol{C}}_{\mathbf{XX}}\tilde{\boldsymbol{\beta}} = \tilde{\boldsymbol{C}}_{\mathbf{XY}}$$

where $\tilde{\boldsymbol{C}}_{\mathbf{XX}} = \boldsymbol{P}\boldsymbol{C}_{\mathbf{XX}}\boldsymbol{P}^{-1}\tilde{\boldsymbol{\beta}} = \boldsymbol{P}\boldsymbol{\beta}$, and $\tilde{\boldsymbol{C}}_{\mathbf{XY}} = \boldsymbol{P}\boldsymbol{C}_{\mathbf{XY}}$ all define a regression problem

$$\hat{Y} = \tilde{\boldsymbol{X}}\, \tilde{\boldsymbol{\beta}}$$

in the new variable $\tilde{\boldsymbol{X}} = \boldsymbol{X}\boldsymbol{P}^{-1}$.

The new covariance matrix $\tilde{\boldsymbol{C}}_{\mathbf{XX}}$ will be diagonal if the columns of the matrix $\boldsymbol{P}$ consist of eigenvectors of the original covariance matrix $\boldsymbol{C}_{\mathbf{XX}}$.

Example 8b - *Multivariate Regression to Examine Relationship of Oil Recovery to Slug Size and Pressure.* For the case here, $\boldsymbol{P}$ and $\boldsymbol{P}^{-1}$ are both

$$\boldsymbol{P} = \begin{bmatrix} 0.7071 & 0.7071 \\ 0.7071 & -0.7071 \end{bmatrix}$$

The new variables are

$$X^{(1')} = 0.7071\, X^{(1)} + 0.7071\, X^{(2)}$$

and

$$X^{(2')} = 0.7071\, X^{(1)} - 0.7071\, X^{(2)}$$

and the new regression equation is

$$\hat{Y} = 0.6847\, X^{(1')} - 0.0289\, X^{(2')}$$

where the variables on the right side are now independent.

The main reason for going through this is that the variance of Y can now be attributed individually to the new variables. Using Eq. (4-3), we have

$$Var\,(\hat{Y}) = (0.6847)^2\, Var(X^{(1')}) + (-0.0289)^2\, Var(X^{(2')})$$

or 63.56 % of the variance of $\hat{Y}$ can be attributed to $X^{(1')}$, only 0.05% to $X^{(2')}$, and the remainder to ε. Variance analysis on redefined variables such as this is called factor analysis, a topic covered in more detail in Carr (1995, pp. 99-102).

10-8 CONCLUDING COMMENTS

Despite the mathematical simplicity of linear least-squares regression, the method has a wealth of statistical features and interpretations that can make it enormously powerful. The material-balance analysis (Example 4 of this chapter) is a good example of this. Simple OLS regression produced estimates of m and N and no statistical analysis was needed to obtain these results. The data, however, have more information that can be interrogated to produce estimated confidence levels for m and N. These precisions have quite an influence on the way we look at the resulting m and N values. Without them, N = 105 million stock tank barrels (MSTB). With them, N is about 105 MSTB, give or take 34 MSTB, and that we need more data to reduce the uncertainty to a tolerable level, dictated by the field economics.

Chapter 9 and this chapter represent some of our experiences with regression. Much more could be said and there are excellent books on the subject, which we have cited. We encourage the reader to explore these aspects further and expand their "regression tool kit." Regression is a fascinating area that seems limited only by the user's ingenuity.

CHAPTER 11

ANALYSIS OF SPATIAL RELATIONSHIPS

We have now covered three chapters on aspects of correlation among different random variables. In addition, in Chap. 6 we discussed heterogeneity and its measures. A more complete exposition of ways to represent spatial structure requires information about both heterogeneity and correlation, except now this is the correlation of a random variable with itself. Such autocorrelation (or *self* correlation) is the subject of this chapter.

Correlation is the study of how two or more random variables are related. In data analysis, there are three types of correlation, as depicted in Fig. 11-1.

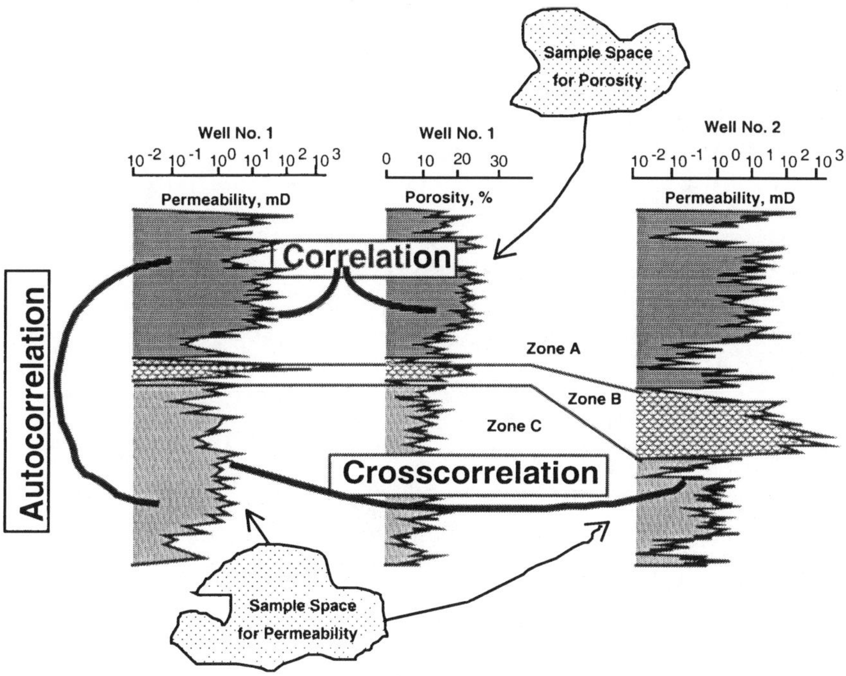

Figure 11-1. Schematic illustration of correlation types for porosity and permeability.

1. *Simple correlation*, such as between permeability and porosity in Well No. 1 in Fig. 11-1, is the relationship between two variables taken from two different sample spaces but representing the same location in the reservoir. The analysis and quantification of this type of relationship were considered in Chap. 8.

2. *Crosscorrelation*, such as of permeability between Well No. 1 and Well No. 2 in Fig. 11-1, is the relationship between two samples taken from the same sample space but at different locations. Examining electrical logs from several wells for characteristic patterns–the generic usage of "correlation" in petroleum engineering–is one example of crosscorrelation.

3. The last type of correlation, *autocorrelation*, is between variables of the same sample space taken at different locations. This is illustrated on the permeability data of Well No. 1 in Fig. 11-1. This correlation addresses, for example, the sequence of lithofacies that appear in a well. The last two types of correlation are both autocorrelation but in different directions.

Because time-dependent data are usually in the form of series of data, sometimes autocorrelation and crosscorrelation fall under the subject of time-series analysis. In what follows, we will make no distinction between data from time-sampled sequences and data from spatially sampled sequences.

11-1 DATA TYPES

Recall from Chap. 2 that there are two types of measurements recorded in series, strings, or chains.

1. Continuous variables recorded on a known or measurable scale, e.g.,
   (a) electric logs measuring resistivity as a continuous function of depth,
   (b) production history or time-dependence of an individual well, and
   (c) core-plug data recorded as a function of specified depth locations.

2. Categorical or discrete variables with/without a known or measurable scale, e.g.,
   (a) lithologic states recorded in a sedimentary succession,
   (b) the occurrence of certain types of microfossils as functions of depth, and
   (c) successions of mineral types encountered at equally spaced points on a transect across a thin-section.

These measurements may be *regular* (equally spaced) or *irregular*. The type of variable and the spacing affect the method and extent of analysis possible, as Fig. 11-2 indicates.

	Regularly Spaced Samples	**Irregularly Spaced Samples**
Interval or Ratio Data	1. Regression 2. Time-trend analysis 3. Autocorrelation or semivariograms 4. Spectral analysis	1. Regularization 2. Regression 3. Autocorrelation or semivariograms
Nominal or Ordinal Data	1. Auto association or cross-correlation 2. Markov chains 3. Runs test 4. Indicator semivariograms	1. Series of events

Figure 11-2. Summary of selected relationship modeling and assessment methods according to the nature and spacing of the sampled variable. Adapted from Davis (1973).

In the following, we discuss a few of these techniques, concentrating primarily on regularly spaced data.

11-2 NOMINAL VARIABLES

A *run* is a successive occurrence of a given discrete variable, which includes a single occurrence. This is the simplest type of sequence because an ordered set of qualitative observations falls under two mutually exclusive states. A test for randomness (independence) of occurrence is formulated from the number of runs of a given state. We can determine the probability of a given sequence or run being created by the random occurrence of two states. We do this by enumerating all possible ways of arranging n_1 items of state 1 and n_2 items of state 2. Taking u to be the total number of runs in a sequence and assuming n_1 and n_2 are greater than 10, u is approximately normally distributed (Feller, 1950, pp. 194-195). The expected value is

$$\mu_u = E(u) = 1 + 2n_1n_2/(n_1 + n_2)$$

with variance

$$\sigma_u^2 = \mathrm{Var}(u) = \frac{2n_1n_2(2n_1n_2 - n_1 - n_2)}{(n_1 + n_2)^2(n_1 + n_2 - 1)}$$

Knowing the mean and variance of u, we can apply a hypothesis test to its normalized version:

$$z = \frac{u - \mu_u}{\sigma_u}$$

Example 1 - *Illustration of Runs Test.* Analysis of an outcrop section revealed the following sequence of rock types sampled every 30 cm along a line:

QQSQQQSQQQSSQQSQQQSSQSQQQSQSS

where S = shale and Q = sandstone. The data are discrete variables that, being geologic features, are usually grouped or autocorrelated. The grouping or autocorrelation could take one of two forms: either there may be fewer transitions (positive autocorrelation) or there may be more transitions (negative autocorrelation or anticorrelation) than the expected number for a random selection of the S's and Q's. We can detect either case using a runs test.

The null and alternative hypotheses in this case are

H_0: $u = \mu_u$ (i.e., no autocorrelation)

H_A: $u \neq \mu_u$ (i.e., positive or negative autocorrelation).

Hence, at a 5% level of significance, $|z| > 1.96$ would lead to rejection of H_0. The statistics suggested by the above are

n_1 = number of shale points = 11	n_2	= number of sandstone points = 18
μ_u = expected number of runs = 14.7	σ_u^2	= variance of u = 6.2
u = observed number of runs = 16	z	$= \dfrac{u - \mu_u}{\sigma_u} = 0.5 < 1.96$

So, we accept H_0 in favor of H_A at the 5% level of confidence. We could analyze this set further to determine the existence of autocorrelation, but the runs test indicates no structure.

The runs test is a simple test for the presence of autocorrelation. In many problems, however, such a test does not provide enough information to permit successful exploitation of the answer it gives. This failing is primarily because of the qualitative nature of discrete variables. For interval or ratio variables, more powerful techniques are possible. We discuss these next.

11-3 MEASURES AND PROPERTIES OF AUTOCORRELATION

Recall from Chap. 8 that the covariance, $Cov(X, Y)$, is a measure of the strength of a relationship between two variables X and Y. In that case, nothing was mentioned about the spatial locations of X and Y because we assumed that both reflected properties at the same location. In a similar way, we can use the covariance to measure the relationship strength between a variable and itself, but at two different locations.

Let Z denote some property (e.g., permeability or porosity) of the rock that has been measured on a regular grid (such as along a core) as sketched in Fig. 11-3. Z_i is the property Z at location x_i, or $Z_i = Z(x_i)$. The location x is a time-like variable (non-negative and always increasing with i). $\Delta h = x_{i+1} - x_i$ is the *spacing* or *class size.*

Number	1	2	3	...	i	i+1	...	...	I
Location	x_1	x_2	x_3	...	x_i	x_{i+1}	...	...	x_I
Value	Z_1	Z_2	Z_3	...	Z_i	Z_{i+1}	...	...	Z_I

Figure 11-3. Schematic and nomenclature of a property Z measured at points along a line.

11-3.1 Autocovariance/Autocorrelation

We can now define the *autocovariance* between data Z_i and Z_j as

$$Cov(Z_i, Z_j) = E\{[Z_i - E(Z_i)][Z_j - E(Z_j)]\} = E(Z_iZ_j) - E(Z_i)\,E(Z_j) \tag{11-1}$$

The quantity $Cov(Z_i, Z_j)$ is a measure of how strongly the property Z at location x_i is related to Z at location x_j. When x_i and x_j are close together, we expect a strong relationship. As x_i and

x_j move farther apart, the relationship should weaken. The properties of the autocovariance are similar to the covariance discussed in Chap. 8, specifically $Cov(Z_i, Z_j) = Cov(Z_j, Z_i)$ and $Cov(Z_i, Z_i) = Var(Z_i)$.

The autocovariance, being a generalized variance, is a measure of both autocorrelation and variability (i.e., heterogeneity). To remove the heterogeneity, we define an *autocorrelation coefficient* between Z_i and Z_j as

$$\rho(Z_i, Z_j) = \frac{Cov(Z_i, Z_j)}{[Var(Z_i)Var(Z_j)]^{1/2}}$$

As in the case of the correlation coefficient discussed in Chap. 8, ρ must be between -1 and +1. Many times both *Cov* and ρ are written in terms of a *lag* or *separation index k,* where $j = i+k$.

The similarities between the definitions of autocovariance and covariance belies a subtlety that has significance throughout the remainder of this discussion. Recall that the basic definition for the expectation operator involved a probability density function (PDF) for the random variable (since we are dealing with two variables here, these are actually joint PDF's). Taken to Eq. (11-1), this means that Z_i and Z_j both must have PDF's associated with them even though they are from the same parameter space. Taken as a single line of data, Fig. 11-3, a PDF at each point is difficult to imagine. If we view the line as only one of several possible lines, however, things begin to make more sense because we can imagine several values of Z existing at a point x_i. If the PDF's for all Z_i are the same, we have assumed *strict stationarity*.

A related hypothesis comes from the need to estimate *Cov*. Equation (11-1) contains $E(Z_i)$, which we would normally estimate with a formula like

$$E(Z) = \frac{1}{I}\sum_{i=1}^{I} Z_i$$

It is not clear that $E(Z)$ estimated in this fashion (the spatial average) would be the same as if we had averaged all the values at x_i from the multiple lines (an ensemble average). Assuming an equality between spatial and ensemble averages is the *ergodic hypothesis*. We assume ergodicity and some form of stationarity in all that we do.

Example 2 - Variance of the Sample Mean (Revisited). We recalculate the variance of the sample mean, as was done in Example 2 of Chap. 4, but now allow the random variable to be dependent. As before, we have

$$Var(\overline{Z}) = Var\left(\frac{1}{I}\sum_{i=1}^{I} Z_i\right) = \frac{1}{I^2}\, Var\left(\sum_{i=1}^{I} Z_i\right)$$

Inserting the definition for variance and multiplying out yields

$$Var(\overline{Z}) = \frac{1}{I^2} \{E[Z_1(Z_1 + \cdots + Z_I) + \cdots + Z_I(Z_1 + \cdots + Z_I)] -$$

$$[E(Z_1)(E(Z_1) + \cdots + E(Z_I) + \cdots + E(Z_I)(E(Z_1) + \cdots + E(Z_I))]\}$$

We can no longer drop the cross terms. Further multiplying and identification with the *Cov* definition yields

$$Var(\overline{Z}) = \frac{1}{I^2} \sum_{i=1}^{I} \sum_{j=1}^{I} Cov(Z_i, Z_j)$$

an equation with a considerably different form from the case with the Z_i independent. The inclusion of autocorrelation may increase or decrease the variance of the sample mean, compared to the independent variable case, since *Cov* can be either positive or negative.

The variance of the sample mean has many different forms. For example,

$$Var(\overline{Z}) = \frac{\sigma^2}{I}\left(1 + \frac{2}{I} \sum_{i=1}^{I} \sum_{j=i+1}^{I} \rho(Z_i, Z_j)\right)$$

where $Var(Z_i) = \sigma^2$. This equation is the starting point for the study of dispersion in an autocorrelated medium or of any similar additive property.

11-3.2 Semivariance

A common tool in spatial statistics is the *semivariance* γ, defined as

$$2\gamma(Z_i, Z_j) = E[(Z_i - Z_j)^2] \qquad (11\text{-}2)$$

The semivariance is also a generalized variance; compare its definition to $Var(Z)=E[(Z- \overline{Z})^2]$. It shares many properties with the variance, most notably that it is always greater than zero. However both *Cov* and γ are limited in utility without further assumptions, which we go into below. Sometimes, a normalized version of the semivariogram is considered:

$$\gamma^*(Z_i, Z_j) = \frac{\gamma(Z_i, Z_j)}{[Var(Z_i) Var(Z_j)]^{1/2}}$$

11-3.3 Stationarity

One of the consequences of strict stationarity (discussed earlier) is that all moments are invariant under translation, that is, they are independent of position. Stationary of order *r* means that moments up to order *r* are independent of position. Since we do not use moments higher than 2, all we require for the moments is second-order stationarity, but even this has some subtle aspects.

We have already used second-order stationarity in Example 2, where we set $Var(Z_i) = \sigma^2$, or

$$Var(Z_1) = E(Z_1Z_1) - E(Z_1)E(Z_1) = E(Z_2Z_2) - E(Z_2)E(Z_2) = \cdots = Var(Z) = \sigma^2$$

which is free of an index on the right side. Also

$$E(Z_1Z_2) = E(Z_2Z_3) = \cdots = E(Z_{I-1}Z_I)$$

from which it follows that $Cov(Z_1, Z_{1+k}) = Cov(Z_2, Z_{2+k})$, etc. (Second-order stationarity implies that first moments are also independent of position.) Thus, second-order stationarity renders the autorrelation measures independent of position:

$$Cov(Z_i, Z_{i+k}) = Cov(k) = Cov_k$$

$$\rho(Z_i, Z_{i+k}) = \rho(k) = \rho_k$$

$$\gamma(Z_i, Z_{i+k}) = \gamma(k) = \gamma_k$$

or, the autocorrelation measures become functions of a single argument. In terms of the line-of-data viewpoint (Fig. 11-3) , these become

$$Cov(k) = Cov(k\Delta h) = Cov(h)$$

$$\rho(k) = \rho(k\Delta h) = \rho(h)$$

$$\gamma(k) = \gamma(k\Delta h) = \gamma(h)$$

where h $(k\Delta h)$ is the *separation distance*, *lag distance*, or sometimes just *lag*.

It is difficult to overstate the simplifications that second-order stationarity brings to the autocorrelation measures. For one thing, there is now a simple relationship between the autocovariance and the semivariance,

$$\gamma(h) = \sigma^2 - Cov(h) \tag{11-3}$$

which holds if the variance is finite. A second advantage is that there is now possible a graphical representation of the autocorrelation measures. A plot of *Cov* versus h is called the *autocovariogram*, γ versus h the *semivariogram* (almost always referred to as the *variogram*), and ρ versus h the *autocorrelogram*. The qualitative behavior of these is discussed in the following paragraphs.

Figure 11-4 illustrates an autocorrelation coefficient ρ decreasing with lag distance (k or h). When ρ approaches zero there is, on the average, no correlation between data at that curve shows a modest amount of autocorrelation, the center curve shows even less, and the right curve shows one with a very long autocorrelation distance. This last figure indicates a trend or a spatially varying mean that would probably be apparent from carefully inspecting the original data set.

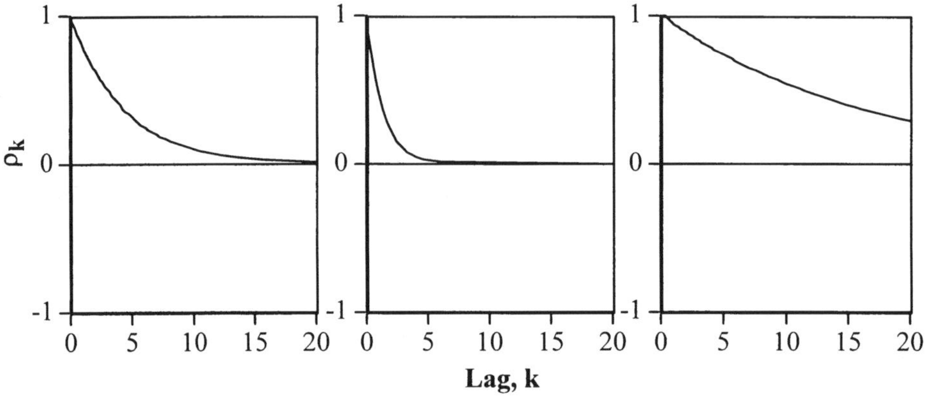

Figure 11-4. Schematic of autocorrelogram with modest (left), little (center), and strong (right) correlation.

Another form of stationarity is sometimes invoked when using the semivariogram. It is slightly weaker than second-order stationarity. If the difference Z_i - Z_{i+k} has a mean and variance that exist and are stationary, then Z satisfies the *intrinsic hypothesis*. It is weaker because $Var(Z)$ may not exist.

It is often *assumed* that second-order stationarity applies. If so, Eq. (11-3) implies that ρ and γ are mirror images of each other. When ρ decreases, γ increases. If $\rho(h) < 0$ for some h, $\gamma(h) > \sigma^2$. Thus, anticorrelation appears as an increase in γ above the variance.

As beneficial as they are, ergodicity and stationarity are difficult to verify in a set of data. Verifying ergodicity would mean measuring multiple lines and seeing if ensemble statistics compare with spatial statistics. However, any such differences would be indistinguishable from heterogeneity. We could, similarly, calculate Cov and γ with successively smaller sets of the data but, at some point, we would be unable to conclude anything because the number of data would be small.

Another subtlety of γ and ρ is that, because they are related to the autocovariance, they must obey certain properties. The most important is that ρ must be a *positive semidefinite function* and, correspondingly, $-\gamma$ must also be a *positive semidefinite function*. Since the mathematical term positive semidefinite may not be familiar to the reader, we will discuss these properties further by examining the implications for the autocorrelation ρ.

The term positive definite is for functions what the requirement to be positive is for scalars ($\lambda > 0$). Semidefiniteness corresponds to the scalar requirement of nonnegativity, $\lambda \geq 0$.

Because we are dealing with functions, however, positive semidefiniteness is more complicated. For the autocorrelation function, ρ, it means that the correlation at one point, x_i, is related to the correlation at any other point, x_{i+k}. For example, we will see later that any interpolation procedure requires that the estimated value at the new location be a linear combination of the values at the I measured locations:

$$\hat{Z} = \sum_{i=1}^{I} \lambda_i Z_i$$

where the weights (λ's) are determined by a specific procedure. If $I = 2$, we have $\hat{Z} = \lambda_1 Z_1 + \lambda_2 Z_2$. Since $\hat{Z}$ is a random variable, it has a variance given by

$$Var(\hat{Z}) = \lambda_1^2\ Var(Z_1) + \lambda_2^2\ Var(Z_2) + 2\lambda_1\lambda_2 Cov(Z_1, Z_2)$$

If Z is stationary,

$$Var(\hat{Z}) = Var(Z)\ [\lambda_1^2 + \lambda_2^2 + 2\lambda_1\lambda_2\rho(1)]$$

Since $Var(\hat{Z}) \geq 0$, we require that $[\lambda_1^2 + \lambda_2^2 + 2\lambda_1\lambda_2\rho(1)] \geq 0$. This can only be achieved for any λ_1 and λ_2 if $\rho(1) \leq 1$. Clearly, when $I > 2$ the situation is analogous but more complicated. This is where the positive semidefiniteness of ρ is needed: the $Var(\hat{Z})$ must be nonnegative for any values the λ's take. Thus, not just any function can be used for ρ; it has to be positive semidefinite. This limits the possibilities quite considerably, and several examples will be considered below.

11-4 ESTIMATING AUTOCORRELATION MEASURES

Like all statistics, *Cov*, ρ, and γ must be estimated. We discuss here only the more common estimators. See Priestley (1981, Sec. 5.3), Cressie (1991, Sec. 2.9), and Chilès and Delfiner (1999, Sec. 2.5) for further estimators and their properties.

11-4.1 Autocovariance/Autocorrelation Coefficient

The most common estimator for the autocovariance between data points Z_i and Z_{i+k} is

$$\hat{C}(Z_i, Z_{i+k}) = \frac{1}{I-k}\sum_{i=1}^{I-k}(Z_i - \bar{Z})(Z_{i+k} - \bar{Z})$$

where $\bar{Z}$ is an estimate for the mean value of Z and $\hat{C}$ is an estimate for *Cov*. Subject to the assumption of second-order stationarity (that *Cov* depends only on the separation index *k*), this becomes

$$\hat{C}(k) = \frac{1}{I-k}\sum_{i=1}^{I-k} Z_i Z_{i+k} - \overline{Z}^2 \tag{11-4}$$

If the separation distance or lag is zero, the above becomes an estimator of the variance that itself becomes an appropriate normalization factor. An estimator for the autocorrelation coefficient is

$$\hat{\rho}(k) = \frac{\frac{1}{I-k}\sum_{i=1}^{I-k} Z_i Z_{i+k} - \overline{Z}^2}{\frac{1}{I}\sum_{i=1}^{I}(Z_i - \overline{Z})^2} \tag{11-5}$$

the denominator being, of course, an estimate of the variance. The estimator in Eq. (11-5), though common, may not be a positive semidefinite function; see Priestly (1981, p. 331).

Example 3 - Autocorrelation of an Ordered Card Deck. Let us calculate the autocorrelogram (Fig. 11-5) for a standard deck of cards (four suites of ace through king each). This is also a concrete way to illustrate autocorrelation, for we can show that "randomness" in card decks really means absence of autocorrelation. Let ace be one, etc., up to king equaling 13. The mean value in the deck is 7 and the standard deviation 3.78. Neither of these statistics expresses the actual spatial ordering, however.

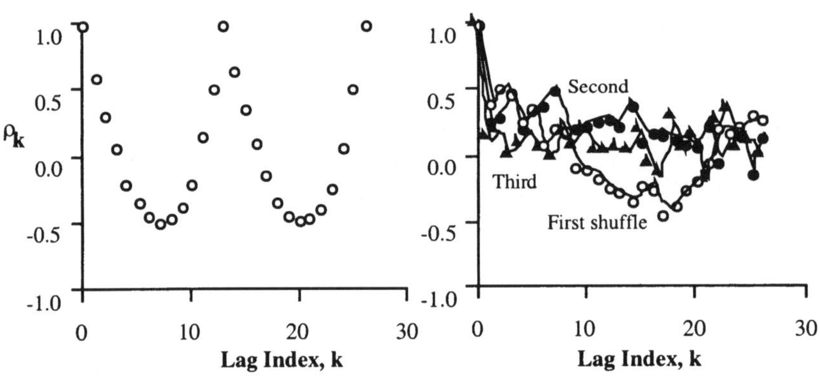

Figure 11-5. The autocorrelogram of an ordered card deck. Originally ordered (left); after one to three shuffles (right).

A completely ordered deck has an autocorrelogram that looks something like Fig. 11-5 (left). There is perfect autocorrelation at every thirteenth card, as expected, and the pattern repeats itself. This means that, given the value and position of any card in the deck, the value of any card that is $13n$ (n an integer) away has the same value. Every seventh card shows very strong negative autocorrelation since these separations are, on average, as far apart as possible in numerical value. In other words, given the position p of a card with value v_p, if $v_p > 7$ (the mean value of the deck), the card at position $p + 7$ is likely to have $v_{p+7} < 7$ (and vice versa).

Figure 11-5 (right) shows the autocorrelograms after several shuffles of the card deck. The autocorrelation is much more erratic (the shuffle was not perfect) but, after the first shuffle, there is a hint of periodicity at every 26th card. More importantly, the autocorrelation coefficient approaches zero only gradually.

After three shuffles, there is very little dependence on adjacent cards. Besides the importance of not shuffling card decks fewer than three times when gambling, this example shows that autocorrelation is a useful tool for evaluating spatial dependencies.

11-4.2 Semivariograms

The standard estimator for the semivariance is

$$2\hat{\gamma}_k = \frac{1}{I-k}\sum_{i=1}^{I-k}(Z_i - Z_{i+k})^2 \qquad (11\text{-}6)$$

This formula expresses the variance of $I - k$ sample increments measured a distance k units apart, again assuming second-order stationarity. The sample semivariogram may look something like Fig. 11-6.

The data points in the sample semivariogram rise with lag, although not necessarily smoothly or monotonically, reflecting greater heterogeneity for larger sample scales. The data may or may not tend to a plateau as the lag increases; similarly, they may or may not tend to zero as the lag decreases. We quantify the shape of these trends in the model curves discussed later in the chapter.

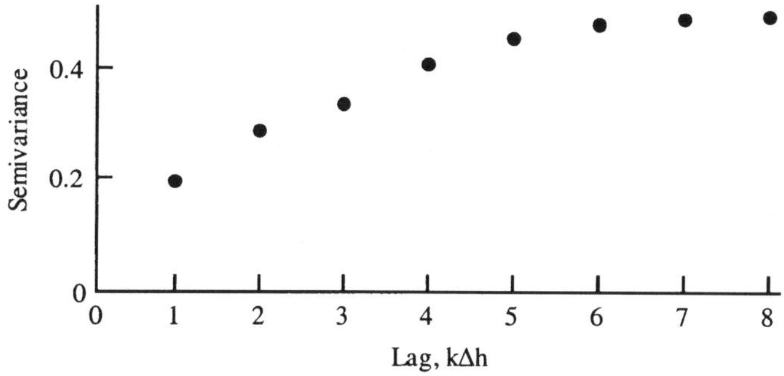

Figure 11-6. Schematic illustration of a sample semivariogram.

Although it is not their prime purpose, both the semivariogram and the autocorrelogram can be used to detect periodicities. As Example 3 showed, large values of ρ for $k > 0$ may reflect a repetitive feature in the data. Similarly, repetitiveness with a semivariogram is shown by a reduction in γ for $k > 0$, This is called a *hole*. Some examples will be considered next.

11-5 SEMIVARIOGRAM BEHAVIOR FOR GEOLOGIC STRUCTURES

Simple hypothetical permeability trends are useful to demonstrate the form of the semivariogram in typical geological scenarios and to demonstrate what can or cannot be extracted in their interpretation.

11-5.1 Large-Scale Trends

Deterministic trends (i.e., with no random component) produce a semivariogram function that is monotonically increasing with increasing lag. Both linear trends (Fig. 11-7a) and step changes (Fig. 11-7b) are common depositional features, although they are rarely as easy to recognize in practice as is suggested by these figures. Cyclic patterns are also common.

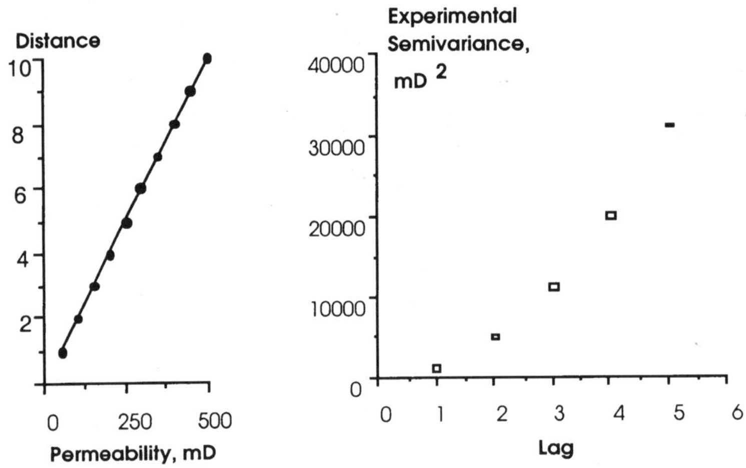

Figure 11-7a. Sample semivariogram for data with a linear trend.

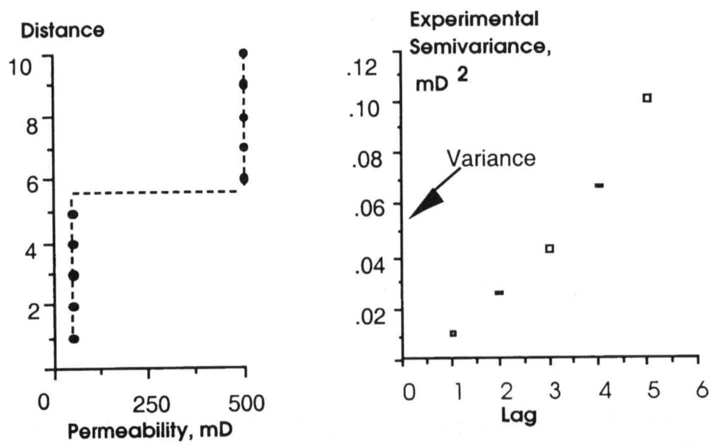

Figure 11-7b. Sample semivariogram for data with a step change.

Both features give similar semivariograms; the form of the semivariogram only indicates an underlying large-scale trend. It provides no insight into the form of that trend. Indeed if the simple trends were inverted (reversed), the resulting semivariograms would be indistinguishable.

11-5.2 Periodicity and Geological Structure

The effects of a periodic component upon the autocorrelation structure of a signal are well-known (e.g., Schwarzacher, 1975, pp. 169-179). Suppose Z has a periodic component with wavelength λ. These components are evident in the sample semivariogram, $\hat{\gamma}(h)$, Fig. 11-8. The depth of the holes will depend upon the size of the periodic component compared to any random component(s) present in Z. The depths of the holes may also diminish with increasing lag, depending upon the nature of the sedimentary system and the nature of its energy source (Schwarzacher, 1975, pp. 268-274). In particular, since even strongly repetitive features in sediments are usually not perfectly periodic, holes may be much shallower than the structure merits. For example, a slightly aperiodic signal (Fig. 11-8 left) has a smaller first hole (Fig. 11-8 right) and much smaller subsequent holes than its periodic equivalent. Yet, the structure represented by each signal is equivalent. So we must be careful using the autocorrelogram or the semivariogram to detect cyclicity. If, however, evidence of cyclicity appears in a sample plot, the cyclicity is probably a significant element in the property variation.

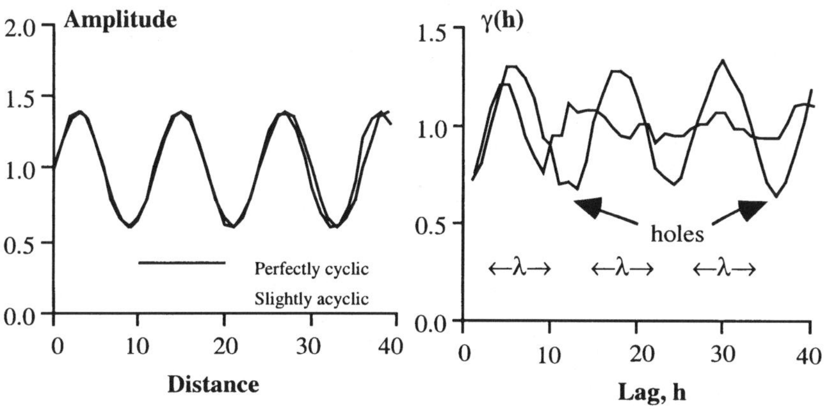

Fig. 11-8. Example permeability plots and semivariograms based on the series $Z(x_i) = 1 + 0.4 \sin(\pi x_i/6) + \varepsilon_i$, where $\varepsilon_i \sim N(0, 1)$ and z_i (= 0, 1, ...) is position, showing evidence of structure (cyclicity with $\lambda = 12$ units). The acyclic series is $Z'(x_i) = 1 + 0.4 \sin(\pi x_i/6 + \delta) + \varepsilon_i$, where $\delta = 0.01[x_i/12]$ and $[x_i/12]$ is the greatest integer less than or equal to $z_i/12$. Left figure shows deterministic components of $Z(x)$ and $Z'(x)$. Adapted from Jensen et al. (1996).

Various nested, graded beds result in semivariograms that reflect underlying trends and periodicity. More complex patterns may not be resolvable (especially real patterns!). See the examples in Fig. 11-9.

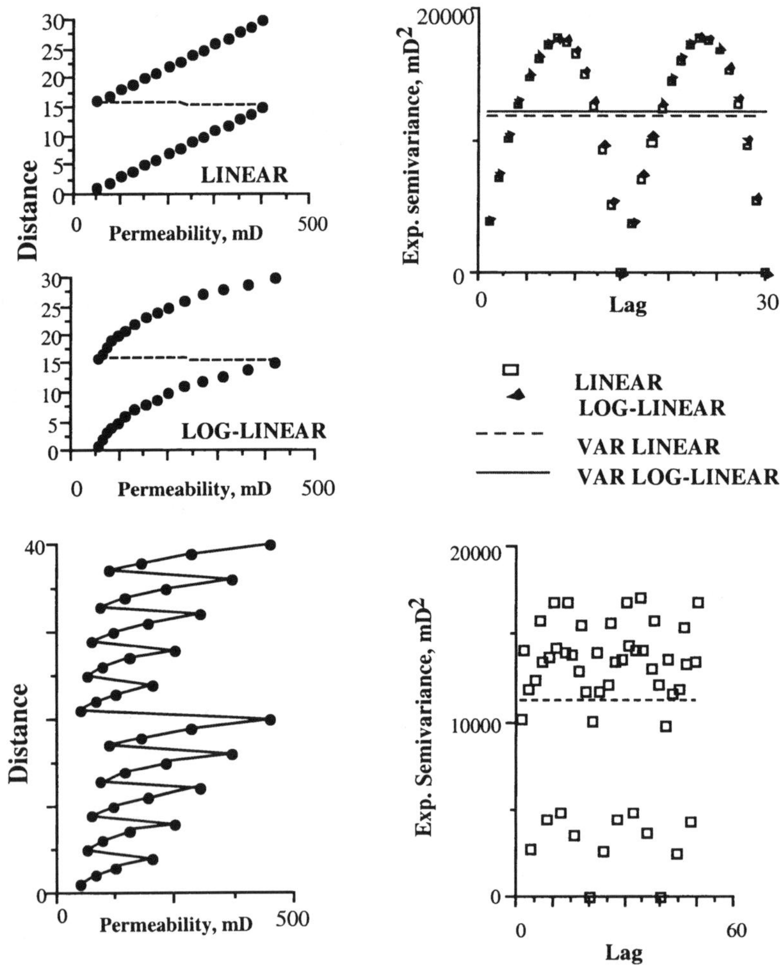

Figure 11-9. Semivariograms for graded and nested patterns.

11-6 SEMIVARIOGRAMS IN REAL MEDIA

Geologic data can be much more complex than shown by certain idealized correlation behavior. We show some examples in this section.

Autocorrelation in clastic formations tends to be scale-dependent, with different scales being associated with laminae, bedforms, and formations. Additionally, autocorrelation tends to be anisotropic; vertical and horizontal autocorrelation structure can be quite different. Cylindrical anisotropy in the horizontal plane is also likely where the deposition is unidirectional or current-generated (e.g., fluvial) in contrast to oscillatory, wave-generated deposits that are more likely to be cylindrically isotropic (e.g., shoreface). The statistical characterization of the properties of geologically recognizable forms is an ongoing area of investigation. As an introduction, we illustrate some of these features in the following items.

11-6.1 Deterministic Structure of Tidal Sediments

In a study of bed-thickness data for an estuarine Upper Carboniferous sequence in Eastern Kentucky (Martino and Sanderson, 1993), autocorrelation analysis was used to determine the spatial relationships of bed thickness. The correlogram shows distinct periodicity with layer periods at 12 and 32. The first periodicity is interpreted to reflect spring-neap tidal variations over 11 to 14 days and the longer range signal as having resulted from monthly variations in tidal range. These results were supported by Fourier analysis.

The semivariogram can sometimes reveal "average" periodicities that are represented by a significant reduction (by 20 to 50%) in variance (holes) at some lags. Two example semivariograms from the Rannoch Formation in two different fields, Fig. 11-10, show a periodicity at 1.2 to 1.4 m (4 to 4.5 ft). This periodicity is similar to that clearly seen in other intervals and is thought to be related to the (hummocky, cross-stratified) bedform thickness.

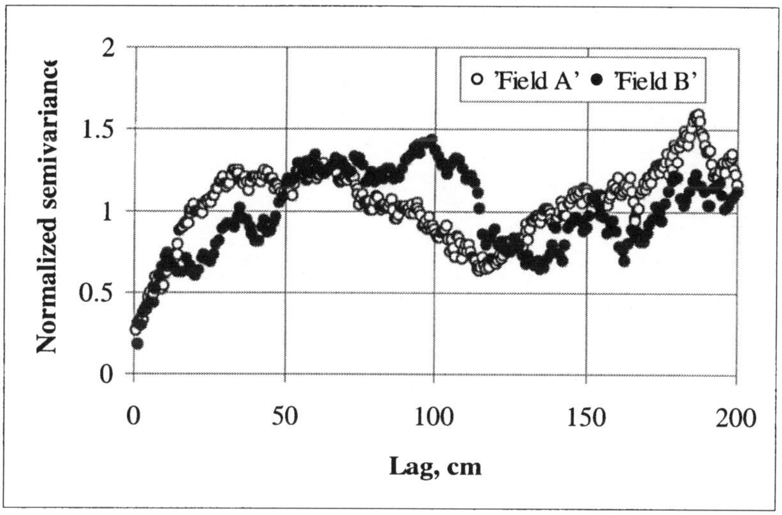

Figure 11-10. Two examples of geologic periodicities in probe permeability data from the Rannoch Formation. The nugget forms about 25% of the variance, while the repetitive bed stacking contributes a further 30% to the variance. The remaining variability is caused by permeability changing within the beds.

11-6.2 Determination of Nested Structure

There is a relationship between sample density, sample length, and the scale of geological variation that one is trying to characterize. The larger-scale variation tends to dominate. Several correlation lengths and scales of periodicity may exist in a formation, and each requires a tailor-made sampling scheme. Figure 11-11 shows this. The general dependency on scale coupled with the cyclic behavior makes it difficult to completely characterize the autocorrelation behavior with even a fractal model.

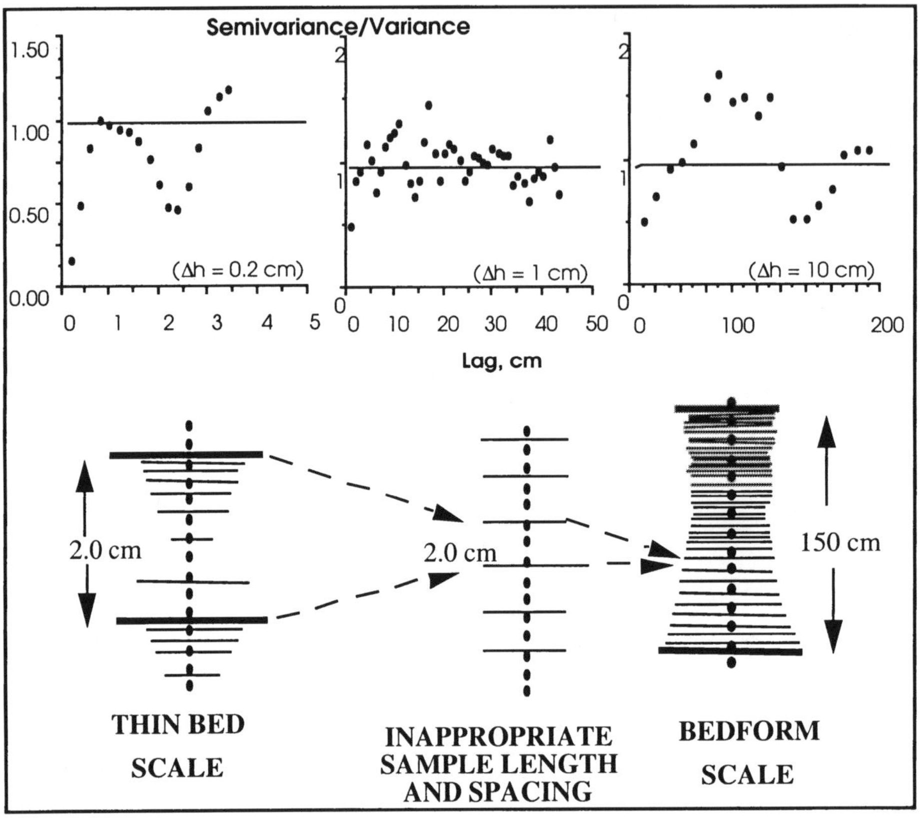

Figure 11-11. Semivariograms at various scales in permeability data from the Rannoch formation. Δh is the distance between measurements.

11-6.3 Semivariograms and Anisotropy

Semivariograms in different directions can be used to demonstrate anisotropy. This could be particularly useful where the trends are not so visually obvious as they are in Fig. 11-12. The filled and open symbols in Fig. 11-12 are taken from data strings perpendicular to each other.

Clastic reservoirs are characterized by relatively weak autocorrelation in the vertical direction and much stronger autocorrelation in the horizontal. In carbonates, the preservation of bedding is not always so clear, but anisotropy may still be present.

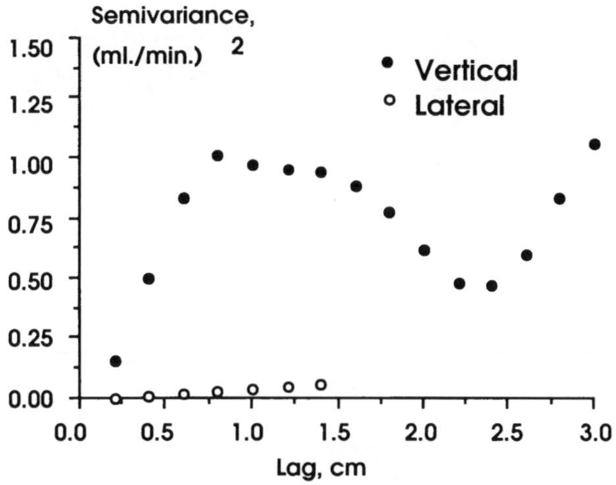

Figure 11-12. Illustration of anisotropy in semivariograms from the Rannoch Formation.

11-7 BIAS AND PRECISION

With the tools now available to estimate autocorrelation, it becomes a natural question as to which is preferable. As is usual, the preference of one measure over another depends on the application, but we can also make some statistical judgments.

Yang and Lake (1988) investigated the bias and precision of the autocovariance and semivariogram. They found that the autocovariance is in general negatively biased, that is, it tends to underestimate the extent of autocorrelation. The source of this bias lies in the estimate of the sample mean, which is required for the estimates of both Cov and ρ. The sample mean does not appear in the definition for the semivariance, Eq. (11-6); it is consequently unbiased. The semivariance is, therefore, the best tool to estimate autocorrelation from this standpoint. These statements are necessarily oversimplifications; the error behavior of the autocorrelogram depends not just on k and I, but also on the underlying structure of the data. Priestley (1981, Sec. 5.3) discusses this and related topics.

Both the autocovariance and the semivariance estimates are relatively imprecise beyond one-half of the sample span, so that fitting models to experimental curves by regression is discouraged. Since it is precisely the autocorrelation at large lags that is of interest in calculating these measures, this can be a serious problem in practice.

The origin behind the lack of precision in the autocorrelation measures lies in the successively smaller pairs available in the estimators, Eqs. (11-5) and (11-6), at large lags. Caution is also required when interpreting estimates based on data sets with significant proportions of missing measurements. In these circumstances, $\hat{\gamma}$ may be as unreliable at some short lags as it is at very long lags. As a simple example, consider the case of six measurement locations ($I = 6$) with two data missing. If $i = 2$ and 4 are missing, there is only one pair to estimate $\gamma(1)$ but two pairs to estimate $\gamma(2)$.

Li and Lake (1994) addressed these problems by an integral estimator that keeps constant the number of pairs throughout the procedure. Maintaining the same number of pairs in the calculation causes the semivariance estimate for the largest lag to be equal to the sample variance. See the original reference for the estimator equation, which is complicated. Figure 11-13 shows some of the results from using this moving-window estimator.

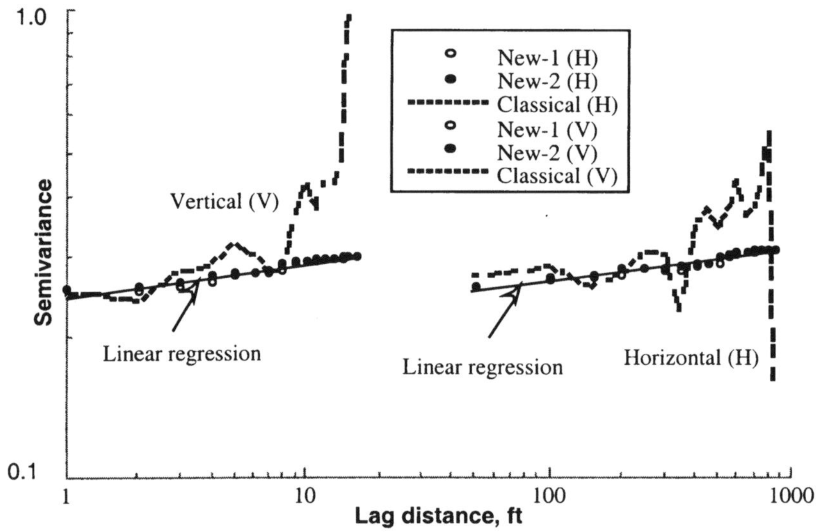

Figure 11-13. Semivariogram estimators for horizontal (H) and vertical (V) data sets from the Algerita outcrop. From Li and Lake (1994).

The plot is on log-log scales to facilitate the interpretation of data with a fractal model. But it is clear that the fluctuations present at large lags in the "classical" (Eq. (11-6)) estimator have been eliminated. The new estimator (New-2) performs better in this regard than all the others. In addition to being more precise, the new semivariance estimator can deal with multidimensional data (rather than along a line as has been the focus of the discussion here) and does not require equally spaced data. The latter aspect means that the new procedure is robust against missing data.

This "moving window" estimator may become an important standard method in the future. Not only does it reduce the semivariance fluctuations at large lags, it also smoothes out fluctuations at small lags, yielding semivariogram behavior that is easier to fit to models near the origin. These effects are important for two reasons: (1) we are able to interpret a model in difficult data situations (for example, with few wells or large well spacing) and (2) because of the smooth near-origin behavior, we can use autofitting methods, such as regression, to obtain a model. Being able to autofit semivariogram functions would eliminate much of the subjectivity in current semivariogram analysis.

11-8 MODELS AND FEATURES

Just as with PDF's and CDF's, we frequently have a need to report the extent of autocorrelation in a fashion that is even more succinct than the estimates obtained from data.

Moreover, some forward applications (see Chap. 12) require negative or positive semi-definite relations that the estimates may not provide. However, just as in the curve fitting to PDF's, there is no strong physical significance to the models apart from the association with the underlying geologic structure discussed above.

11-8.1 Finite Autocorrelation Models

We cover only only a few of the models most commonly used in reservoir characterization applications. See, for example, Cressie (1991, Sec. 2.5), Chilès and Delfiner (1999, Sec. 2.5), or Kitanidis (1997, Sec. 3.4) for others. All are, however, positive or negative semi-definite, as appropriate, and can be used in three-dimensional characterizations.

Exponential model of the autocorrelogram:

$$\rho(h) = exp(-\frac{h}{\lambda_L}) \qquad (11\text{-}7)$$

This one-parameter model of autocorrelation is popular for its simplicity. The parameter λ_L is the *autocorrelation length*. $\rho(h)$ in this model is always positive and monotonically decreasing. Consequently, it should not be used for cyclic or periodic structures. Clearly, as λ_L approaches zero, autocorrelation vanishes.

Spherical model of the semivariance:

$$\gamma(h) = \begin{cases} \left(\sigma^2 - \sigma_o^2\right)\left[\frac{3}{2}\left(\frac{h}{\lambda_R}\right) - \frac{1}{2}\left(\frac{h}{\lambda_R}\right)^3\right] + \sigma_o^2 & h \le \lambda_R \\ \sigma^2 & h > \lambda_R \end{cases} \qquad (11\text{-}8)$$

This most popular of all autocorrelation models is the three-parameter spherical model. σ^2 is the *sill*, σ_0^2 the *nugget*, and λ_R is the *range*. These quantities are illustrated in Fig. 11-14, which shows the spherical model fit to a small set of data.

The sill should approximate the sample variance, and the range the extent of the autocorrelation. The nugget is an inferred quantity, since it can never be measured directly. A semivariogram that has a nugget approximately equal to its sill exhibits a *pure nugget effect* and, similarly to having a small range, has no autocorrelation in the underlying data. The pervasiveness of the spherical model has led to the incorporation of range, sill, and nugget into the more general statistical terminology. For example, a data set with a "large range" is strongly autocorrelated even though the spherical model may not apply at all.

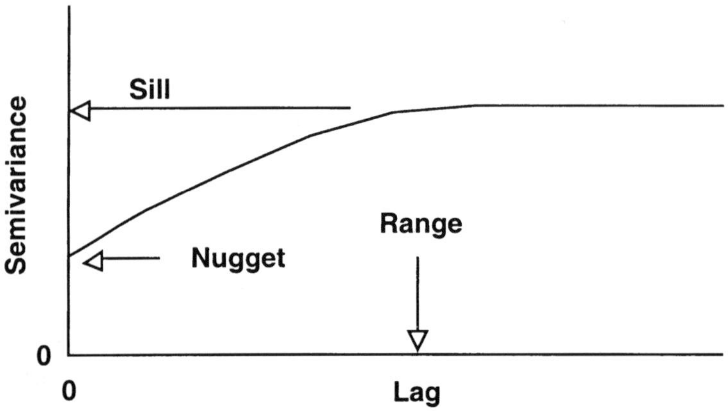

Figure 11-14. Illustration of spherical semivariogram model.

A semivariogram model that does include a degree of cyclicity and, therefore, has some physical significance attached to its behavior is the *hole effect* model given by

$$\gamma(h) = \sigma_0^2 + (\sigma^2 - \sigma_0^2)[1 - \lambda_H sin(h/\lambda_H)/h]$$

The peaks and troughs (the holes) of this model are at solutions to the equation $h/\lambda_H = \tan(h/\lambda_H)$. The first peak is at $h/\lambda_H \cong 3\pi/2$ and the first hole is at $h/\lambda_H \cong 5\pi/2$. The first hole has depth 0.13 of the nugget— sill distance, $\sigma^2 - \sigma_0^2$. This is the deepest hole of any of the known models for 3-D correlation and is the minimum possible for the model to be negative semidefinite. As seen earlier, deeper holes in some directions are common in many sedimentary systems. Therefore, an anisotropic covariance is needed to model this behavior with a model of the form

$$\gamma(h) = \sigma_0^2 + (\sigma^2 - \sigma_0^2)[1 - cos(h/\lambda_H)]$$

in the direction where the cyclicity is strongest.

11-8.2 Integral Scale

The correlation length and range are often used interchangeably, probably because they both express the same general idea. However, they are not the same, each being unique to the respective model. A quantity that also expresses the extent of autocorrelation is the *integral scale*, defined as

$$\lambda_I = \int_0^\infty \rho(x)dx \qquad (11\text{-}9)$$

λ_I is defined for any autocorrelation coefficient with a finite correlation, but, apart from this, it is independent of the model chosen. Of course, $\lambda_I = \lambda_L$ if the model is exponential.

11-8.3 Infinite Autocorrelation Models

A growing body of literature suggests that many geological data sets do not have finite correlation. Indeed, the data in Fig. 11-13 can be so interpreted. In this eventuality, there is a class of models abstracted from fractal terminology.

The fractional Brownian motion (fBm) semivariogram,

$$\gamma(h) = \gamma_o \, h^{2H} \tag{11-10}$$

is a two-parameter model that contains a variance at a reference (h = 1) scale, γ_o, and the Hurst or intermittency coefficient, H. H takes on the role of the various lengths in Eqs. (11-7) to (11-9), but, of course, γ is unbounded in general. Figure 11-13 illustrates one of the ways to estimate H by plotting γ versus h on log-log coordinates. See Hewett (1986) for discussion of other methods.

A related model is the fractional Gaussian noise (fGn), which has a much more complicated model equation. However, Li and Lake (1993) have shown that, under most practical circumstances, fGn with $0 < H < 1$ is equivalent to fBm for $1 < H < 2$ as long as H+1 is used instead of H in Eq. (11-10).

The unbounded nature of γ (and the simplicity of the model) makes fractal models particularly appealing for geological characterization (e.g., Burrough, 1983a,b). But, it is conventional to truncate the model at some upper and lower scales simply in order not to be extrapolating too far outside the range of the data (Li and Lake, 1993).

11-8.4 Fitting Covariance Models to Estimates

The choice of appropriate autocovariance model based on data is often not straightforward. A number of models might "almost" fit a sample semivariogram or correlogram. Combinations of models can also be used, since the sum of covariance models will still have the appropriate positive- or negative-semidefiniteness property. Carr (1995, Chap. 6) and Journel and Huijbregts (1978, Chap. 4) give examples.

The fitting of a model must account for what behavior in the sample covariance plot is "real" and what aspects arise from sampling variability. If the range or correlation length can be related to a physical aspect of the system, then the model is better tied to the geosystem being assessed. For example, the range has been observed to coincide with the thickness of tidal channels (Jensen et al., 1994). Holes in the sample semivariogram have been found to frequently reflect sedimentary cyclicities, and the nugget may represent undersampled smaller-scale variability (Jensen et al., 1996). Therefore, an iterative procedure may be needed where features observed in the sample covariance plot are investigated for any relationship with sedimentary aspects. Another helpful test is to apply a jack-knife procedure to the sample covariance to determine what behavior might be unexplained by sampling variability. See Shafer and Varljen (1990) for an example of this approach.

11-9 AUTOCORRELATION AND THE INFORMATION CONTENT OF DATA

Until this chapter, we have ignored the fact that measurements made at nearby locations in the reservoir may be correlated. All of the theory presented in Chap. 5 concerning variability assessments of estimates implicitly assumes that all the data are independent and makes no allowance for autocorrelation; a sample of size I is assumed to consist of I uncorrelated measurements. We know, however, that this may not be correct. For example, measurements within a facie may be highly autocorrelated, whereas measurements from different facies may well be uncorrelated. How does correlation affect the theory presented in previous chapters? Some idea about this was presented through Example 2; we discuss the implications in the following paragraph.

Autocorrelation reduces the information content of data. That is, I correlated measurements will only be worth I_{eff} uncorrelated measurements, where $I_{eff} < I$. The greater the degree of correlation, the smaller is I_{eff}/I. For example, we have for the arithmetic average

$$\bar{X} \sim N(\mu, \sigma^2/I)$$

if the I data are uncorrelated and come from a population with mean μ and variance σ^2. However, if the data have a first-lag correlation of $\rho = \rho(1) \neq 0$, then (Bras and Rodriquez-Iturbe, 1985)

$$\bar{X} \sim N\left\{\mu, \frac{\sigma^2}{I}\left[1 + \frac{2}{I}\,\frac{I\rho(1-\rho) - \rho(1-\rho^I)}{(1-\rho)^2}\right]\right\}$$

Consequently, the variance of the estimate $\bar{X}$ may increase.

Correlation of data can arise also when samples are bunched. For example, consider the situation discussed in Sec. 5-9, where a reservoir has been drilled as in Fig. 11-15.

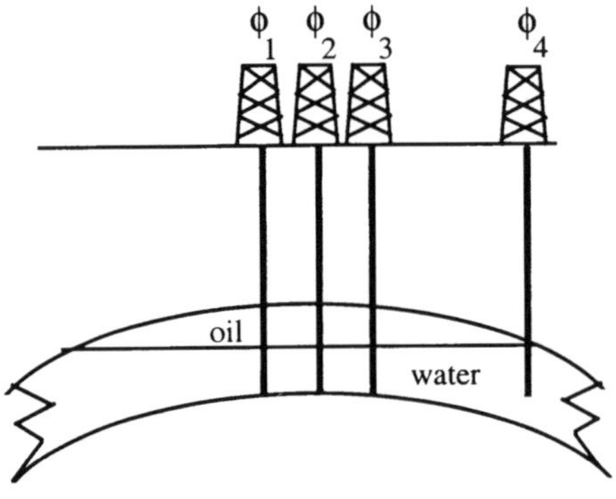

Figure 11-15. Schematic of variable sampling density for average porosities.

Would a representative porosity in the oil-saturated medium be given by $(\phi_1 + \phi_2 + \phi_3 + \phi_4)/4$? Clearly, the answer is no; a simple average would probably be optimistic since three of the four wells penetrate the largest part of the oil region. The ϕ values for wells 1, 2, and 3 do not represent the same volume of material as does ϕ_4. A better average porosity would be obtained by declustering the data–weighting each ϕ by the volume it represents in the sum.

Dealing with correlated data can be difficult. It is an area of continuing research for which there are as yet more questions than answers. Barnes (1988) offers a recent and helpful discussion of the problem. It is clear, however, that the first problem is to become aware that the problem exists.

11-10 CONCLUDING REMARKS

The statistical study of autocorrelation is still in a state of relative infancy. We have developed a fairly substantial number of tools to do it, some with a strong theoretical base, but we are still learning how to use them. Of particular interest is how geologic features can best be conveyed through the statistics. One thing seems obvious: autocorrelation is very important to reservoir performance and we, therefore, are highly motivated to make the connection between it and fluid flow. Chapter 12 describes a means of doing this.

CHAPTER 12

MODELLING GEOLOGICAL MEDIA

Thus far, our discussion has been to develop and illustrate tools for the analysis of data, primarily to characterize them using particular models and for diagnostic purposes. An extraordinarily powerful use of statistics is to apply the models to construct synthetic data sets for use in subsequent calculations.

This procedure was mentioned in Chap. 1, and we have implicitly used it several times already. The idea is to use the data and any geological and physical information to produce statistically plausible reproductions (*realizations*). This approach underlies regression, where we take a model, $\hat{Y} = f(X, \beta) + \varepsilon$, and data to establish the unknown parameters (β) of the model. Once determined, the model can be applied to produce estimates of Y. These estimates will usually differ from the true values of Y but, if we have modelled well, the errors will be small.

For regression, every time we set $X = X_0$, the model will produce the same estimated value for Y. Thus, regression is a method for producing identical realizations. Other methods will produce different estimates each time the estimator is applied. Each realization conforms to the statistical, geological, and physical information we incorporate into the model. The different estimates, however, result because we have not included sufficient information to uniquely predict the reservoir property at each desired location in the domain.

The general area of using statistical methods to create replicas of variable distributions in the earth sciences has been called *geostatistics*. In this chapter we discuss the use of statistical techniques and statistics to generate sets of data. We have already done this in a simple way using the Monte Carlo technique for reserves estimates and stochastic shale models in Chap. 3. Usually these sets will be on arrays or statistical fields that can be used in a variety of calculations: reserve calculations (hydrocarbons in place), locating the top of structures, or as numerical simulation input. Before becoming this specific, however, we give a systematic summary of the basic techniques for generating statistical fields based on only a limited amount of measured data, a problem that is among the most basic in earth sciences.

12-1 STATISTICAL POINT OF VIEW

To begin with, we return to Fig. 3-15 to discuss a point of view that holds throughout this chapter. The following remarks apply to any spatial dimensionality, but they are easiest to visualize in two dimensions. Likewise, the remarks hold for all parameters relevant to a given application, even though we discuss the figure as a univariate field.

The statistical point of view is that every point in the field (and each random variable) has a probability distribution function (PDF). For measured points, the PDF consists of a single value (a "spike" or zero-variance PDF, assuming no measurement error). For the unmeasured points, the PDF has a nonzero variance and, without further restrictions, can take on an arbitrary shape. For example, if we assume second-order stationarity, the PDF must have the same first and second moments everywhere in the domain.

This point of view is central to the statistical techniques discussed in this chapter. But we must remember that it is only a point of view; there is in reality a single value for each variable at each point. If we sample one of the unknown points, assuming no measurement error, it becomes known and its PDF converts to a spike (with a value, it is presumed, that is consistent with the imagined PDF).

12-2 KRIGING

There are several statistical procedures for interpolating data. These procedures form the basis for sophisticated techniques, so it is important that we understand the fundamental ideas.

Kriging is a basic statistical estimation technique. Named after its originator, D. Krige (1951), it was first used to estimate ore grade in gold mines. It was then theoretically developed by researchers, primarily in France (e.g., Matheron, 1965). As we shall see, there are now many variations of Kriging, but all start with the same fundamentals and all are related to regression.

Consider a three-dimensional field as illustrated schematically (in two dimensions) in Fig. 3-15. We seek an estimate of a property $\hat{Z}$ at an unmeasured location (point o) that is based on the Z_i, $i = 1,\ldots, I$ known values.

$$\hat{Z} = \sum_{i=1}^{I} \lambda_i Z_i \tag{12-1}$$

Equation (12-1) is a common interpolator. For the specific way we calculate the λ_i described below, it is the *Kriging estimator.* It is linear in the Z_i. The values λ_i, $i = 1,\ldots, I,$ are the Kriging *weights*. Clearly, if we determine the λ_i, we can calculate $\hat{Z}$.

12-2.1 BLUE Estimators

Statistical estimation techniques are best linear unbiased estimators (BLUE). "Best," in a statistical sense, means that the estimators have been arrived at through minimization of some type of variance. "Linear" means that the estimator is a linear combination of data values at known points, and, as we have discussed in Chap. 5, "unbiased" means that the expectation of the estimator will return the true (or population) value. How these criteria are imposed is evident in the following development of the Kriging estimator.

The linearity of Eq. (12-1) simplifies the procedure significantly. However, it also means that the $\hat{Z}$ will be approximately normally distributed (Gaussian). Recall from the discussion of the Central Limit Theorem in Chap. 4 that any random variable that is the sum of a large number of independent, additive events will become Gaussian, regardless of the underlying PDF of the events. If we view the $\lambda_i Z_i$ as the independent events, the estimator in Eq. (12-1) satisfies these criteria if I is large. Unfortunately, $\hat{Z}$ will be Gaussian even if the Z_i are not; hence, statistical additivity (Chap. 5) requires us to transform data (the Z_i) to normality before analysis and field generation. If the Z_i are normally distributed, the Kriging estimator will give the means of the imagined distributions in Fig. 3-15, which are now themselves normal. A nice feature of the normal PDF is that the $\hat{Z}$ will also be the most likely (the mode) value at the point o.

12-2.2 Simple Kriging

To determine the weights λ_i, we first define and then minimize the simple Kriging (SK) variance σ_{SK}^2:

$$\sigma_{SK}^2 = E[(Z - \hat{Z})^2]$$

In this definition, $\hat{Z}$ is the estimate of Z at a single point. The Z itself is unknown, but we can proceed without this knowledge because of the stationarity assumptions, which we implicitly have invoked. σ_{SK}^2 is interpreted as the variance of the distribution at the point being estimated (Fig. 3-15). Substitution of Eq. (12-1) and rearrangement gives

$$\sigma_{SK}^2 = \sigma_o^2 - 2\sum_{i=1}^{I} \lambda_i \sigma_{oi}^2 + \sum_{i=1}^{I}\sum_{j=1}^{I} \lambda_i \lambda_j \sigma_{ij}^2 \qquad (12\text{-}2)$$

where we have employed the convention that $\sigma_o^2 = Var(Z)$, $\sigma_{oi}^2 = Cov(Z, Z_i)$, and $\sigma_{ij}^2 = Cov(Z_i, Z_j)$ are known quantities. Written in this form, σ_{SK}^2 is a function of the weights, since the autocovariance model alone determines the autocovariances.

To minimize σ_{SK}^2, we employ the now familiar argument (Chap. 9) that we seek a point in λ_i space such that $d\sigma_{SK}^2 = 0$. If the weights are independent, the minimal σ_{SK}^2 will occur when

$$\frac{\partial \sigma_{SK}^2}{\partial \lambda_j} = 0 \text{ for } j = 1, 2, \ldots, I$$

Because of the use of "dummy" subscripts in Eq. (12-2), taking the derivatives requires some explanation. Differentiating the first term gives

$$-2\frac{\partial\left(\sum_{i=1}^{I}\lambda_i\sigma_{oi}^2\right)}{\partial\lambda_j} = -2\lambda_j\sigma_{oj}^2$$

where we have assumed that the weights are independent of each other. Since by stationarity the covariances are independent of the weights, there remains only one term from the original sum. Differentiating the second term gives

$$\frac{\partial\left(\sum_{i=1}^{I}\sum_{m=1}^{I}\lambda_i\lambda_m\sigma_{im}^2\right)}{\partial\lambda_j} = \frac{\partial}{\partial\lambda_j}\left(\sum_{i=1}^{I}\lambda_i\left(\lambda_1\sigma_{i1}^2 + .. + \lambda_j\sigma_{ij}^2 + .. + \lambda_I\sigma_{iI}^2\right)\right)$$

upon use of m as a dummy subscript and partial expansion of the inner sum. The chain rule yields

$$\frac{\partial\left(\sum_{i=1}^{I}\sum_{m=1}^{I}\lambda_i\lambda_m\sigma_{im}^2\right)}{\partial\lambda_j} = \left(\sum_{i=1}^{I}\frac{\partial\lambda_i}{\partial\lambda_j}\left(\lambda_1\sigma_{i1}^2 + .. + \lambda_j\sigma_{ij}^2 + .. + \lambda_I\sigma_{iI}^2\right) + \sum_{i=1}^{I}\lambda_i\frac{\partial\left(\lambda_1\sigma_{i1}^2 + .. + \lambda_j\sigma_{ij}^2 + .. + \lambda_I\sigma_{iI}^2\right)}{\partial\lambda_j}\right)$$

which leads to

$$\frac{\partial\left(\sum_{i=1}^{I}\sum_{m=1}^{I}\lambda_i\lambda_m\sigma_{im}^2\right)}{\partial\lambda_j} = \left(\sum_{j=1}^{I}\lambda_j\sigma_{ij}^2 + \sum_{i=1}^{I}\lambda_j\sigma_{ij}^2\right) = 2\sum_{i=1}^{I}\lambda_j\sigma_{ij}^2$$

Adding these terms and setting them equal to zero gives the following equations:

$$\sum_{i=1}^{I}\sigma_{ij}^2\lambda_i = \sigma_{oj}^2 \quad j=1,\ldots,I \tag{12-3a}$$

which can be written in matrix form as

$$\begin{bmatrix} \sigma_{11}^2 & \cdot & \cdot & \cdot & \sigma_{1I}^2 \\ \cdot & \cdot & \cdot & \cdot & \cdot \\ \cdot & \cdot & \cdot & \cdot & \cdot \\ \cdot & \cdot & \cdot & \cdot & \cdot \\ \sigma_{I1}^2 & \cdot & \cdot & \cdot & \sigma_{II}^2 \end{bmatrix} \begin{bmatrix} \lambda_1 \\ \cdot \\ \cdot \\ \cdot \\ \lambda_I \end{bmatrix} = \begin{bmatrix} \sigma_{o1}^2 \\ \cdot \\ \cdot \\ \cdot \\ \sigma_{oI}^2 \end{bmatrix} \tag{12-3b}$$

or in vector notation as

$$\Sigma^2 \cdot \Lambda = \Sigma_o^2 \tag{12-3c}$$

The SK λ_i's are therefore the solution to these *normal equations*. Equation (12-3b) is similar to the regression matrix Eq. (10-7).

Strictly speaking, the requirement $d\sigma_{SK}^2 = 0$ locates only extrema in the λ_i space, not necessarily a minimum. We should, therefore, interrogate the second derivative of σ_{SK}^2 to see that we have arrived at a maximum or a minimum. This is left as an exercise to the reader; however, the linearity of Eq. (12-3), caused originally by the linearity of the Kriging estimator, means that the solution to these equations will always be a minimum σ_{SK}^2.

The elements in the above matrix and in the vector on the right side are autocovariances. As will be seen below, these can all be obtained from the semivariogram. The right side deals with autocorrelations among the known points and the point *o* at which Z is to be estimated; the matrix elements account for autocorrelations among the known points. The autocorrelation matrix contains variances along its diagonal and is symmetric because of the reciprocal property of the autocovariance. The symmetry means that the inversion of Eq. (12-3) is amenable to various specialized techniques. Although the autocorrelation matrix contains nonnegative elements along its diagonal, it is not, in general, positive definite, although this is true for non-oscillatory semivariogram models. Finally, if we invoke weak stationarity (Chap. 11), all of the autocovariances become functions only of their separation distances (or of the separation vector if the autocorrelation is anisotropic). The locations of the Z_i being fixed, the matrix of the σ_{ij}^2 in Eq. (12-3b) need be inverted only once to estimate $\hat{Z}$ at other points.

We can eliminate the inner sum of the last term of Eq. (12-2) with Eq. (12-3a) to give a short form for the minimized SK variance,

$$(\sigma_{SK}^2)_{MIN} = \sigma_o^2 - \sum_{i=1}^{I} \lambda_i \sigma_{oi}^2 \tag{12-4a}$$

for the point *o*. Unlike the formula for the weights, this equation requires a variance of the unknown *Z*. By convention, we evaluate σ_o^2 from the sample variance or from the semivariogram.

Many applications require estimates $\hat{Z}_1, \hat{Z}_2, \ldots, \hat{Z}_M$ on a grid of *M* values. The above equations nicely generalize with matrix notation in this case. The Kriging estimator is

$$\hat{\mathbf{Z}} = \Lambda^T \mathbf{Z} \tag{12-5}$$

where $\hat{\mathbf{Z}}$ is an *M*-element vector of the estimated values, $\mathbf{Z}$ an *I*-element vector of the known values, and Λ^T is an $I \times M$ matrix of the Kriging weights λ_{ij}:

$$\Lambda^T = \begin{bmatrix} \lambda_{11} & . & . & . & \lambda_{1M} \\ . & . & . & . & . \\ . & . & . & . & . \\ . & . & . & . & . \\ \lambda_{I1} & . & . & . & \lambda_{IM} \end{bmatrix}$$

The subscripts on the λ_{ij} refer to the measurement position and estimation location, respectively. Solving for the weights from Eq. (12-3c) and combining with Eq. (12-5) gives an expression for the vector of estimates,

$$\hat{\mathbf{Z}} = [(\Sigma^2)^{-1}\,\Sigma_o^2\,]^T\,\mathbf{Z} \qquad (12\text{-}6)$$

where the elements of Σ^2 and Σ_o^2 are the autocovariances between the known values and between these values and the estimated points, respectively. Inverting Σ^2 can be computationally intensive, if the number of data is large, since it is an *I* by *I* matrix. But, as suggested above, the inversion need be performed only once for each point estimated.

Example 1 - Simple Kriging as an Exact Estimator. An estimator is *exact* if it reproduces the sampled data points. Show that, under some fairly easy conditions, the SK estimator is exact.

We begin with Eq. (12-3a) to estimate the $j = k^{th}$ known point:

$$\sum_{i=1}^{I} \sigma_{ik}^2 \lambda_i = \sigma_{kk}^2$$

Upon rearranging,

$$\sigma_{1k}^2 \lambda_1 + \sigma_{2k}^2 \lambda_2 + \cdots + \sigma_{kk}^2\,(\lambda_k - 1) + \cdots + \sigma_{Ik}^2 \lambda_I = 0$$

If the σ_{ik}^2 are independent of each other (they depend only on separation distance) and not all zero, we must have

$$\lambda_k - 1 = 0 \text{ and } \lambda_j = 0 \quad \text{for } j \neq k$$

Therefore, the weight for the kth point is one and all others are zero. As would be expected, $\sigma_{SK}^2 = 0$ at this point from Eq. (12-4a). Since the point *k* was chosen arbitrarily, this result applies to all the known points. Hence, the Kriging estimator reproduces the values at the known points.

Exactness is a property not shared between Kriging and regression, and, in fact, exactness will not apply here if the σ_{ik}^2 are dependent, as might occur if the semivariogram model used had a nonzero nugget. Indeed, you may show that the Kriging estimator reduces to the inexact arithmetic average in the limit of the semivariogram model approaching a pure nugget.

12-2.3 Kriging with Constraints

Many applications require additional constraints on the Kriging estimates, constraints that arise from statistical grounds, from geology, from physical limits (nonnegative porosity, for example), or from other data. There are several ways to constrain Kriging estimates, but the one most consistent with the linearity of the Kriging estimator is the method of Lagrange multipliers. We illustrate this method through an example. See Beveridge and Schechter (1970, Chap. 7) for an exposition of Lagrange multipliers.

12-2.4 Ordinary Kriging

Unfortunately, the $\hat{Z}$ from simple Kriging are biased (BLE, not BLUE), because we did not constrain SK to estimate the true mean at unsampled locations. The SK bias can be determined as

$$b_{\hat{Z}} = E(\hat{\mathbf{Z}}) - E(Z)$$

$$= \mathrm{E}\left(\sum_{i=1}^{I} \lambda_i Z_i\right) - \mathrm{E}(Z) = \sum_{i=1}^{I} \lambda_i \mathrm{E}(Z_i) - \mathrm{E}(Z) = \left(\sum_{i=1}^{I} \lambda_i - 1\right)\mathrm{E}(Z)$$

Now, if we presume that the original data are themselves unbiased, or

$E(Z_i) = E(Z) = \mu_Z$ so that

$$b_{\hat{Z}} = 0 = \mu_Z\left(\sum_{i=1}^{I} \lambda_i - 1\right)$$

Hence, the bias can be removed by forcing the λ_i to sum to one. The problem of determining the Kriging weights now evolves to a constrained minimization, that is, minimizing σ_{SK}^2 subject to the statistical constraint

$$\sum_{i=1}^{I} \lambda_i = 1 \tag{12-7}$$

To apply the Lagrange multiplier technique to minimize the Kriging variance, Eq. (12-2), subject to constraint Eq. (12-7), we form the following objective function:

$$L(\lambda_1, \ldots, \lambda_I, \mu) = \sigma_o^2 - 2\sum_{i=1}^{I}\lambda_i\sigma_{oi}^2 + \sum_{i=1}^{I}\sum_{j=1}^{I}\lambda_i\lambda_j\sigma_{ij}^2 + 2\mu\left[\sum_{i=1}^{I}\lambda_i - 1\right]$$

The factor 2 before the last term is for subsequent mathematical simplicity. We can either add or subtract the last term since it will be zero ultimately, but the form indicated above has a slight mathematical advantage in that it will lead to a symmetric autocovariance matrix below. μ is the *Lagrange multiplier*, a new variable in the problem.

Minimizing the objective function in the same way as for Simple Kriging above (now setting $dL = 0$) leads to the following equations for *Ordinary Kriging* (OK):

$$\sum_{i=1}^{I}\lambda_i\sigma_{ij}^2 = \sigma_{oj}^2 + \mu \qquad j = 1,\ldots,I \tag{12-8}$$

which can be written in the following matrix form:

$$\begin{bmatrix} \sigma_{11}^2 & . & . & \sigma_{1I}^2 & 1 \\ . & . & . & . & 1 \\ . & . & . & . & . \\ \sigma_{I1}^2 & . & . & \sigma_{II}^2 & 1 \\ 1 & . & . & 1 & 0 \end{bmatrix} \begin{bmatrix} \lambda_1 \\ . \\ . \\ \lambda_I \\ \mu \end{bmatrix} = \begin{bmatrix} \sigma_{o1}^2 \\ . \\ . \\ \sigma_{oI}^2 \\ 1 \end{bmatrix}$$

The problem to be solved is nearly the same as for SK; there is an additional column in the autocovariance matrix to account for the Lagrange multiplier and there is an additional row to account for the constraint. Of course, there is now an additional unknown, the Lagrange multiplier itself.

For large problems, say $I > 10$, there is a negligible difference in effect between solving the SK equations and the OK equations, since the latter just involve a slightly expanded matrix. However, the matrix, while still symmetrical in the form shown, contains a zero on the diagonal that restricts the applicable inversion techniques. Just as in SK, the matrix is invariant as the point to be estimated moves; hence, the autocorrelation matrix need be inverted only once for multiple estimations.

We can combine the equation for the OK objective function and Eq. (12-8) to give the minimized OK variance,

$$(\sigma_{OK}^2)_{MIN} = \sigma_o^2 - \sum_{i=1}^{I}\lambda_i\sigma_{oi}^2 + \mu \tag{12-4b}$$

Being constrained, the OK variance is usually greater than the SK variance because the additional constraint does not allow the attainment of as deep a minimum.

Example 2 - Ordinary Kriging with Three Known Points. Consider the arrangement of points shown below (Knutsen and Kim, 1978).

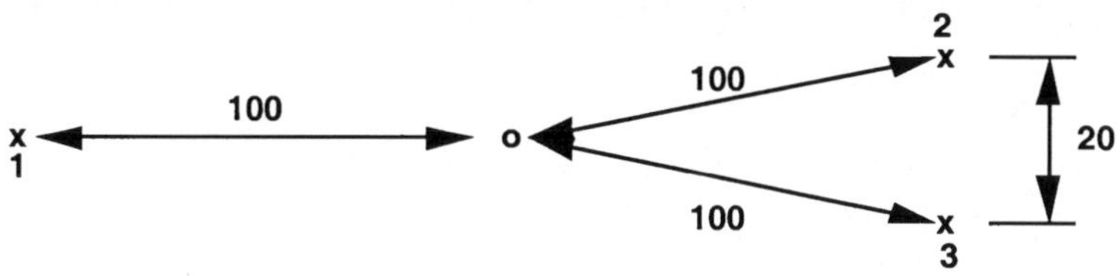

Figure 12-1. Layout of known points (x) and point (o) to be estimated.

Estimate the value of $\hat{Z}$ at point o. The theoretical semivariogram is given as

$$\gamma(h) = \begin{cases} 0.01h & h < 400 \\ 4 & h > 400 \end{cases}$$

Since the semivariance is written as a function of separation distance only, we have assumed weak stationarity. Hence, the autocovariances and semivariances are related as (Chap. 11) $Cov(h) = Cov(0) - \gamma(h)$. We now calculate the elements of the OK equations as

$$\sigma_{11}^2 = \sigma_{22}^2 = \sigma_{33}^2 = Cov(0) = 4$$

$$\sigma_{12}^2 = \sigma_{21}^2 = \sigma_{13}^2 = \sigma_{31}^2 = Cov(199) = Cov(0) - \gamma(199) = 2.01$$

$$\sigma_{23}^2 = \sigma_{32}^2 = Cov(20) = Cov(0) - \gamma(20) = 3.8$$

$$\sigma_{o1}^2 = \sigma_{o2}^2 = \sigma_{o3}^2 = Cov(100) = Cov(0) - \gamma(100) = 3$$

and the OK equations in matrix form are

$$\begin{bmatrix} 4 & 2.01 & 2.01 & 1 \\ 2.01 & 4 & 3.8 & 1 \\ 2.01 & 3.8 & 4 & 1 \\ 1 & 1 & 1 & 0 \end{bmatrix} \begin{bmatrix} \lambda_1 \\ \lambda_2 \\ \lambda_3 \\ \mu \end{bmatrix} = \begin{bmatrix} 3 \\ 3 \\ 3 \\ 1 \end{bmatrix}$$

Solution of this set of equations yields $\lambda_1 = 0.487$, $\lambda_2 = 0.256$, $\lambda_3 = 0.256$, and $\mu = 0.026$, from which we take the estimate to be

$$\hat{Z} = (0.487)(10) + (0.256)(5) + (0.256)(15) = 9.99$$

where $Z_1 = 10$, $Z_2 = 5$, and $Z_3 = 15$. The OK variance can be obtained from Eq. (12-4b) as $\sigma^2_{OK} = 1.026$. The $\hat{Z}$ is nearly the same as the arithmetic average, to which OK degenerates (all weights equal) in the limit of no autocorrelation.

As simple as this exercise is, it illustrates several points. First, note that $\lambda_2 = \lambda_3$ and that λ_1 is nearly equal to $\lambda_2 + \lambda_3$. This near equality occurs because points 2 and 3 are so close together that they are essentially acting as a single point to the estimate. If points 2 and 3 were spaced farther apart, λ_2 would not equal λ_3. The sharing of weights for clustered known points is known as a *transfer of influence* (Knutsen and Kim, 1978). Fields with clustered points will have larger Kriging variances than fields with widely separated points, a fact that obviously indicates the most efficient way to sample a property in a field (but ignores scales of variation in geology and the financial aspects of the sampling strategy).

Now suppose point 2 is moved so that it is directly behind point 3 from point *o*. We would find, on solution, that λ_2 becomes negative and λ_3 exceeds one. (Negative weights are permissible as long as the weights continue to sum to one.) The negative weight for λ_2 overcomes an undue influence from λ_3. By the same token, point 3 is *shielding* point 2.

Another general property of OK is that if all the points, both known and to be estimated, are so far apart from each other that they are uncorrelated, the Kriging estimator becomes the arithmetic average. This property is consistent with the notion, first discussed in Chap. 5, that the arithmetic average is the best estimator of the mean for a set of uncorrelated data–as the points would certainly be in this case. If this happens for the SK estimator, the estimates all become zero.

This discussion of moving sampled points around might lead one to believe that the semivariogram and the data are independent. In reality, the semivariogram is estimated from the data so that any shift in the sampling points will cause a change in the semivariance. However, the change will be small if the semivariogram is based on a large number of points and only a few points are shifted. Such is presumed to be the case with the points in Example 2. Note that it is possible to perform Kriging with fewer sampling points than are in the entire set.

Recall from Chap. 6 that the N_0 approach was described for obtaining enough samples to allow the arithmetic average to be predicted to within 20% of the true mean 95% of the time. These Kriging results indicate that the N_0 method is a simplification and has to be modified to account for autocorrelation. When the sample locations are such that the measurements are autocorrelated, more measurements must be taken to keep the variability of the estimated mean within 20%. See Chap. 11 and Barnes (1988) for a discussion of the problem.

Finally, you may wonder why we resort to Lagrange multipliers to enforce the unbiased condition in OK instead of renormalizing the SK weights by their sum. Such a procedure does not result in a minimized σ^2_{SK} (the estimate would be LUE, not BLUE). To see this, let β be the sum of the weights. We multiply and divide the second term in Eq. (12-2) by β and the third term by β^2. The minimization proceeds as before to result in the following minimized SK variance:

$$\sigma_{SK}^2 = \sigma_o^2 + (\beta^2 - 2\beta) \sum_{i=1}^{I} \lambda_i \sigma_{oi}^2$$

This equation shows that σ_{SK}^2 is the smallest only when $\beta = 1$, the same result as Eq. (12-4a). Hence, such a renormalization will lead to a larger Kriging variance than SK. This variance may or may not be larger than the corresponding OK variance, but the Lagrange multiplier technique always leads to the deeper minimum.

12-2.5 Nonlinear Constraints

One of the more interesting, and largely unexplored, avenues in Kriging is the possibility of adding more constraints. In the case of Kriging applied to permeability, in particular, the linear unbiased constraint is weak because of the lack of additivity of this variable for many situations (see Chap. 5). However, this can be modified somewhat using the approach suggested in Chap. 10 (e.g., Kriging log(Z) or Z^{-1}).

Well testing is a useful means of inferring the larger-scale flow properties of an aquifer or reservoir. In this class of procedures, the permeability in the vicinity of a well is inferred, usually by imposing changes in the rate of the well and measuring changes in pressure. The permeability so obtained (we will call it k_{wt}) is a good reflection of the average flow properties of the reservoir, but it says little about its variability because it is a global (that is, it applies to a large region) rather than a point measure. Moreover, it is not clear how to average the point values within the test region to arrive at k_{wt}.

Campozana (1996) has added the nonlinear constraint $k_{wt} = f(k_1, k_2, \ldots, k_M)$ to Kriging through the Lagrange multiplier procedure. He deals with the global nature of k_{wt} by minimizing the sum of the Kriging variances at all points (recall that the Kriging variances are all positive) and with the discrepancy in scales by allowing k_{wt} to be related to the point values through the following power-averaging formula:

$$k_{wt} = \left(\sum_{m=1}^{M} k_m^{\omega} \right)^{1/\omega}$$

The exponent ω in this formula can be obtained through minimization, just like the weights. Of course, the point-wise autocorrelation is maintained through the autocovariances in the Kriging procedure.

Figure 12-2 shows the results of adding the k_{wt} constraint to the permeability field. The upper-left plot is a two-dimensional, fine-scale permeability field that is the test case for this procedure. k_{wt} for this field is known through a simulation of right-to-left flow. The upper-right plot in Fig. 12-2 shows a reconstruction of the field using OK based on data known at nine equally spaced points within the field; the lower-left plot shows the result of the Kriging with the k_{wt} constraint. The heterogeneity of the field is reproduced much better with the k_{wt} constraint. The example illustrates how Kriging can be used to apply global information to improve arrays of point information.

To be sure, the improvement shown in Fig. 12-2 comes with a price. The nonlinearity of the constraint means that iteration is required to attain a minimized solution and larger matrices must be inverted. More subtly, the lack of additivity may mean there is a basic inconsistency between the Kriging estimate and the power average. However, the results of Fig. 12-2 are encouraging. Other nonlinear constraints are possible, of course. See Barnes and You (1992) and Journel (1992) for an illustration of how bounds may be added to Kriging.

12-2.6 Indicator Kriging

Kriging is a desirable estimator: it is linear, unbiased, and exact and has minimal variance. Despite these attractive features, it has three other aspects that are less desirable.

1. Kriging tends to generate a synthetic field that is too smooth; it under-represents the variability. This is a property that has been established from application; however, it is also evident from the filtering aspects of the estimator itself.

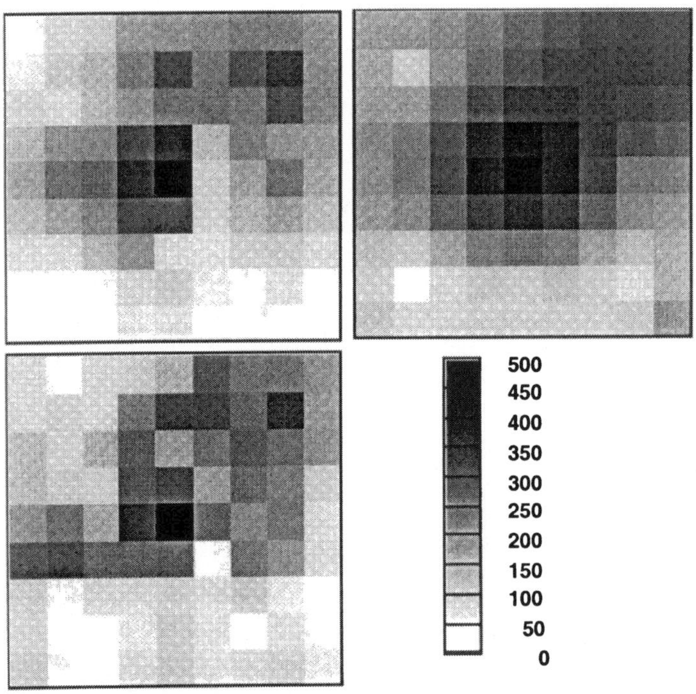

Figure 12-2. Results of Kriging constrained to well-test data. Original field is in upper left; ordinary Kriging reconstruction in upper right; simple Kriging with well-test constraint is in lower left. The scale is in mD. From Campozana (1996).

2. Like regression, Kriging is deterministic; the same data values Z_i will yield the same estimates. Whether or not this is viewed as an advantage depends on the application, but such a procedure cannot be used for uncertainty estimation.

3. The fields generated are Gaussian. The restriction to the generation of Gaussian fields (and the requirement of ensuring consistency) can also be bothersome.

We can get around the deterministic nature of Kriging by regarding $\hat{Z}$ and σ_{OK}^2 (for example) as the mean and variance of locally Gaussian PDF's (see below). Remarkably, Kriging can be easily modified to eliminate all three of these problems by generating the entire CDF of the estimates (Fig. 3-15) at each point. This procedure is called *indicator Kriging*.

Before proceeding, you should review the material in Chap. 3 on CDF's. There we discovered that a CDF consists of a plot of the frequency of occurrence $F(z)$ of a random variable Z less than some limit z. $F(z)$ was nondecreasing and consisted of two types: continuous CDF's, appropriate for continuous variables, and discrete CDF's, appropriate for discrete variables. Recalling the definition of probability (Chap. 2), for a given *cutoff value* z_c, $F(z_c)$ represents

$$F(z_c) = \frac{\text{Number of samples with } Z \leq z_c}{\text{Total number of samples}}$$

Other terms for z_c are the *critical* or *threshold value*. z_c varies between a minimum z_{min} and a maximum z_{max} and can take on several values.

Let $I(z_{cn})$ for $n = 1,\ldots,N$ represent an indicator variable defined as

$$I(z_{c_n}) = \begin{cases} 1 & Z_i \leq z_{c_n} \\ 0 & Z_i > z_{c_n} \end{cases} \qquad (12\text{-}9)$$

$I(z_{c_n})$ is an on-off (Boolean) variable because it can take on only two values. N is less than or equal to the number of data values. The estimated CDF follows from

$$\hat{F}(z_c) = \frac{1}{N}\sum_{n=1}^{N} I(z_{c_n})$$

and letting z_c run from z_{min} to z_{max}. $\hat{F}(z_c)$ is the sample or empirical CDF of $\hat{Z}$

We form the indicator semivariogram γ_I exactly as we did for the semivariograms discussed in Chap. 11. These devices have the same meaning as they did there except they are now expressing the continuity (extent of autocorrelation) of $\hat{Z}$ values less than z_c. The indicator semivariogram γ_I can therefore be used to concentrate on a particular range of the Z that is known to be most significant to the application, e.g., the connected large-permeability range in flow simulations. γ_I, of course, is always less than one because $0 \leq I(z_{c_n}) \leq 1$ since $I(z_{c_n})$ is either zero or one.

Perhaps the most desirable feature of the indicator semivariogram is that it is robust–that is, γ_I does not depend strongly on the nature of the distribution of the Z_i nor is it affected by the presence of extreme values. The latter factor makes the indicator semivariogram very useful in assessing autocorrelation in noisy data.

It is also clear that γ_I will depend on the value of z_c chosen, a decided increase in complexity compared to previous semivariogram analyses. However, the semivariogram is frequently robust to the z_c, in which case a single γ_I is satisfactory for the subsequent construction. If z_c is the median value in a data set, the subsequent Kriging is known as *median indicator Kriging.*

The actual indicator Kriging proceeds exactly as in Kriging, except γ_I is used in place of γ and the data values Z_i are replaced by their respective indicators I_i. (Replacing the data values with indicators and using the indicator semivariogram is a form of nonparametric statistics.) The following example indicates how to use them.

Example 3 - Indicator Kriging for Three Known Points. Rework Example 2 to generate a CDF for $\hat{Z}$ at the point o. Take the indicator semivariogram to be

$$\gamma(h) = \begin{cases} h/1600 & h < 400 \\ 0.25 & h > 400 \end{cases}$$

and independent of the value of $I(z_{c_n})$ chosen. As before, we use this to form the Kriging equations.

$$\begin{bmatrix} 0.25 & 0.125 & 0.125 & 1 \\ 0.125 & 0.25 & 0.238 & 1 \\ 0.125 & 0.238 & 0.25 & 1 \\ 1 & 1 & 1 & 0 \end{bmatrix} \begin{bmatrix} \lambda_1 \\ \lambda_2 \\ \lambda_3 \\ \mu \end{bmatrix} = \begin{bmatrix} 0.188 \\ 0.188 \\ 0.188 \\ 1 \end{bmatrix}$$

where we have now used $Cov(h) = Cov(0) - \gamma_I(h)$. The solution to these equations yields the same weights as before: $\lambda_1 = 0.487$ and $\lambda_2 = \lambda_3 = 0.256$. Since γ_I is independent of z_c, these weights will remain the same throughout the example. For $z_{c1} = 5$, the indicators for points 1, 2, and 3 are 0, 1, and 0, respectively, from which we calculate

$\hat{F}\,(5) = 0.487(0) + 0.256(1) + 0.256(0) = 0.256$

for the CDF value corresponding to $z_{c1} = 5$. Similarly, for $z_{c2} = 10$ we have $I = \{1, 1, 0\}$ and for $z_{c3} = 15$, $I = \{1, 1, 1\}$. These give $\hat{F}\,(10) = 0.743$ and $\hat{F}\,(15) = 1$, which give the discrete CDF at the point o (Fig. 12-3). To estimate uncertainty, we sample the CDF in the same fashion as was done in earlier chapters.

The discreteness of the $\hat{F}$ depends on the number of known points. In Example 3, the CDF has only three steps because there are only three points. Clearly, the more data there are, the better is the approximation to the population CDF. As in the other types of Kriging, the autocovariance matrix does not change from point to point (if γ_I is truly independent of z_c). Even so, generating $\hat{F}$ can become time-consuming if the full CDF must be constructed at every point in a field that has a large amount of data. But, unlike ordinary Kriging, the weights λ_i here must be nonnegative, otherwise the properties of the $\hat{F}$ (Chap. 3) will not, in general, be satisfied.

Finally, and most interestingly, the indicator variable may be associated with some geologic feature (e.g., $I = 0$ is shale, $I = 1$ is sand). When this occurs, indicator Kriging is estimating the probability that the particular feature exists at the point of estimation. Such a procedure, expanded to include several different types of indicators, offers a direct way to convert geologic classifications into a synthetic field. We discuss this further in Section 12-4.

When applying indicator Kriging, users will typically choose from one to five cutoffs based on geologic or engineering criteria, e.g., net pay, lithologies indicated from wireline logs, flow units > 100 mD, etc. In this fashion, indicator semivariograms will have geologic and/or engineering significance. However, such few cutoffs will generate a local CDF that is segmented and not fully representative of the range and shape of the actual CDF. Such CDF's will give a poor representation of the local statistics; the median in Fig. 12-3, for example, is estimated to be about 7, whereas it is actually 10 from the data and from the nonsegmented CDF's.

Figure 12-4a shows that segmentation produces artifacts in regions of the CDF where the curvature of the CDF is large, where the estimated CDF "short hops" across the empirical CDF. Short hopping can be corrected by selecting additional cutoffs (increase the number of data or categories) in the large-curvature areas (Fig. 12-4b).

Additional semivariogram modeling is required for these nongeological/engineering cutoffs; usually only one to three additional cutoff points are needed.

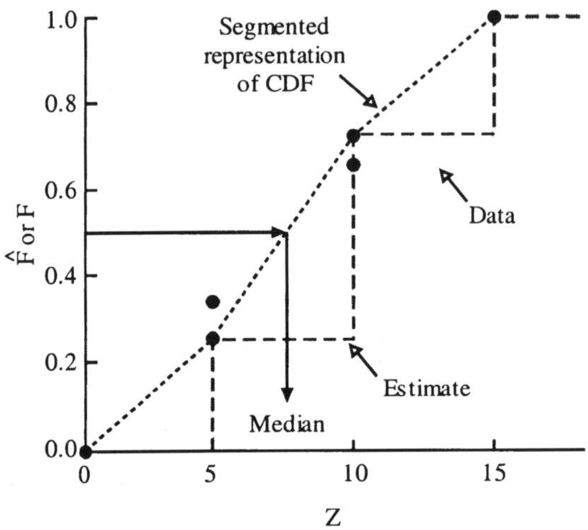

Figure 12-3. Discrete CDF of Z at the point *o* (Fig. 12-1) contrasted with the CDF of the Z_i from Example 3. The indicator CDF is different from the empirical CDF (values between 5 and 10 are less probable) because it is now conditioned on the data.

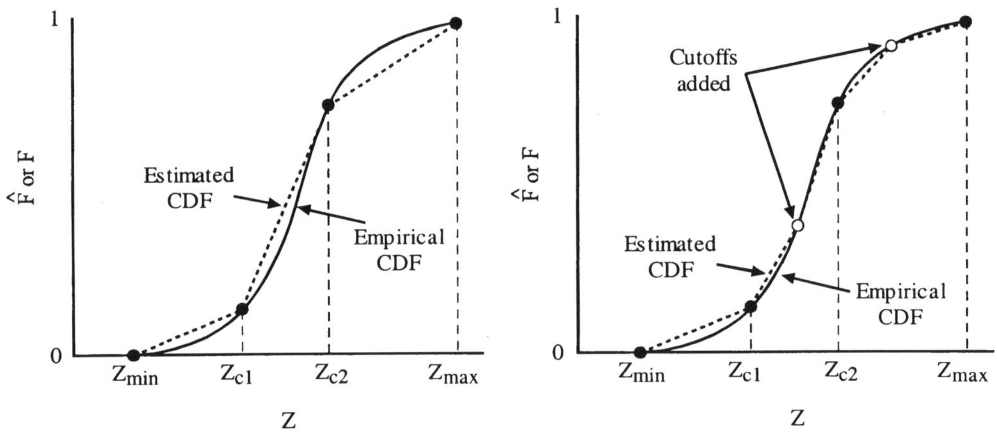

Figure 12-4. Illustration of the effect of adding additional cutoff points to eliminate "short hopping."

12-3 OTHER KRIGING TYPES

This discussion has far from exhausted the different possibilities of Kriging. Other forms are briefly described below.

12-3.1 Universal Kriging

This form seeks to account for drift or a long-range autocorrelation in the data by subjecting the data to a preprocessing step. The preprocessing can be as basic as fitting a low-order polynomial to the original data or more sophisticated, such as segmenting the data according to the geologic patterns.

12-3.2 Disjunctive Kriging

This is another way to deal with the non-Gaussian nature of Kriged data. Here the data are first rendered Gaussian by the appropriate transform before Kriging proceeds.

The ideas behind universal (e.g., stratigraphic coordinates) and disjunctive (e.g., the *p*-normal transform) Kriging have become common practice. The remaining form of Kriging is more involved.

12-3.3 Co-Kriging

One of the most useful generalizations is to define the original Kriging estimator, Eq. (12-1), in terms of more than one data set. For example, if the need is to estimate $\hat{Z}$ based on known Z_i and Y_i, as might be the case where we would estimate permeability from permeability and porosity data, the estimator would be

$$\hat{Z} = \sum_{i=1}^{I} \lambda_i Z_i + \sum_{i=1}^{I} \nu_i Y_i$$

The weights, ν_i, account for any change of units between Y and Z. The ensuing Kriging manipulations are exactly the same as those above, but with an expanded set of equations to be solved since terms involving the correlation between the Z_i and the Y_i will appear. Co-Kriging is a powerful tool that can incorporate a broad range of reservoir data in its most general form.

12-4 CONDITIONAL SIMULATION

As we mentioned above, Kriging fields are deterministic and cannot be used to quantify uncertainty. In the following, we review some of the more common procedures, collectively termed *conditional simulation* (CS), to generate a stochastic random field. The term conditional arises because, as we shall see, the fields so generated are exact, just as with Kriging. Nearly all techniques, which are more algorithms than mathematical developments, use some form of Kriging. In all cases we are to generate a random field consisting of values in $j = 1, \ldots, J$ cells based on $i = 1, \ldots, I$ sampled data points. There are therefore $(J - I)$ points to be estimated or simulated initially based on I conditioning points. We have previously determined the appropriate semivariogram model either from the I data points or from geology and we assume the values at the conditioning points are free of error.

12-4.1 Sequential Gaussian Simulation

Sequential Gaussian simulation (SGS) is a procedure that uses the Kriging mean variance to generate a Gaussian field. The entire procedure is the following:

1. Transform the sampled data to be Gaussian. The most common technique is the normal scores technique (Chap. 4). This is a natural precursor to any Gaussian technique.
2. Assign each of the (J - I) unconditioned cell values to be equal to those at the nearest conditioned cells. The values in the conditioned cells do not change in the following.
3. Define a random path through the field such that each unconditioned cell is visited once and only once.
4. For each cell along the random path, locate a prespecified number of surrounding conditioning data. This local neighborhood, which may contain data from previously simulated cells, is selected to roughly conform to the ellipse range on the semivariogram model.
5. Perform ordinary Kriging at the cell using data in the local neighborhood as conditioning points. This determines the mean of the Gaussian distribution $\hat{Z}$ and the variance σ_{OK}^2. The local CDF is now known since the mean and variance completely determine a Gaussian distribution.
6. Draw a random number in the interval [0, 1] from a uniform distribution. Use this value to sample the Gaussian distribution in step 5. The corresponding transformed value is the simulated value at that cell.
7. Add the newly simulated value to the set of "known" data, increment I by 1, and proceed to the next cell as in step 4.

Repeat steps 4 to 7 until all cells have been visited. This constitutes a single realization of the procedure. Multiple realizations are effected by repeating steps 3 to 7. All the values in all realizations are back-transformed to their original distribution before subsequent use.

At each step in the random path through the grid, a new conditioning data point is added to the list of original data points. Use of the local neighborhood is to avert the problem of inverting an ever-increasing Kriging matrix (Eq. (12-6)) as more and more points are estimated. It is also possible to use a prespecified maximum number of nearest conditioning data, including previously simulated values, in the estimate. In this way the Kriging matrix will never be more than this number of points within the ellipse and the SGS method will be able to handle large fields. Of course, ordinary Kriging in this procedure can be replaced by simple Kriging or co-Kriging as the need dictates.

The local neighborhood restriction, however, can not preserve large-scale trends. This can be modified to some extent by increasing the local neighborhood at the expense of greater effort. Obviously, some care is required in using fractal semivariograms. Alternatively, we can select a set of conditioning points that are a mix of points inside and outside of the neighborhood. The estimated points in SGS act as "seed points" for new stochastic features (e.g., low-, medium-, or high-value regions) in the random field. Since the SGS procedure ensures continuity in the random field according to the semivariogram, these stochastic features can persist and will be geologically acceptable if the semivariogram is consistent with the geologic structure. Cells far from conditioning points can take on values that are equiprobable over the full range of the empirical CDF; hence, SGS maximizes uncertainty in this sense.

12-4.2 Random Residual Additions

In the following, we discuss a procedure that is computationally more intensive but does not involve the subjectivity of determining a local neighborhood. We call it the procedure of *random residual additions* (RRA).

Figure 12-5 illustrates the procedure; the data points are indicated with *x*'s, and we are to fill in a random surface in this one-dimensional example. Five steps are required to produce the desired field.

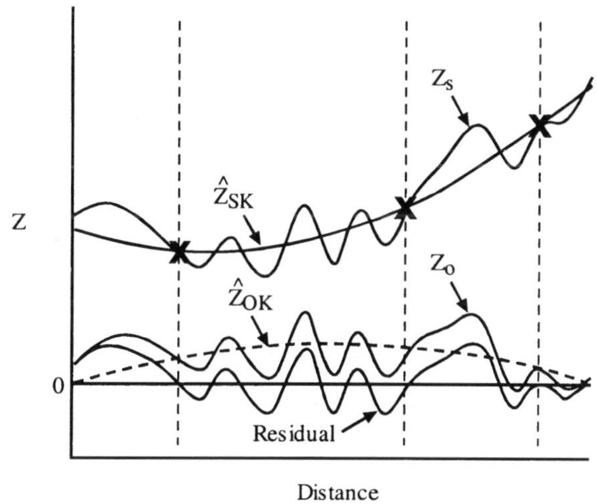

Figure 12-5. Schematic of random residual addition procedure. $\hat{Z}_{OK}$ is the first Kriging surface (deterministic); Z_s is the unconditioned surface; $\hat{Z}_{SK}$ is the second Kriging surface; Z_o is the final surface.

1. Generate a Kriging surface. This surface, shown as the dashed line $\hat{Z}_{OK}$ in Fig. 12-5, has the usual attributes of Kriging, the most important here being the exactness and the determinism.

2. Generate an unconditioned surface. This surface, shown as the erratic solid line Z_s in Fig. 12-5, has the desired statistical properties of the final surface, the appropriate second moments as conveyed through the semivariogram. But, of course, being unconditioned, this surface does not pass through the measured data. We discuss ways to generate this surface later. In general, many of the statistical tools used in this generation are the same tools used in the Kriging step. The unconditioned surface is not deterministic.

3. Generate a second Kriging surface. This step is the most important of all. The second Kriging surface, shown as $\hat{Z}_{SK}$ in Fig. 12-5, passes through the unconditioned surface at spatial coordinates that coincide with those of the measured data in Fig. 12-5. The $\hat{Z}_{SK}$ surface, therefore, treats $i = 1,2,\ldots,I$ points on the Z_s surface as known.

4. Subtract the unconditioned surface from the second Kriged surface. These differences are known as *residuals*. The residuals are zero at the measured data locations because of the coordinate conditioning in step 3.

5. Add the residual surface to the original Kriged surface (step 1). Because the residuals are zero at the measured points, the original data will be reproduced in the synthetic surface.

The final surface Z_0 in Fig. 12-5 has many desirable properties: it matches the measured data, has the appropriate degree of heterogeneity (if there is consistency between the statistical devices used in generating the various surfaces), is unbiased, and remains minimal. Many of these attributes carry through from the ordinary Kriging step because of the linearity of the operations. Most importantly, the final surface is stochastic; steps 2 to 5 will be different for each execution and constitute a realization for the entire process.

Throughout this book, we have been making distinctions between deterministic and random processes (although both may be treated statistically). The RRA procedure offers a clear demonstration of the combination of the two (i.e., it is a hybrid method). The determinism comes from step 1 and the randomness from steps 2 to 5. Indeed, many applications may be so deterministic that only the ordinary Kriging step is necessary. Of course, other surfaces may be so erratic that the determinism is negligible, but these limits may not be so apparent in the original data. RRA, then, offers a way of blending these two aspects in, it is hoped, the appropriate amounts.

12-4.3 Generating an Unconditioned Surface

There are two basic techniques for generating an unconditioned surface: spectral methods and averaging methods.

Spectral methods rely on the ability to represent a synthetic field with a Fourier series (Shinozuka and Jan, 1972). The key information here is a plot of the amplitude of each term in that series versus its frequency, known as a spectral density plot, which can be analytically related to autocovariance functions (Jenkins and Watts, 1968). This means that spectral methods are slightly limited in their ability to use a large number of autocovariance functions, especially those related to fractal semivariograms (Bruining, 1991). Spectral methods are difficult to implement in more than one dimension; however, a version of spectral methods, the turning-bands method (Mantoglou and Wilson, 1982), is quite efficient. This method relies on projecting a one-dimensional data set throughout a two-dimensional field by a series of geometrical constructions around a rotating line. The price for this efficiency is a slight, spurious periodicity in the generated field (Ghori et al. 1990).

Averaging methods exhibit more flexibility than spectral methods. They range from the relatively simple source-point method (Heller, 1972) to the most complicated matrix-decomposition method. The source-point method is simple because it does not require a matrix inversion. It can generate a wide variety of autocovariance functions, but fields with a prespecified function must be developed by trial and error (Bhosale, 1992).

The matrix decomposition method (MDM) is the most computationally intensive of the averaging methods (Yang, 1990), and its advantages are to be balanced against the computation cost. The computation overhead results from the repeated inversion of an autocovariance matrix. Even though the many inversion procedures are quite efficient, this task is still challenging even on the most advanced computers. Various approximations and the use of parallel computing seem to be of great advantage here (de Lima, 1995).

Both the foregoing CS procedures generate Gaussian fields that are frequently at odds with actual (geologic) distribution of properties. For example, the procedure can generate a stochastic feature that a geologist, based on sedimentary principles, knows is deterministic. These difficulties can be avoided with a careful geologic study that precedes or, better yet, is done in conjunction with the field generation. With this in mind, CS can be done entirely within the geologic framework or the geology can be superimposed through a multilevel CS. Chapter 13 shows some examples. The principal difficulty with these CS procedures is that they tend to be time-consuming, especially if multiple realizations are desired. Of course, being able to generate multiple realizations is the principle justification for CS. However, we can use indicator mathematics to generate geologically based realizations, as the following paragraphs describe.

12-4.4 Generating Facies Distributions

This method is called the *sequential indicator simulation-probability distribution function* (SIS-PDF) method. The objective is to generate a random field that classifies each cell into a facies category. Once each cell is so assigned, we can associate a property to the cell from the PDF of that facie.

We begin by assuming that the spatial distribution of facies categories is mutually exclusive, that is, two facies cannot exist at the same cell. The spatial pattern of each facie is governed only by its overall probability of occurrence and its semivariogram (e.g., structural type, range, orientation, and anisotropy). The overall facies PDF represents the field-wide occurrence of each category; it is usually obtained from well data or a facies map. Indicator simulation is ideal for modeling such "discrete" variables as these (Deutsch and Journel, 1992; Alabert and Massonnat, 1990). The SIS-PDF procedure for categorical indicators has the following steps (Jordan and Goggin, 1995).

1. Define a random path through the reservoir model such that each unconditioned cell is visited once and only once. Conditioned cells contain wells that are initially given known facies categorical values.

2. For each cell along the random path, locate a prespecified number of conditioning facies data, including facies data from wells, pseudowells, and previously simulated cells.

3. Using indicator Kriging (above), estimate the conditional probability for each facies category. Each facie will be estimated independently using individual facies semivariograms.

4. Normalize each facies' probability by the sum of probabilities over all facies and build a corresponding local conditional CDF from the normalized probabilities.

5. Draw a uniform random number in [0, 1] and determine the simulated facies category in the current cell by sampling the local CDF generated in step 4.

6. Repeat steps 2 to 5 for each cell in the random path, paying attention to the local availability of conditioning facies data in the previously simulated cells. New facies "realizations" are obtained by reinitializing the random path through the model, beginning at step 1.

Jordan and Goggin (1995) used this procedure for modeling facies patterns and their uncertainties in sand-rich turbidite systems. Facies categories, identified from interpreted seismic-amplitude map extractions, calibrated using core and log data, are (a) channels, (b) overbank deposits, (c) turbidites, and (d) distal lobes. Major facies trends were preserved by sampling the interpreted facies maps in "high probability" areas (e.g., pseudowell picks). Therefore, the SIS-PDF technique can model local facies uncertainty at the scale of cells in a full-field flow simulation.

12-5 SIMULATED ANNEALING

Simulated annealing (SA) is a method for producing a field with known properties. The procedures attempt to simulate physical processes, such as melting, freezing, and genetic reproduction, through a variety of algorithms. By the same token, the number of applications has also been diverse. Its use in reservoir characterization is fairly recent (Farmer, 1989) although it shows great promise (Datta Gupta, 1992; Panda and Lake, 1993).

Simulated annealing overcomes many of the difficulties with the methods described above. The data do not have to be normally distributed (in fact, any type of distribution is acceptable), stationarity is not necessary, conditioning is easy, and there are virtually no limits on the type of autocorrelation or heterogeneity a given field may possess. Indeed, SA can entirely ignore autocorrelation if necessary. The single largest advantage of SA is that it can incorporate several constraints, each from a different source, as long as these do not conflict. Hence, it is possible to generate a field that, in principle, satisfies all of the data gathered in a specific application. Finally, SA, as a class of techniques, is very simple and direct. Balanced against these advantages are the facts that SA tends to be quite computationally intensive and it has an uncertain theoretical pedigree. The trade-offs among these items should become clear as we discuss the main SA techniques, but first we cover some background items.

12-5.1 Background

If a metal or a glass is heated to near its melting point and then cooled to a given temperature T, the properties of the material depend on the rate of the cooling. For example, the material might be stronger, less brittle, or more transparent when cooled at a slower rate. The explanation for the different properties is usually based on diffusion arguments; if cooling is too fast, there is not enough time for physical defects or entrained gas to diffuse from the material. The process of slow cooling is called *annealing*.

From a thermodynamic viewpoint, each step in the cooling results in an equilibrium state or a minimum in the free energy surface. This is idealization because true equilibrium at any temperature T is attained only for an infinitely slow cooling rate. The end of each cooling step, then, must represent a local equilibrium or a local minimum in the free energy surface.

A central tool in the application of SA is the Boltzman or Gibbs cumulative density function given by

$$F(\Delta E) = \exp(-\Delta E/T) \tag{12-10}$$

where ΔE is the change in the energy, T is the temperature, and F is the frequency of occurrence of a given ΔE. Both T and the energy change are positive. The CDF in Eq. (12-10) is the complement of the one-parameter exponential distribution function first discussed in Chap. 4. Based on this equation, we see that the most likely energy change at a given T is zero and that nonzero energy changes become less likely as T decreases.

It is important to remember that SA is essentially an argument by analogy; the quantities we have been calling *energy* and *temperature* are really just thermodynamic names for variables that behave like them. The energy, in fact, plays the role of the objective function in previous work. A typical form for an objective function is

$$E = \sum_{m=1}^{M} \omega_m (\tilde{z}_m - \tilde{z}_{c_m})^2$$

where there are $m = 1,2,\ldots,M$ constraints, and ω_m is a weighting factor, selected in part to make the units consistent in this equation. The $\tilde{z}_M$ and $\tilde{z}_{c_m}$ are actual and computed statistics of the field, respectvely; usually they are a function of the Z_i, $i = 1,2,\ldots, I$ points to be generated. This equation is desirably vague for it is here that all constraints on a given field apply. The $\tilde{z}$ may be anything; we will show examples of $\tilde{z}$ as points on a semivariogram, global averages, and tracer data. The key property of E is that it be nonnegative, but this can be satisfied by other forms. For example,

$$E = \sum_{m=1}^{M} \omega_m \,|\tilde{z}_m - \tilde{z}_{c_m}|\,|$$

also satisfies the nonnegative requirement. As we shall see, we take only differences of E in the SA procedures, rather than derivatives as in conditional simulation; hence, the absolute value form is as convenient as the squared form.

Simulated annealing can be implemented in several ways. We cover two basic types: the Metropolis and the heat-bath algorithms. As is usual, we generate a grid of $j = 1, 2, \ldots, J$ properties Z_j that satisfy the constraints and that are also drawn from a specific empirical (experimental) CDF.

12-5.2 Metropolis Algorithm

This algorithm, first proposed by Metropolis et al. (1953), is the simplest SA algorithm. It has the following steps.

1. Generate a random field on the grid of J values by drawing some initial Z_j from the desired empirical CDF.

2. Based on these values, calculate $\tilde{z}_m$ for the M constraints.

3. Calculate the objective function E_{old}. If this is the first outer iteration, set the initial temperature to E_{old}.

4. Choose a gridblock randomly and draw a new value Z_j^{new} for this cell from the empirical distribution.

5. Recalculate the objective function E_{new}.

6. If E_{new} is less than E_{old}, accept the new value (that is, replace the original Z_j with Z_j^{new}), set $E_{old} = E_{new}$, and return to step 4.

7. If E_{new} is greater than E_{old}, calculate $\Delta E = E_{new} - E_{old}$ and the Gibbs probability F from Eq. (12-10).

8. Pick a random number r from a uniform distribution between 0 and 1. If $r > F$, reject the Z_j^{new} and return to step 4. Otherwise, go to step 9.

9. If $r < F$, accept the new value (even though ΔE has increased), let $E_{old} = E_{new,}$ and return to step 4.

   Repeat steps 4 to 9 until a suitable number of acceptances has occurred. These steps constitute the search for a local minimum at a given temperature. Usually the number of accepts should be somewhat greater than J, but it is not necessary to be more specific since we will ultimately be seeking a global minimum after the temperature has been lowered.

10. Test for convergence. Check to see if ΔE is below some preset tolerance. If it is, terminate the procedure; if not, continue to the next step.

11. Lower the temperature according to some cooling schedule such as $T_{new} = r_f T_{old}$ and return to step 4. r_f is a reduction factor ($r_f < 1$) that is typically around 0.8.

The Metropolis procedure involves two levels of cycling; the inner level, steps 4 to 9, at constant temperature, is noniterative and the outer level, steps 4 to 10, is iterative. Figure 12-6, a plot of E versus the cumulative cycling counts of three outer cycles, illustrates how convergence is attained.

In this figure, T is decreasing to the right with inflections intended to show when a new temperature is adopted. E decreases at each fixed temperature, but not necessarily monotonically; the cycle for T_1 illustrates a local minimum for E there. Steps 7 to 9 are intended to boost E over such local extrema. Since the algorithm uses E at two levels only, it is a form of a Markov process.

The generality of this procedure is somewhat offset by the *ad hoc* nature of many of the steps, for example, the number of inner cycles and the rate of cooling. To set these values for a particular application normally requires some trial and error. From a rigorous standpoint, the most serious problem with SA is that there is no guarantee that the process will converge. However, the procedure is quite flexible, as the following simple example shows.

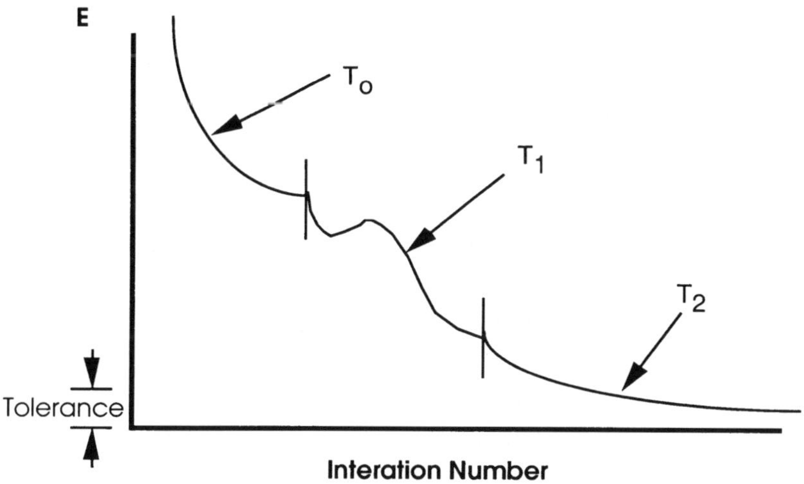

Figure 12-6. Schematic convergence of simulated annealing.

Example 4 - Simulated Annealing with the Metropolis Algorithm (from M. N. Panda, personal communication, 1994). Use the Metropolis algorithm to generate a four-cell ($J = 4$) grid with a geometric mean $\bar{z} = 340$. Use the absolute-value form of the energy definition.

For later comparison, let the true distribution be

$$\begin{bmatrix} 100_1 & 500_2 \\ 300_3 & 900_4 \end{bmatrix}$$

where the subscripts indicate the cell number. Of course, we know only the value of the geometric mean before we do the annealing.

In this case there is only $M = 1$ constraint and we perform the inner cycle seven times at each temperature. The discrete experimental PDF is the set of nine equally likely values (100, 200, 300, 400, 500, 600, 700, 800, 900). This means that if a random number falls between 0 and 0.111, the selected value is 100; between 0.444 and 0.555, it is 500; etc. Table 12-1 shows the steps for the results for the first temperature $T_o = 106$.

Table 12-1. Inner cycle for Metropolis example.

1 Move Cell	2 Rand. No.	3 Cell 1	4 Cell 2	5 Cell 3	6 Cell 4	7 Geo. Mean	8 E	9 Delta E	10 Gibbs Prob.	11 Rand. No.	12 Deci-sion
	Initial	400	500	400	500	447	106				
1	0.82	800	500	400	500	532	191	85	0.45	0.57	Reject
1	0.59	600	500	400	500	495	154	48	0.64	0.51	Accept
3	0.99	600	500	900	500	606	265	111	0.35	0.78	Reject
4	0.96	600	500	400	900	573	232	78	0.48	0.44	Accept
2	0.01	600	100	400	900	383	42	-190			Accept
4	0.61	600	100	400	500	331	10	-33			Accept
3	0.13	600	100	200	500	278	63	53	0.61	0.38	Accept

Column 1 is the randomly selected cell, and column 2 is the random number for selecting the trial value z_i^{new}. Column 7 shows the current geometric mean. Columns 8 and 9 are the energy (E) and the energy change (ΔE). The Gibbs probability and the corresponding random number are not needed unless the energy change is positive. Three trials were accepted even though the energy change was positive. The energy has decreased from 106 to 63 (not monotonically) during this cycle.

Table 12-2 further illustrates the entire process by showing only the accept steps until convergence. The horizontal lines show when the temperature has been lowered (we have used $r_f = 0.8$ here). Again, several trials were accepted even though the energy increased; indeed, three trials resulted in no decrease of the energy whatsoever. We can illustrate the nature of the convergence in this simple problem with Fig. 12-7.

As illustrated, the convergence is not monotonic because of the possibility of accepting a positive energy change. The final answer,

$$\begin{bmatrix} 900_1 & 500_2 \\ 300_3 & 100_4 \end{bmatrix}$$

is not the same as the actual one. The converged solution, which may be regarded as a realization of the process, differs from the true distribution that we gave at the first of the example because the objective function contained no information regarding the spatial arrangement of the values. However, the final solution has the desired geometric mean.

Table 12-2. Acceptance steps for Metropolis example.

Temp.	Move Cell	Rand. No.	Cell 1	Cell 2	Cell 3	Cell 4	Trial Geomean	E	Delta E	Gibbs Prob.	Rand. No.
106	1	0.59	600	500	400	500	495	154	48	0.64	0.51
	4	0.96	600	500	400	900	573	232	78	0.48	0.44
	2	0.01	600	100	400	900	383	42	-190		
	4	0.61	600	100	400	500	331	10	-33		
	3	0.13	600	100	200	500	278	63	53	0.61	0.38
85	3	0.06	600	100	200	500	278	63	0		
	2	0.66	600	500	200	500	416	75	13	0.86	0.53
	4	0.35	600	500	200	400	394	53	-23		
	1	0.18	200	500	200	400	299	42	-11		
	4	0.09	200	500	200	100	211	129	88	0.36	0.18
	3	0.11	200	500	200	100	211	129	0		
	1	0.65	500	500	200	100	266	75	-54		
68	3	0.58	500	500	600	100	350	9	-66		
	1	0.63	500	500	600	100	350	9	0		
	3	0.58	500	500	600	100	350	9	0		
	3	0.47	500	500	500	100	334	6	-3		
54	1	0.31	900	500	500	100	387	46	40	0.48	0.21
	3	0.44	900	500	300	100	341	0	-6		

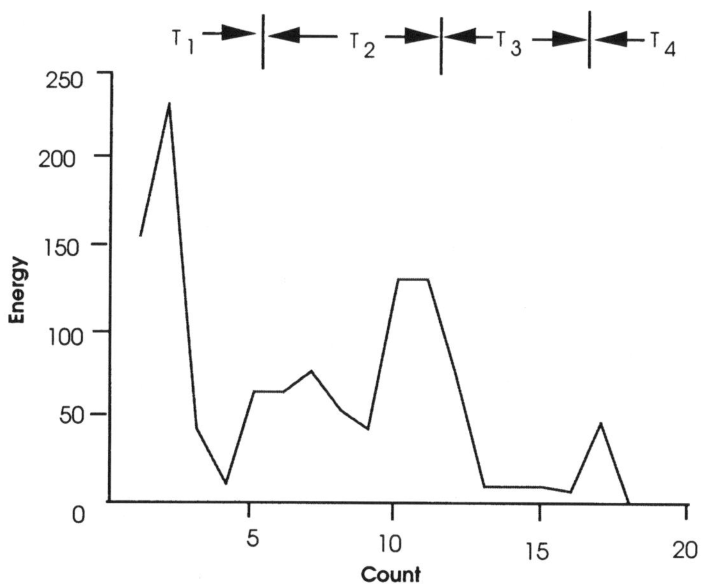

Figure 12-7. Convergence rate for Example 4.

This example illustrates both the strength and weaknesses of SA. The objective function is statistical but it does not deal with spatial arrangement, although this could have been included. The procedure converges, but it converges slowly, especially toward the end of the process where most of the Gibbs trials are rejects. The multiple rejections can significantly increase the computation effort of the process.

12-6 MODELLING EXAMPLES

As a tool for generating the input to reservoir flow simulations, statistical techniques offer an unprecedented means of generating insights into reservoir processes and improving efficiency. We give a few illustrations in this section. In general, we present results of two types: results of fluid-flow simulations that have used statistical modeling to generate their input permeabilities or maps of the input itself. All of the cases to be discussed have used some form of conditional simulation (CS), and, except where noted, all have used random field generation in conjunction with geological information.

12-6.1 History Matching

Figure 12-8 shows calculated (from a numerical simulator) and actual results of water cut from a waterflood in a U.S. oilfield.

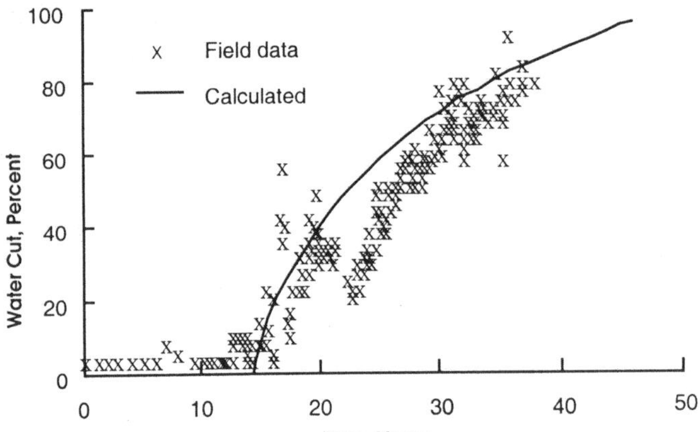

Figure 12-8. Actual and simulated production from a waterflood. From Emanuel et al. (1989).

The calculated results used a hybrid streamline approach in which the results of several two-dimensional cross sections were combined into a three-dimensional representation. The permeability distributions in the cross sections were generated by CS, using, in this case, a fractal representation that was derived from analysis of wireline logs and permeability-porosity transforms. The calculated and actual results agree well.

The process of bringing the output of a numerical simulator into agreement with actual performance is known as history matching. History matching is a way of calibrating a simulator before using it for predictions. In most cases, history matching requires more than

50% of a user's time, time that could be better spent on analysis and prediction. For the case illustrated in Fig. 12-8 (and in several other cases in the same reference), the indicated history match was attained in only one computer run. Historical data are best used to test and diagnose problems in the model rather than to determine the model.

Subsequent experience has shown that such agreement is not so easily attained for individual wells and, even for entire field performance, some adjustment may be needed. However, the point of Fig. 12-8 is that CS significantly improved the efficiency of the analyst.

12-6.2 Geological Realism

Figure 12-9a shows a three-dimensional permeability distribution from a carbonate reservoir in east Texas generated using CS and fractal semivariograms. The dark shading indicates low-permeability regions and the small circles indicate well locations. This realization was generated with substantial geological insight, the most important of which was the identification of six separate facies, indicated in the figure as layers. Only the first two facies (layers 1 and 2) were statistically distinct from the others (Yang, 1990), but the properties within each were nevertheless generated to be consistent with the respective permeability PDF.

Subsequent analysis revealed the existence of a pinchout (a deterioration in reservoir quality) to the southeast in this field. The pinchout is represented by the solid lines in Fig. 12-9b. Such features are not a standard in field-generation procedures and modeling; however, with some effort they can be incorporated.

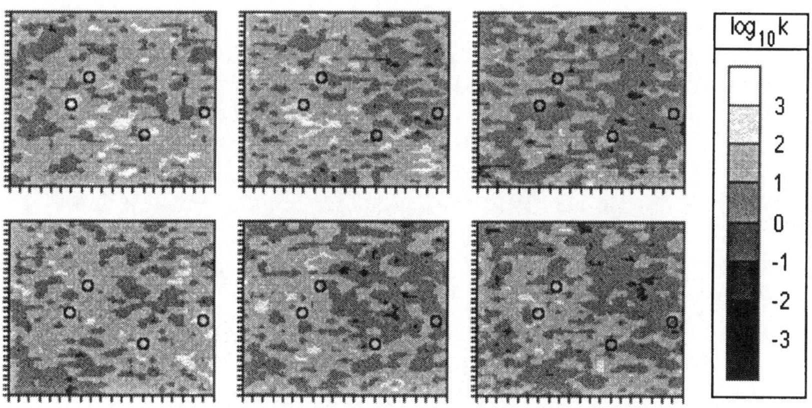

Figure 12-9a. Three-dimensional permeability realization of an east Texas field. From Ravnaas et al. (1992). (Top row contains layers 1, 2, and 3).

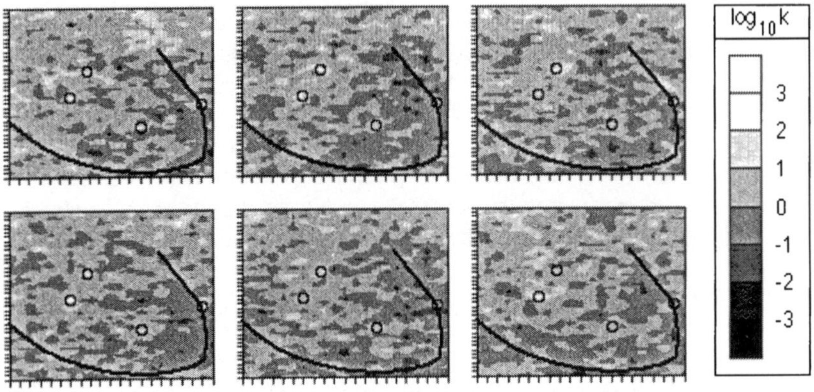

Figure 12-9b. Three-dimensional permeability realizations of an east Texas field with pinchout. From Ravnaas et al. (1992).

As discussed in Yang (1990), the pinchout is introduced by a two-level CS. The first level constructs the pinchout along the given line using data along the vertical line from the right-most well in the figure. This well, a poor producer, has properties that are representative of those along the pinchout. After the first level, the second-level CS proceeds, now using data from the first level as conditioning points (some call these points "pseudowells"). The result is a gradual but sporadic deterioration of reservoir quality to the southeast in all layers, as well as a general alteration of permeability in all layers.

Another way to impart geological realism is through the use of pseudoproperties. These properties are the usual transport properties—permeability, relative permeability, and capillary pressure—that have been adjusted so that Darcy's law applies at the scale of interest. The adjustment takes into account the underlying geology of the field, the flow properties, and the prevailing boundary conditions.

The main point from these exercises is that geostatistics and CS procedures are extremely flexible; if geology is known, it can be honored with suitable manipulation both as a preprocessing step (previous section) or in postprocessing.

12-6.3 Fluid-Flow Insights

Numerical simulation has been in common practice for several years, and we have come to believe in the correctness of its results if we are confident in its input. This procedure requires careful geological, engineering, and statistical assessment. The history-matching process often requires post-scaleup adjustments to layer or near-wellbore properties in a three-dimensional model. Model adjustments are the result of discussions between the development geologist and reservoir engineer assigned to the project team about remaining reservoir property uncertainties.

Now, we take a different approach and, using hypothetical models, examine the sensitivities of recovery processes to changes in the statistical modeling. Statistical modeling through CS offers a way to control both the heterogeneity and continuity of an input flow field with sufficient generality to gain novel insights into reservoir processes.

We have known for several decades that miscible displacements in which the solvent is more mobile than the phase being displaced form an erratic displacement front, bypassing, and as a consequence, inefficiently recovering the resident fluid. The ratio of mobilities of such a displacement (the mobility ratio) being greater than one, the displacement is termed *adverse*. This phenomenon, among the most interesting in all of fluid dynamics, has been modeled in the laboratory by numerical simulation and through analytic methods. Nearly all of this work has been in homogeneous media. The question then arises: what happens to adverse miscible displacements in the presence of realistic heterogeneity?

This was the issue in the work of Waggoner et al. (1992) extended by Sorbie et al. (1994), who studied adverse miscible displacements through two-dimensional cross sections of specific heterogeneity and continuity. This work used the Dykstra-Parsons coefficient V_{DP} (Chap. 6) to summarize heterogeneity and the spherical semivariogram range (Chap. 11) in the main direction of flow to summarize continuity. The displacements fell into three categories.

1. Fingering: displacements in which bypassing occurred, caused by the adverse mobility ratio.

2. Channeling: displacements in which bypassing occurred, caused by the permeability distribution.

3. Dispersive: displacements in which no bypassing occurred.

Both fingering and channeling displacements manifest inefficient recovery and early breakthrough of the solvent. The dispersive displacements showed efficient recovery and late breakthrough. The principal distinction between fingering and channeling is that a fingering displacement becomes dispersive when the mobility ratio becomes less than one; a channeling displacement continues bypassing.

These classifications are of little use without some means of saying when a given type will occur. Figure 12-10 shows this on a flow-regime diagram.

The figure shows some interesting features. For example, it appears possible to completely defeat fingering with uncorrelated heterogeneity as, for example, might occur within a carbonate bed (Grant et al., 1994). However, the overwhelming impression is of the ubiquitousness of channeling. In fact, for typical values of V_{DP}, fingering does not occur and there are dispersive displacements only in essentially random fields. It appears that fingering occurs only in regimes where V_{DP} is less than 0.2. Such low values are normally possible only in laboratory displacements. The transition to channeling flow normally occurs at a heterogeneity index I_H of around 0.2. For typical field V_{DP} (>0.6), geologic "flow units" need only extend and interconnect over 30% of the interwell spacing to support channeling. Although the minimum channeling criteria changes for different mobility ratios, there are many mature fields developed in stacked-sand or carbonate (Goggin et al.,1993) environments where flow units are equal or greater than well spacing (dimensionless range > 0.7).

The above observations render a remarkable change in the point of view of miscible displacements: rather than being dominated by viscous fingering, as had been thought, they appear to be channeling. Indeed, channeling appears to be the principal displacement mode for most field displacements. These insights have operational significance. For example, it seems obvious that decreasing well spacing (increasing the dimensionless range) will always lead to more channeling and, consequently, less sweep and ultimate recovery. Such a loss might be compensated by an increased rate; however, smaller well spacings will always lead to more pressure and flow continuity between wells. The decision to infill-drill or pattern realign a waterflood for improved or accelerated oil recovery is always an economic one, often driven by how well any remaining geologic uncertainties have been addressed at a field's mid-life decision point.

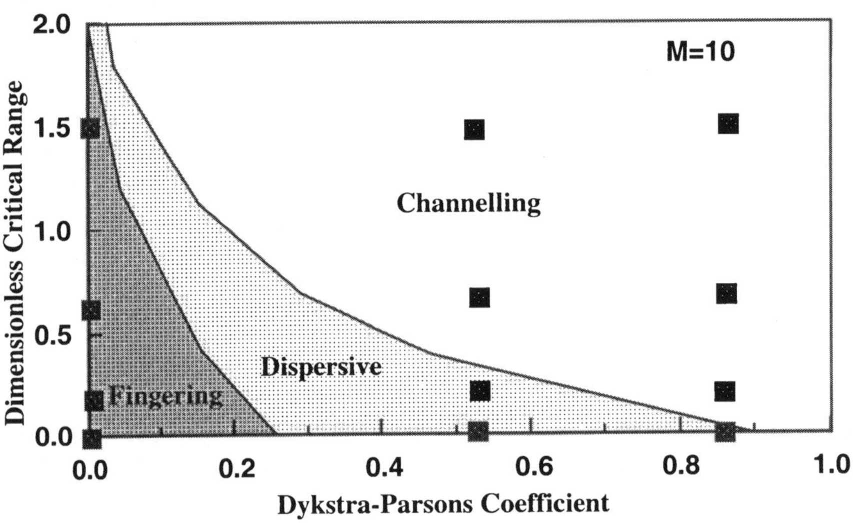

Figure 12-10. The various types of flow regimes for a matched-density miscible displacement. From Waggoner et al. (1992). Filled boxes indicate computer runs.

The Waggoner et al. (1992) conclusions have been extended to unmatched density displacements by Chang et al. (1994), to more general cases by Sorbie et al. (1994), and to immiscible displacements by Li and Lake (1994). In the context of this discussion, we emphasize that such insight can only be possible with a suitably general representation of heterogeneity afforded by CS.

12-6.4 Wellbore Placement Models

We have up until this point concentrated on reservoir models that have been built to understand or predict reservoir performance (directly or by analogy). Another area in which the models can be used is well planning.

In a study of a mixed aeolian and fluvial reservoir, Siefert et al. (1996) undertook a well-placement study. The question they addressed was whether, in a depositionally complicated gas field, there is an optimum length and orientation to drill a horizontal producing well. The question was answered by assuming that a certain proportion of the best facies was needed for a successful well.

Many thousand synthetic wells were drilled in various geostatistical realizations of the model (Fig. 12-11). Each well was analyzed for the frequency and number of aeolian sandbodies intesected (the best reservoir units) along each borehole (Fig. 12-12). Azimuth G turned out to be oblique to the geological grain. Paralleling the geological grain might have encountered the maximum aeolian sand interval intersected, but also with a significant chance of missing the aeolian sandbodies altogether. Conceptual reservoir models like these show how geostatistical models can be used to assess the uncertainty in well planning, in addition to their use in reservoir simulation, and provide results that might not be immediately apparent.

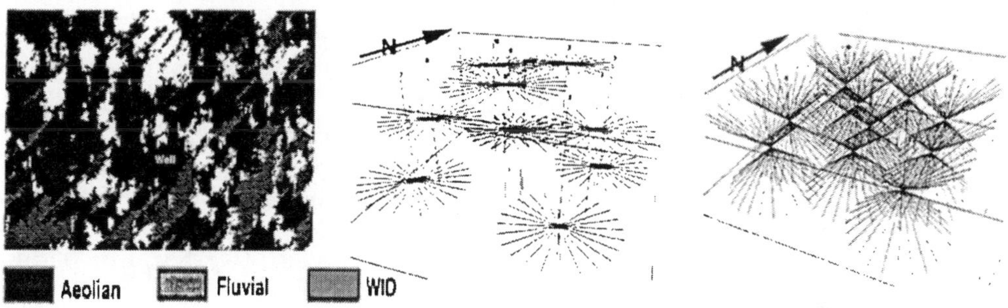

Figure 12-11. Geostatistical model of a complex reservoir and the synthetic wells placed at a wide variety of locations, lengths, and orientations (from Siefert et al., 1996).

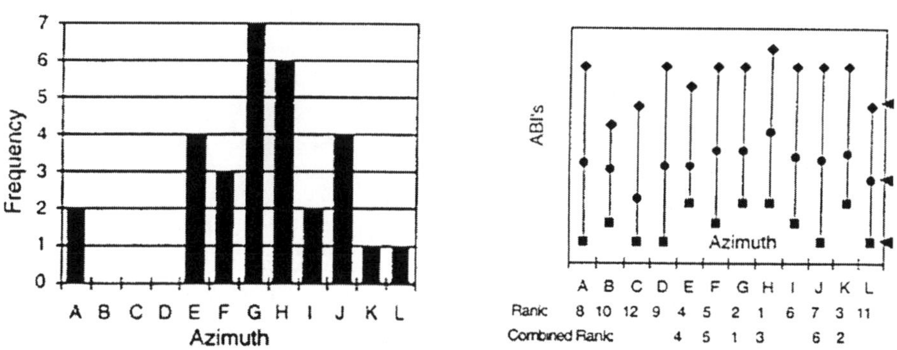

Figure 12-12. Geostatistical model results showing that the optimum orientation will be azimuth G based on the frequency (left plot) of sandbodies intersected and the number of aeolian bodies intersected (ABI's). After Seifert et al. (1996).

Another approach to well planning is given by Hoye et al. (1994). In this study of the Troll West field in Norwegian waters of the North Sea, there was uncertainty in the reservoir structure as interpreted from seismic data in undrilled areas. Stochastic modeling was used to generate depth and thickness distributions for specific well locations. These distributions could be used to provide average estimates of reservoir thickness likely to be encountered in a well or the probability of finding greater than a specified minimum required thickness. As an example, a probability map for 9 m of pay sand in the oil leg of the field was presented. Probability maps are important for well planning.

12-7 OTHER CONDITIONING

As has been the thrust of this entire book, the above section has concentrated on geologic and well data, the so-called static data. Other data sources abound in engineering practice and should, as time and expense permit, also be part of the reservoir model. The most common of these are the following.

1. *Pressure-transient data.* In these types of tests, one or more wells are allowed to flow and the pressure at the bottom of the wells observed as a function of time. In some cases, the pressure is recorded at nonflowing observation wells. Pressure-transient data have the decided advantage that they capture exactly how a region of a reservoir around a wellbore is performing. But this regional property is measured on a different scale than point properties such as cores and logs. Resolving the disparity in scales between the regional and point measurements and choosing the appropriate interpretation model remain significant challenges in using pressure-transient data. (See Sec. 12-2.5 for the results of attempts to incorporate pressure-transient data into Kriging.) However, this type of data is quite common and may be inexpensive to acquire, making it an important tool in the tool box. New offshore fields are increasingly being developed with reliable continuous, down-hole pressure and water-cut monitoring gauges. These monitoring devices, along with appropriate data analysis software and down-hole flow controls, should allow reservoir and production engineers to quickly modify how waterfloods are managed, thereby maximizing oil rate and overall field recovery.

2. *Seismic data.* The ultimate reservoir characterization tool would be a device that could image all the relevant petrophysical properties over all the relevant scales and over a volume larger than interwell spacing. Such a tool eludes us at the moment, especially if we consider cost and time. However, seismic data come the closest to the goal with current technology.

   There are two basic types of seismic information: three-dimensional, depth-corrected traces "slabbed" from a full cubic volume of a reservoir and two-dimensional maps of various seismic attributes. Seismic data are generally considered as soft constraints on model building because of the as-yet limited vertical resolution. However, because of the high areal sampling rate, seismic data can provide excellent lateral constraints on properties away from wells. Seismic data integration into statistical models, through using impedance information as a facies or porosity predictor, has become common in large projects (e.g., Bashore et al., 1994; Chambers et al., 1994; Wolf et al., 1994; Jordan and Goggin, 1995; Tjolsen et al., 1996; MacLeod et al., 1996).

3. *Production data.* Like pressure-transient data, production data (rates and amounts of produced fluids versus time) reflect directly on how a reservoir is performing. In mature fields, spinner or temperature surveys are commonly used to capture zonal fluid injection or production performance. Pressure profiles are collected in new infill wells to monitor zonal pressure depletion and to estimate pressure connectivity with nearby producers. Consequently, such data form the most powerful conditioning data available. Like the seismic integration, incorporating production data is a subject of active research (see Datta-Gupta et al., 1995, for use of tracer data) because it is currently very expensive. The expense derives from the need to run a flow simulation for each perturbation of the stochastic field. Furthermore, production data will have most impact on a description only when there is a fairly large quantity of it, and, of course, we become less and less interested in reservoir description as a field ages.

12-8 CONCLUDING REMARKS

The technology of statistics, particularly as applied to fluid-flow modeling, has madc remarkable advancements. We know that using statistics to assign the input to a simulator improves the efficiency of the modeler. We must verify, however, that the procedures generate realistic heterogeneity; our discussion above implies that the statistical assignments can be made to agree with geology As has been the theme throughout this text, the statistics are adjuncts to, not replacements for, geology. (The next chapter has further discussion.) These considerations make it likely that the practice of using statistics to assign flow simulation input will be standard technology in the near future. But there is one other argument for the use of statistical assignments, not yet realized, that may supersede all of the previous ones in importance.

We continually observe (and occasionally contribute to) the ongoing revolution in computer technology. What is possible today–a few million cells in a numerical simulation–was only a dream a few years ago, and with this capability comes a significant burden in knowledge management. A typical numerical simulation requires, at a minimum, three scalar permeabilities, a porosity, a fluid saturation, and a pressure for each cell. For a one-million-cell simulation, then, six million numbers are required. Simply assigning these numbers for the simulation is a significant task, and adjusting specific ones for the purpose (say) of history matching becomes impossible.

On the other hand, each parameter field can be efficiently controlled with an appropriate semivariogram; thus, instead of six million numbers, we control the simulation with only 12, if a two-parameter semivariogram is used and the entire field is one geologic classification. Twelve parameters represent a nice balance between flexibility and what can be realistically achieved. Thus, it is far more efficient to assign fields with statistics than to assign the numbers directly. This benefit may be the most important one for future application, especially if the trend in increasing computer power continues.

The keys to managing reservoir variability are to recognize, quantify, and rank all geologic uncertainties that impact both the size of the oil resource (original oil in place) and its producibility (predicted oil, water, and gas profiles). These should be done as early in the field development as possible. Each new stage of field development offers an opportunity to collect data, sometimes costly, that will help reduce the highest-ranking uncertainties, and thereby cause a re-evaluation of remaining geologic factors.

Reservoir management plans (RMP's) must be flexible enough to loop back through the model-updating process at each major field development stage. More importantly, the RMP must document and rank remaining geologic uncertainties, so that the subsurface teams know the range of allowable model adjustments for history match and prediction of reservoir performance. The economic impact of well-designed RMP's is only now being recognized as a high priority, especially in development areas where long-term, high-capital expenditures are required, e.g., hostile climates, deep-water, new oil pipelines infrastructure, new gas processing and transport infrastructures.

Flow simulation models will increase in sophistication equal to available data at each stage, and both deterministic and stochastic methods will be used. As we've seen throughout this book, model adjustments should be made within this mixed framework as well as within an evolving geologic understanding. Starting at the largest volumetric scale, geometric and reservoir properties will be adjusted by new time-to-depth structures, by fault blocks, by stratigraphic zones, by channel or facies regions, and by petrophysical PDF's, correlations, and variograms.

Earth modeling tools will also increase in sophistication to meet the ever-increasing data requirements of very large (>100 million cells) geologic models with hundreds of regions and associated reservoir properties, including measures of uncertainty. Modeling frameworks will take full advantage of all interpreted and measured data by merging topological, geometric, and property modeling tools tuned to an ever-increasing understanding of geologic processes.

CHAPTER 13

THE USE OF STATISTICS IN RESERVOIR MODELLING

In this chapter, we apply some of the concepts introduced in previous chapters for evaluation and modelling of reservoir, or potential reservoir, performance. These case studies show how the statistical concepts have been used as basic tools in the geoengineering toolkit for evaluation and management. The chapter ends with some guidelines for practical reservoir modelling.

Remember that the appropriateness of a statistical or engineering model depends on the ultimate application; more sophistication and complexity do not always result in a "better" model. For example, a tank-type material-balance model of a field can yield a fairly accurate estimate of reserves and may, in certain circumstances, even provide as much insight as a more sophisticated model (Dake, 1994, p. 4). Moreover, a crude but easy-to-use model will often fit better into time and budget constraints. The question in these cases is not whether a detailed model is better than a simple one, but rather how *much* better. However, more detail always yields a better answer, time and budget notwithstanding as long as it is competently done. For detailed models, how well they capture the essential physics and geology determines their accuracy. Modelling in this chapter means the distribution of reservoir properties according to statistical and/or geologic principles.

For most properties, both recognizable (essentially deterministic) geological elements at some scale(s) and random elements are present in reservoir data. The random element can represent natural variability, sparse sampling, noisy data, or a combination of all. Stochastic modelling can be applied to media that are neither entirely deterministic nor random.

Chapter 1 introduced three types/descriptions of reservoir structures:

Strictly deterministic: Correlatable and mappable reservoir units with well-understood internal sedimentary architecture and predictable petrophysical properties from a primary textural control.

Mixed deterministic and random (i.e., stochastic): Correlatable and mappable reservoir units containing an internal sedimentary architecture and/or petrophysical properties defined by vertical and lateral spatial statistics (e.g., facies PDF's, porosity CDF's, and associated semivariograms). The alternative could also occur: well-defined property distributions within reservoir units whose stratigraphic boundaries are poorly defined.

Purely random: No apparent geological explanation for the distribution of petrophysical properties in a heterogeneous to very heterogeneous reservoir defined by very uncertain boundaries.

These categories are in order of the amount of information required for modelling. The deterministic style requires the most information; measurements at all geologically relevant scales are needed and the geological units must be pieced together. On the other hand, the purely random style needs only estimates of the mean and variance of the important flow properties to produce a model. The result of the lack of information in the random and stochastic cases is that many different but statistically similar realizations can be produced. The deterministic approach produces only one realization.

Reservoir and aquifer properties are deterministic only in the most qualitative sense; on the other hand, strict randomness is extremely rare in nature. However, as the cases below show, statistics has a role in all three categories. In fact statistics, used in tandem with geoscientific concepts for reservoir modelling, provides power and authenticity not achievable by either method alone.

The degrees of either determinism and/or randomness, which can be assessed by geological description and sampling, will determine the best modelling strategy. While a reservoir or aquifer may fall into different categories at different scales, our case studies tend to concentrate on the flow unit (in the Hearn et al., 1984 definition) or interwell scale (which ranges from around 0.1 km in an onshore setting, to about 1 km, in an offshore location).

13-1 USING STATISTICS FOR RESERVOIR DIAGNOSIS

A range of statistical tools can be used to identify and prove the influence of geological structure on fluid flow. Two very useful diagnostic tools in analysis of reservoir properties are the variability (as measured by the C_V) and the structure (as measured by the semivariogram). If a preliminary analysis of the reservoir data showed a C_V less than 0.5 (i.e., homogeneous), and an ideal spherical semivariogram (i.e., small nugget, well defined sill, no periodicity), then the reservoir unit may be considered homogeneous. This assumption is valid provided the sampling has been sufficient. Any other complexity (e.g., structural geological complexity such as faulting) must be accounted for separately in the model. If, on the other hand, the $C_V > 0.5$ (i.e., heterogeneous, which is usually more likely), then the sampling issues should be considered carefully, the structure carefully appraised, and models thoughtfully constructed.

One of the most difficult properties to estimate is the vertical permeability. Variation in petrophysical properties (e.g., porosity and permeability) are usually related to the manner in which the medium was deposited. In fact, textural characteristics of sediments (e.g., grain size and sorting) often play an important role in determining levels and distributions of petrophysical properties (Panda and Lake, 1995). The hierarchical nature of sediments is such that reservoirs often comprise systematic groupings and arrangements of depositional elements. Sedimentology uses a knowledge of this organization to interpret sediments in terms of depositional environment. The environments contain a great deal of variability in all aspects, but "typical" elements exist and the architecture (stacking patterns) of those elements reveals the processes at work during deposition. The degree of stacking of geological elements will influence the vertical flow paths (and vertical permeability) of the formation.

The three cases included below illustrate how the estimation of this property might be approached.

13-1.1 Rannoch Formation North Sea

Permeability measurements at various scales can reveal various scales of variability within a geological formation. The Rannoch Formation, an important North Sea reservoir unit, shows variations in permeability structure that are related to variations in the geological structures (Fig. 13-1). At the smallest scales (30 cm, Fig. 13-1 right), the wavy-bedded material shows permeability variation in excess of one order of magnitude, whereas the low-contrast mica-laminated interval is relatively uniform. A permeability semivariogram at this scale in the wavy bedded material reveals small-scale repetitive laminaset structures (Fig. 13-2 left) reflected in the zero-nugget, periodic semivariogram.

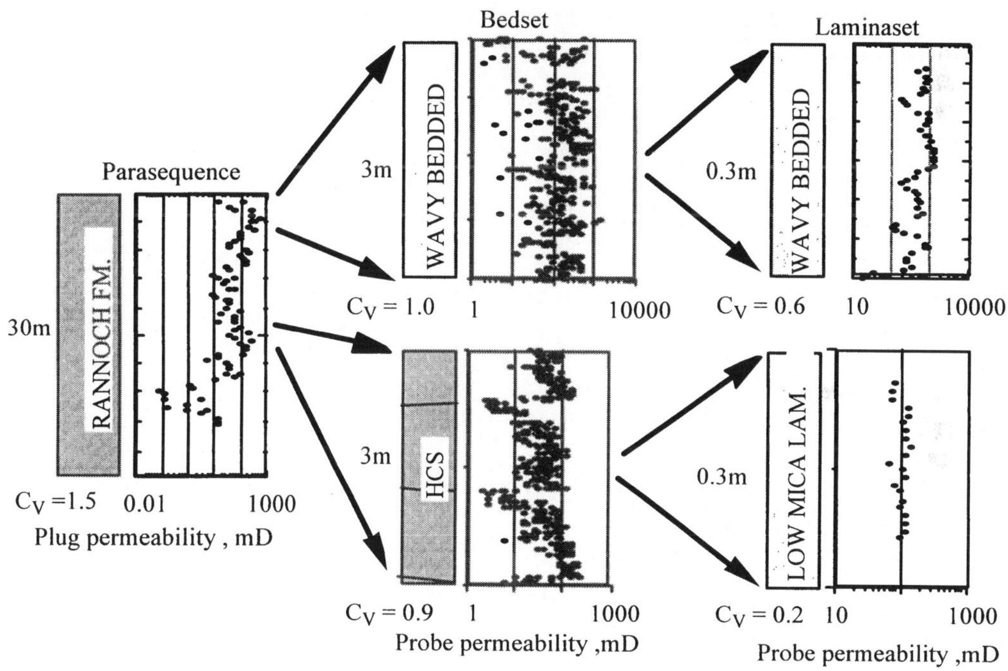

Figure 13-1. Heterogeneity at different scales in the Rannoch formation. The lamina-set packages that occur in the wavy-bedded and HCS beds have different degrees of heterogeneity and vertical lengths. For the purposes of illustration, only certain intervals are shown; for more details and discussion, see Corbett (1993).

In the same formation, for a series of measurements at the bed scale (i.e., over 3 m as in the center figures in Fig. 13-1), there is only small scale structure in the wavy bedded interval (i.e., no larger-scale bed features such as trends or cycles). In the hummocky cross-stratified (HCS) interval on the other hand, we can clearly see cycles in the permeability profile. These are also apparent in the semivariogram for this interval (Fig. 13-2 center).

Permeability measured on a set of core plugs over the formation (30 m, Fig. 13-1 left) show a larger scale variation that increases upward, which is reflected by the no-sill semivariogram (Fig. 13-2 right). The nugget in the core plug semivariograms is reflecting the small scale variability seen by the probe data at various scales in the formation (Fig. 13-1) and is therefore diagnostic of small scale (sub-core plug spacing) variability not fully captured by the plug sampling strategy. The suggestion of some structure (periodicity) in the plug semivariogram (Fig. 13-2 right) may also be identifying some 10 m cyclical event worthy of further geological consideration.

Figure 13-1 illustrates very clearly the hierarchical or nested autocorrelation structure within a complex reservoir. These scales should be considered separately in an engineering appraisal of the data. The small-scale structure in the wavy-bedded rock type may have a strong influence on capillary-dominated sweep efficiency. The beds seen in the HCS interval will influence the viscous-dominated inflow performance into a well. The large-scale trend will influence the gravity-dominated sweep of this interval. Models at various scales are needed to make these assessments (Corbett and Jensen, 1993b).

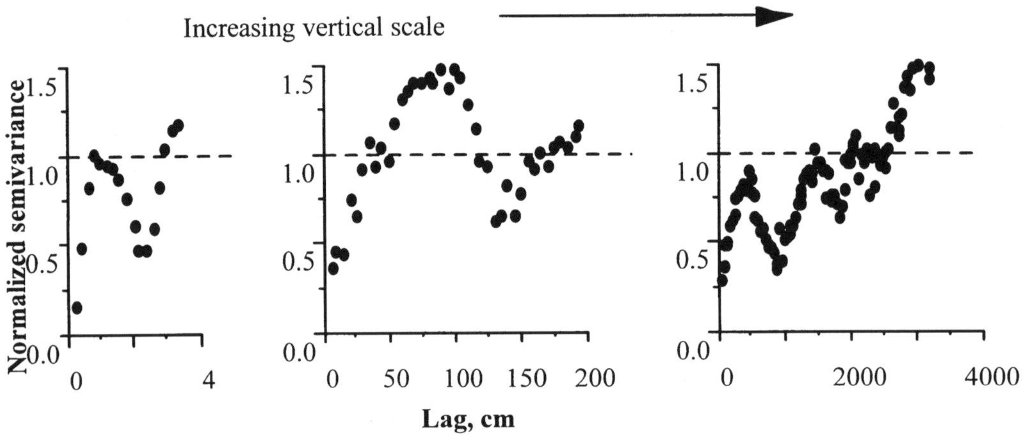

Figure 13-2. Spatial patterns of permeability in the Rannoch Formation from representative semivariograms over three length scales. Note the nested hierarchy of spatial structure and the absence of nugget for the finest-scale data. The non-zero nugget at larger scales represents the poor resolution of the lamina by larger sampling programs. The semivariance is normalized by the sample variance.

The C_V 's for the various intervals are also diagnositic. The C_V of the low-mica lamination is 0.2 (Fig. 13-1). Using our $(10\ C_V)^2$ criterion from Chap. 6, four samples would suffice where we have acquired 21. In the wavy-bedded interval, the 48 samples satisfy the 36 required by the I_0 criterion. For the 3 m intervals of wavy-bedded and HCS, the 300 probe measurements are in excess of the 100 and 84, respectively, required. The probe data in this study adequately capture the small-scale variation. In fact, there is even scope, in a repeat study, for increasing the sample spacing to 3 cm for the interval shown, without losing significant definition. It may be practical to alter the spacing for different rock types or to run with the spacing required by the most variable facies.

At the parasequence scale (30 m) the core-plug profile contains 76 core plugs (approximately one per foot or 30 cm). The sampling criterion requires 225 samples for the measured variability. The plugs are insufficiently sampled, as was suggested by the nugget in the semivariogram and the resolution by the probe of small-scale geological structure. The core-plug spacing equates to one sample for each of the intervals shown in Fig. 13-1 (right).

What might be the effect on the vertical permeability of all this geological structure seen in Figs. 13-1 and 13-2? We can estimate the permeability anisotropy by taking the harmonic average for the vertical permeability (k_v) and the arithmetic average for the horizontal permeability (k_h) and calculating the ratio (k_v/k_h).

We have seen in Chap. 5 how the respective averages can be used for estimation of vertical and horizontal flow. Probe data can usually be assumed to be isotropic; plug data tend to show more anisotropy. The estimator assumes that the system is perfectly and continuously layered and the layers are of equal thickness and measured by the individual points. Small-scale experiments have been conducted (Fig. 13-3) to evaluate these assumptions. The arithmetic average of the probe data was found to be a good estimator of the horizontal permeability. The harmonic average tends to overestimate the vertical permeability. There is, in the latter case, a tendency of thin low-permeability layers in rocks to escape detection.

Similar experiments have also been conducted in carbonates where the variability is greater (Corbett et al., 1999). In this latter set of experiments, the horizontal permeabilities were estimated by the geometric to harmonic averages. Similar results for the vertical permeability suggested carbonates are more isotropic and that effective flow is very dependent on the degree of pore/vug connectivity.

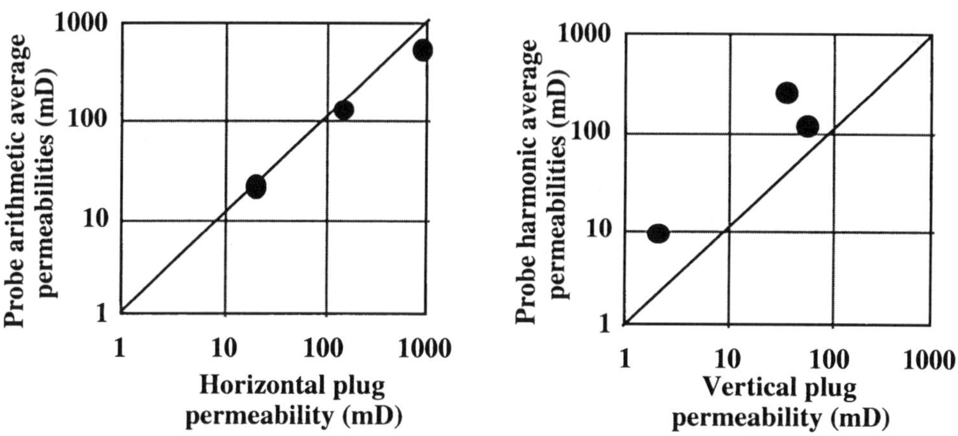

Figure 13-3. Plug-scale experiments in the Rannoch Formation. Comparisons of arithmetic-average probe measurements with horizontal plugs (left), and harmonic-average probe measurements with plug vertical permeabilities (right).

Extending the use of arithmetic and harmonic averaging over different vertical distances (i.e., increasing scale) illustrates how the ratio varies with scale in a formation. The k_v/k_h ratio decreases with scale (Fig. 13-4) but also decreases differently in different rock types within the same formation. An estimate of the average k_v/k_h ratios from vertical and horizontal core-plug pairs is unlikely to be *directly* applicable to larger-scale models. The Rannoch Formation as a whole unit may be more anisotopic (with a k_v/k_h ratio of 0.01 in Fig. 13-4) than is suggested by the average of the vertical plug to horizontal plug ratio (0.1 in this case).

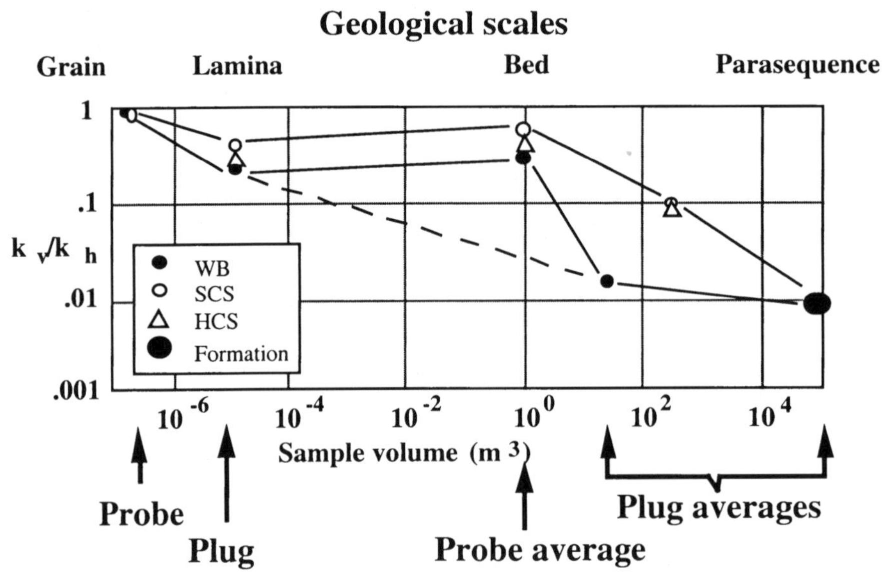

Figure 13-4. Estimates of permeability anisotropy in the Rannoch Formation based on harmonic and arithmetic averaging of probe and plug data (Corbett, 1993). WB, SCS and HCS refer to different rock types; wavy-bedded, hummocky cross-stratified and swaley cross-stratified lithofacies. The wavy bedded rock type may be more anisotropic at the bed scale than the other two. The geological scales refer to the stratal elements discussed in Chap. 1.

A simulation model at the large, interwell, scale was able to match the well performance once the variability at the small scale was accounted for (Corbett et al., 1992). This was achieved by imposing a hierarchical series of deterministic models of the geological structure (at each of the various scales shown in Fig. 13-2) to determine the effective property for use in successively larger-scale models. Although the approach used could (and probably should) be extended by generating several realizations at each stage (to assess the sensitivity of the result), the resulting match to flow was considered acceptable.

13-1.2 Sherwood Sandstone, Irish Sea Basin

Vertical permeability can be directly measured in the subsurface by a wireline tester (WT) device. This device can be used to test use of the harmonic and arithmetic average of small-scale measurements as estimators of vertical and horizontal permeability, respectively. For the Sherwood Sandstone example, discussed above and in Chap. 6, we will see how the vertical permeability measurement compares to the estimates from upscaled probe data over increasing vertical intervals.

Figure 13-5 is a schematic of the geological core description shown alongside the permeability measurements. Channel abandonments are thin shaly intervals clearly seen on the graphic core log. The clean sands and clay drapes represent the high- and low-permeability intervals, respectively, within the stacked fluvial interval. The probe permeability data detect variation not seen by the core plugs (insufficient sampling by the latter). The WT values of vertical and horizontal permeability are shown over the interval for which the measurement can be applied. The low-permeability abandonment interval at 4242.5 ft is clearly influencing the WT vertical permeability estimate. The core plugs did not detect the abandonment channel interval because they are too widely spaced. The vertical-to-horizontal permeability ratio estimated from the WT of about 0.01 agrees well with the largest scale ratios in Fig. 13-4.

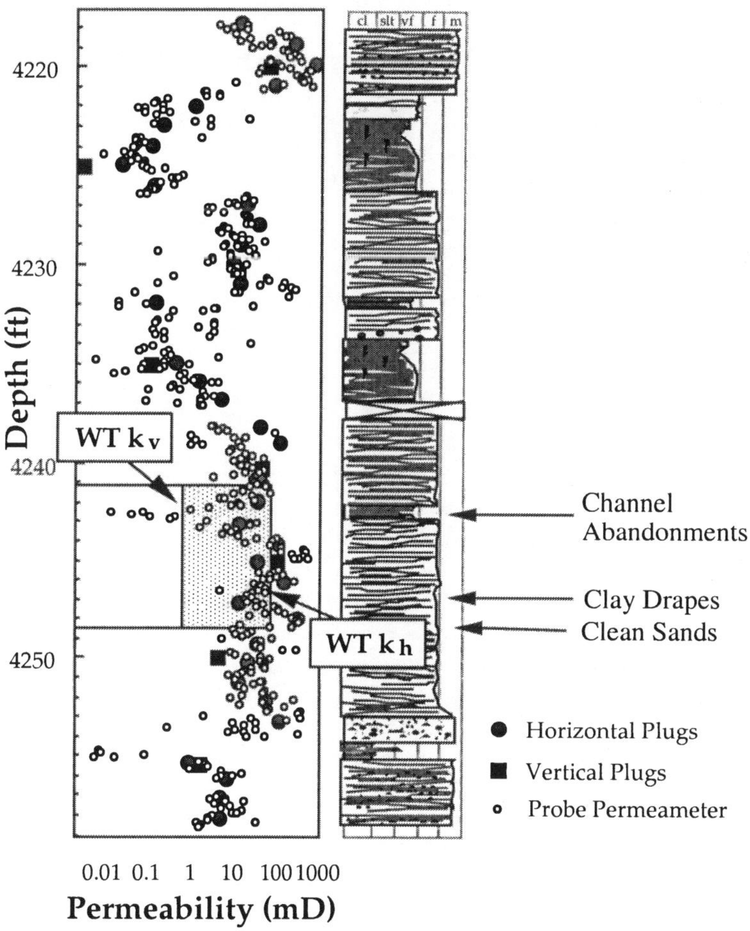

Figure 13-5. Interval of Sherwood Sandstone from the Irish Sea Basin (Thomas et al., 1997). Where vertical plugs are taken, the core-plug k_v/k_h ratio ranges from 0.1 to 3.

The disagreement of core-plug and WT k_v/k_h ratios might have been anticipated by reviewing the C_V over the interval. The plug C_V is 1.2 and therefore the 16 plugs are not very representative. The probe C_V for the same interval is 1.5, slightly higher than the plug C_V, since higher and lower permeability intervals have now been detected. The 160 probe measurements in the interval, while still not sufficient (225 suggested from I_0 criterion), are an improvement. The low permeability intervals can be barriers to vertical flow and the occurrence of these can be detected by an unordered Lorenz Plot (see Fig. 6-9)--but only if the sampling is sufficiently dense. In this study, a porosity measurement at the scale of the probe permeability measurements was not available. With such measurements, a probe scale, unordered Lorenz Plot would show more thief and barrier zones.

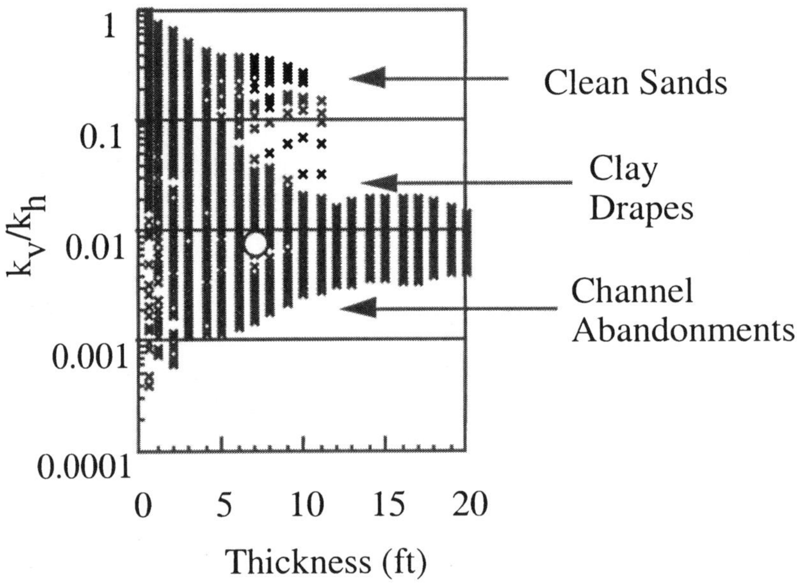

Figure 13-6. Estimates of permeability anisotropy in the Sherwood Sandstone. Thickness in this plot is the scale over which the respective measurements have been averaged. Large circle is the WT estimate; crosses are the probe permeameter estimates. Refer to Thomas et al. (1997) for more details.

In Fig. 13-6, the effects of clay drapes, channel abandonments, and clean sand properties and scale are highlighted. At a small scale, each element is relatively isotropic. With increasing scale of interval investigated, the presence of clay drapes and abandonments cause a significant anisotropy within the formation. The dynamic WT measurement in this example is used as a calibration of the statistical estimation model, which, in this case, leads to an acceptable prediction. There is a dramatic drop in vertical permeability (and consequent decrease in the k_v/k_h ratio) at an interval of approximately 3 ft. This spacing coincides with the average thickness of the fluvial sandstone beds and is reflected in the semivariogram (Fig 13-7 left). At 10 ft, the average bedset thickness, the k_v/k_h ratio has stabilized. This distance is also evident in the semivariograms (Fig. 13-7).

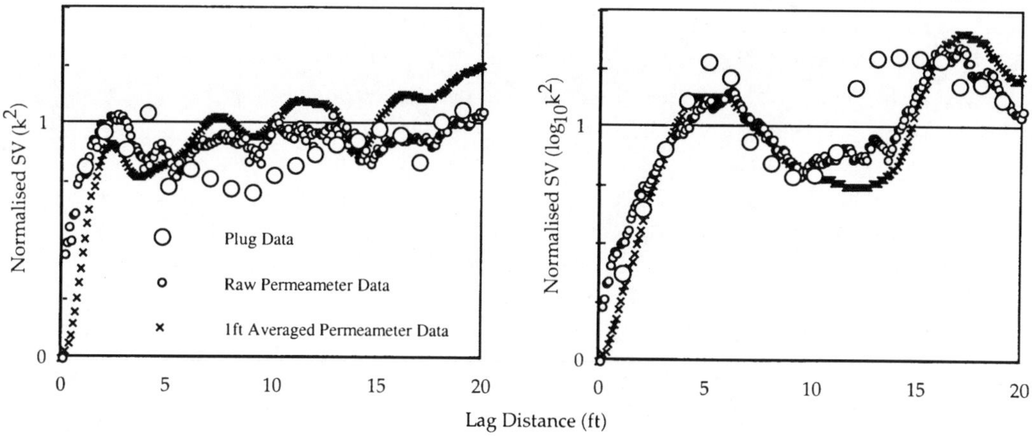

Figure 13-7. Sherwood Sandstone probe and core plug semivariograms (Thomas et al., 1997).

13-1.3 San Andres Formation Shelf Carbonates Permian Basin, Texas

This example is of a carbonate reservoir on which a statistical study has been carried out (Lucia and Fogg, 1990; Fogg et al., 1991). Flow units extend between wells but no internal structure can be easily recognized or correlated. The San Andres formation carbonates are very heterogeneous with a C_V of 2.0 to 3.5 (Kittridge et al., 1990; Eisenberg et al., 1994; Grant et al., 1994). A vertical profile of horizontal permeability (Fig. 13-8) from a wireline-log-derived permeability estimator shows two scales of bedding. With very large heterogeneity within the reservoir, it is important to consider how that heterogeneity is distributed/organized and to exploit statistical tools and geological inference to build appropriate models.

A sample semivariogram along a vertical profile (Fig. 13-9) shows a spherical behavior out to a range of about 8 ft. This range appears to be between the scales of the two bedding types (A and B) shown in Fig. 13-8. The spherical model fit to the sample semivariogram overlooks the hole (at around 24 ft) that may be reflecting Type B scale cyclicity. The structure of Type A beds may not be reflected in the sample semivariogram because of the effect of the larger "dominating" Type B cycles.

Horizontal semivariograms were derived from permeability predictions generated from initial well production data (Fig. 13-10). Because of the sparse data, these semivariograms are indistinguishable. But there is a mapped high-permeability grainstone trend associated with reef development that is consistent with an anisotropic horizontal autocorrelation structure. The geology is critical to interpreting these semivariograms (refer to Fogg et al., 1991, for further discussion). For further discussion of semivariogram anisotropy (also termed geoanistotropy) refer to Kitanidis (1997).

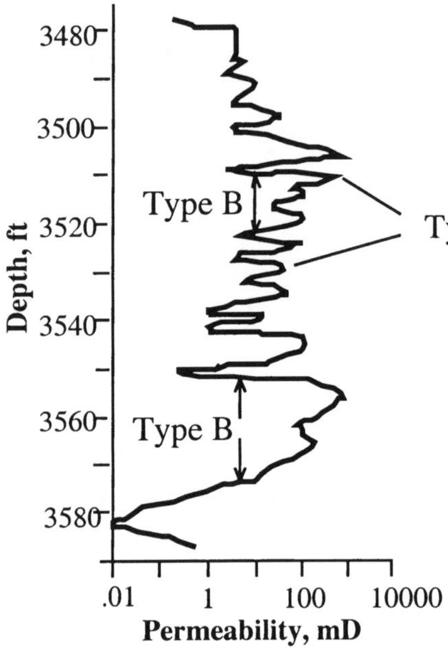

Figure 13-8. Horizontal permeability profile for the San Andres carbonate generated from a porosity-based predictor. There are two types (A and B) of heterogeneity at the bed scale. From Fogg et al. (1991).

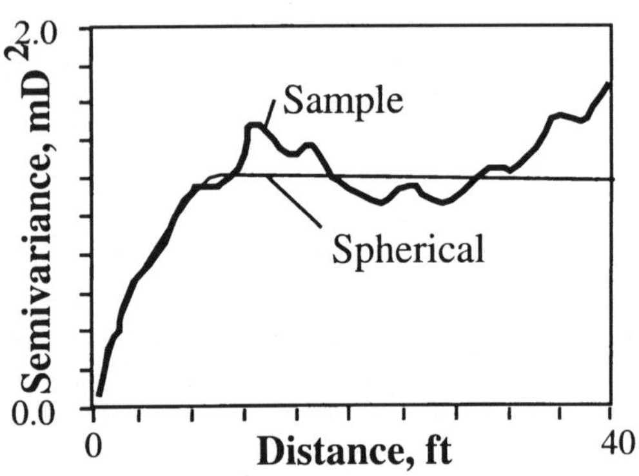

Figure 13-9. Spherical model fitted to a vertical sample semivariogram. The spherical model ignores the hole at 24 ft that indicates vertical cyclicity (i.e., periodicity). From Fogg et al. (1991).

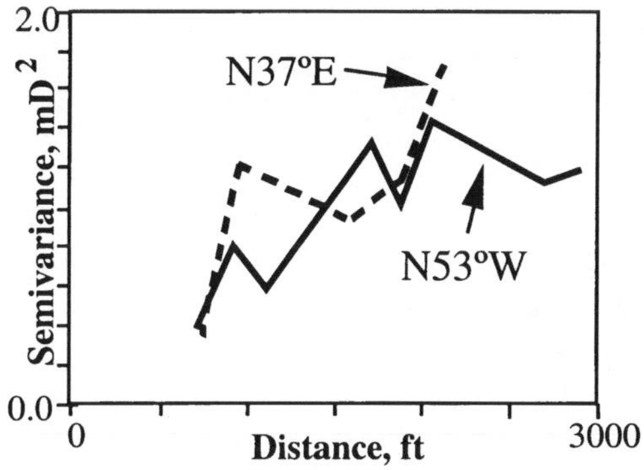

Figure 13-10. Sample semivariograms in two horizontal directions for the San Andres carbonate. Semivariograms with different ranges can express the anisotropy suggested by the geological model. From Fogg et al. (1991).

Conditional simulations (conditioned on the well data) of the permeability field in the San Andres formation were generated to investigate the influence of infill drilling (Fig. 13-11). In this case, the permeability field was modeled as an autocorrelated random field using single semivariograms in the vertical and horizontal directions. Both realizations show degrees of heterogeneity and continuity. However, there are still deficiencies in the characterization that might prove important.

1. The profile in Fig. 13-8 suggests a reservoir dominated by two scales of beds. In the realizations in Fig. 13-11, it is difficult to see that this structure has been reproduced.

2. The autocorrelated random field model does not explicitly represent the baffles caused by the bed bounding surfaces; these may have a different flow response from that shown by the models in Fig. 13-11. The surfaces have been incorporated in subsequent models (Grant et al., 1994; Jennings et al., 1988) which show important effects.

The panels in Fig. 13-11 also illustrate a generic problem with pixel-based stochastic modelling: there is no good *a priori* way to impose reservoir architecture. Stratigraphic grids, composed of three-dimensional cells with regular x-y coordinates but variable z-coordinates, are commonly used to capture vertical variations in reservoir-layer thickness (Rachmawati et al., 1997; Chambers et al., 1999). Conditional simulations are generated within these pixel-based but geologically aligned coordinate systems. The stratigraphic grid has two major advantages over purely pixel-based models: (1) the cells are vertically aligned along main depositional surfaces, thereby aligning conditional simulations with petrophysical trends and, (2) scaled-up, flow simulation will be aligned with geologically-constrained flow streamlines. If the geometry of internal bedding units or channels were determined to be important, other modelling techniques (e.g., object-based) could provide additional constraint within this stratigraphic framework.

The San Andres study illustrates how semivariograms can be used to generate autocorrelated random fields that are equiprobable and conditioned on well data. These realizations can form the basis of further engineering studies; however, it is unlikely that all 200 realizations will be used in flow simulation. Usually only the extreme cases or representative cases, often selected by a geologist's visual inspection, will be used. The floods of each of the realizations shown in Fig. 13-11 give horizontally isotropic behavior: flooding left-to-right and right-to-left produced the same flow result.

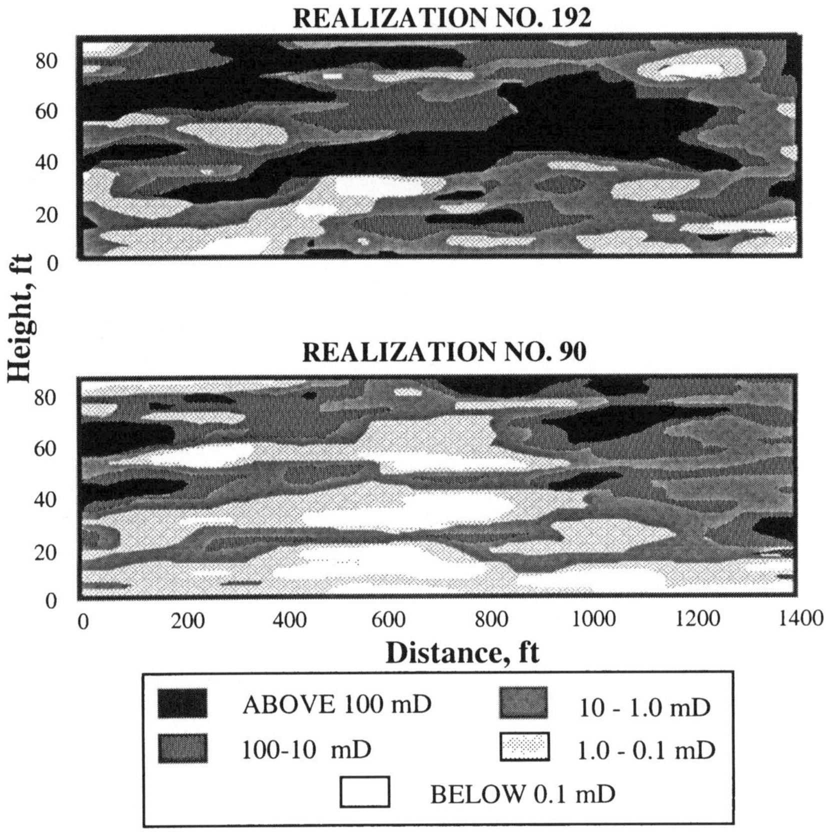

Figure 13-11. Two realizations of the spatial distribution of the permeability field for the San Andres generated by theoretical semivariograms fit to the sample semivariograms shown in Figs. 13-9 and 13-10. The permeability fields are the same in the two realizations for the extreme left and right ends of the model where the conditioning wells are located. From Fogg et al. (1991).

In extensions to this study (Kerans et al., 1994; Goggin et al., 1993) showed the effects of incorporating the layer (sequence) boundaries into the model. These stratigraphic markers have the effect of compartmentalizing the reservoir vertically and produce significantly poorer oil recovery performance (Fig. 13-12). The flood was also now anisotropic, left-to-right flooding being different from right-to-left flooding (Kerans et al. 1994).

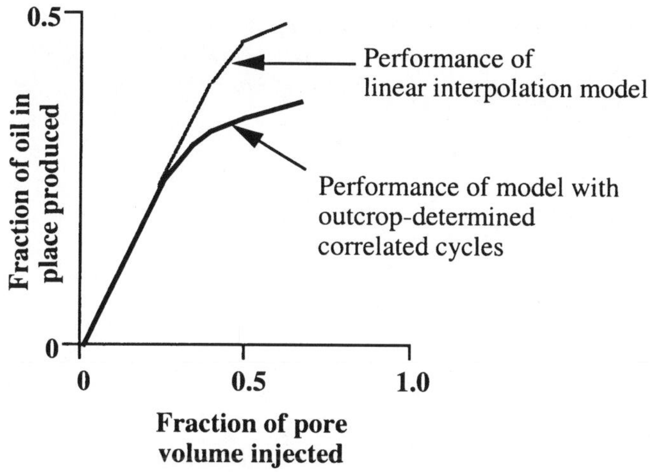

Figure 13-12. Flow performance for a model using simple linear interpolation compared to same for a model incorporating geological layering for the San Andres Carbonate (Kerans et al., 1994).

Further analysis was carried out on the San Andres outcrop data by Jennings et al. (1988), who identified long-range periodic structure in the semivariograms with a wavelength of 140-180 ft (42-54 m). The geological reasons for this periodicity are unknown, but the periodic structure did influence the flow response in a 15-layer, 3-D model with a quarter 5-spot well arrangement. The model was isotropic in the areal plane as the models were based on 1-D permeability transects and/or 2-D vertical permeability grids. The study concluded that the relatively low interlayer variability was more important in the flow modeling than the much greater variability within the layers with very short correlation lengths (less than 4.5 ft or 1.5 m). The small-scale structure was essentially washed out by the long-range flow characteristics. The lack of causal definition on the larger scale periodicity means that the geology was not able to suggest areal anisotropy, which might be expected in such a system. The importance of the layered geological structure, detected in the semivariograms (Fig. 13-8), was illustrated by this study.

The importance of establishing the effect of reservoir layers is widely recognized, either by modeling discrete layers with separate properties (e.g., Goggin et al., 1993; Grant et al., 1994; Sweet et al., 1996; Rachmawati et al., 1997) and/or by incorporating the appropriate upscaled properties (e.g., Corbett and Jensen, 1993b; Sweet et al., 1996). The layers and the properties may be deterministic or stochastic; commonly a mixture is used.

13-2 USING STATISTICS WITHOUT GEOLOGY

Questions concerning the role of geology often arise during modelling studies. Does the geology matter and, if so, by how much? While the interaction of heterogeneity and flow processes is still being researched, there are increasing indications that geological information may be quite important for accurate flow behavior models. The Page Sandstone is a case of particular interest, as it addresses this question.

A succession of works has detailed the geology (Chandler et al., 1989), the permeability distribution (Goggin et al., 1988; Goggin et al., 1992) and fluid flow (Kasap and Lake, 1990) through the Page Sandstone, a Jurassic eolian sand that outcrops in the extreme of northern Arizona near Glen Canyon Dam.

The geological section selected for the flow simulation was the northeast wall of the Page Knob (Goggin, 1988). The data were derived from probe permeameter measurements taken on various grids and transects surveyed on the outcrop surfaces, together with profiles along a core. The individual stratification types comprising the dunes, i.e., grainfall and windripple, have low variabilities (C_V's of 0.21). The interdune element has a C_V of 0.81. The interdune material is less well-sorted than the dune elements. Grainfall and windripple elements are normally distributed, the interdune log-normally. The individual stratigraphic elements in this eolian system are well-defined by their means, C_V's, and PDF's.

The vertical outcrop transects are more variable ($C_V = 0.91$) than the horizontal transects ($C_V = 0.55$), an anisotropy that seems to be typical for most bedded sedimentary rocks. The global level of heterogeneity for the Page Knob is probably best represented by the transect along the core, which has a $C_V = 0.6$. Semivariograms were calculated for the grids and core profiles. The grids allowed spherical semivariogram ranges to be determined for various orientations. These ranges indicate the dip of the crossbeds; the ranges are 17 m along the bed and 5 m across the bed (Goggin, 1988). Hole structures are present in most of the semivariograms, indicating significant permeability cyclicity that corresponds to dune cross-bed set thickness.

For our purposes, the most important aspect of this work is the modelling of a matched-density, adverse-mobility-ratio miscible displacement through a two-dimensional cross section of the Page Sandstone. Figure 13-13 shows a typical fluid distribution just after breakthrough of the solvent to the producing well on the right. The dark band is a mixing zone of solvent concentrations between 30% and 70%. The region to the left of this band contains solvent in concentrations greater than 70%, to the right in concentrations less than 30%. The impact of the variability ($C_V = 0.6$) and continuity can be seen in the character of the flood front in the large fluid channel occurring within the upper portion of the panel. There is a smaller second channel that forms midway in the section. As we shall see, both features are important to the efficiency of the displacement.

The simulation in Fig. 13-13 represents one of the most geologically realistic, deterministic, numerical flow simulations, because it attempts to account for every geologic detail established through prior work. Specifically,

1. permeabilities were assigned according to the known stratification types at every grid block;

2. the permeability PDF for each stratification type is well-known;

3. a random component was assigned to the permeability of each grid block to account for variance of the local PDF's;

4. cross-bedding was accounted for through the assignment of a full tensorial permeability in about one-third of the grid blocks;

5. the model accounted for the specific geometry of the features through the use of finite elements; and

6. each bounding surface feature (low permeability) was explicitly accounted for with at least one cell in each surface.

In all, over 12,000 finite-element cells were required to account for all of the geologic detail in this relatively small cross section. Indeed, one of the purposes of the simulation was to assess the importance of this detail through successively degraded computations. See Kasap (1990) for more details.

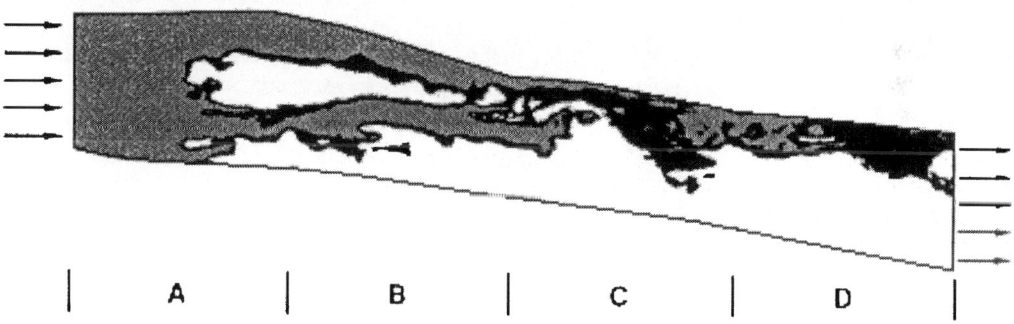

Figure 13-13. Distribution of solvent during a miscible displacement in the detailed simulation of the Page Sandstone outcrop. From Kasap (1990). The flow is from left to right.

Another purpose of the deterministic simulation was to provide a standard against which to measure the success of conditional simulation (CS), discussed in Chap. 12. Figure 13-14a shows two CS-produced realizations of solvent flowing through the Page cross section. Compared to Fig. 13-13, the permeability distribution was conditioned to a significantly degraded data set that used only data at the two vertical wells (imagined to be on the edges of the panel) and information about a horizontal semivariogram. The field on which the simulation of Fig. 13-14a was constructed used a spherical semivariogram. Figure 13-14b shows the same field constructed with a fractal semivariogram. Note that both flow simulations show a dependency on the long-range structure as noted by Jennings et al. (1988).

We compare both figures to the distribution in Fig. 13-13. Qualitatively, the fractal representation (Fig. 13-14b) seems to better represent the actual distribution of solvent; it captures the predominant channel and seems to have the correct degree of mixing. The spherical representation (Fig. 13-14a) shows far less channeling and too much mixing (that is, too much of the panel contains solvent of intermediate concentrations). However, a quantitative comparison of the two cases shows that this impression is in error (Fig. 13-15). The distribution that gave the best qualitative agreement (Fig. 13-14b) gives the poorest quantitative agreement. Such paradoxes should cause us concern when performing visual comparisons; however, there is clearly something incorrect about these comparisons. The key to resolving the discrepancy lies in reviewing the geology.

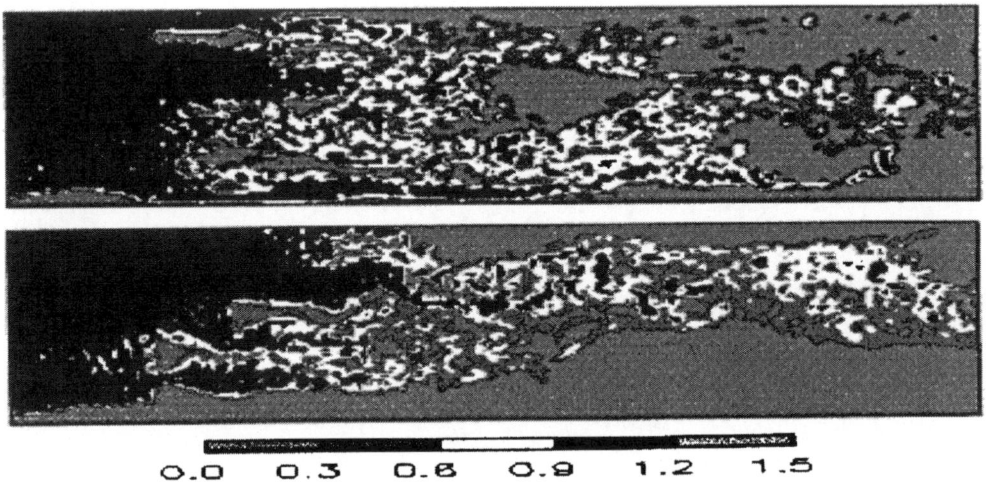

Figure 13-14a. Simulated solvent distribution through cross section using a spherical semivariogram. From Lake and Malik (1993). The mobility ratio is 10. The scale refers to the fractional (at the injected) solvent concentration. Flow is from left to right.

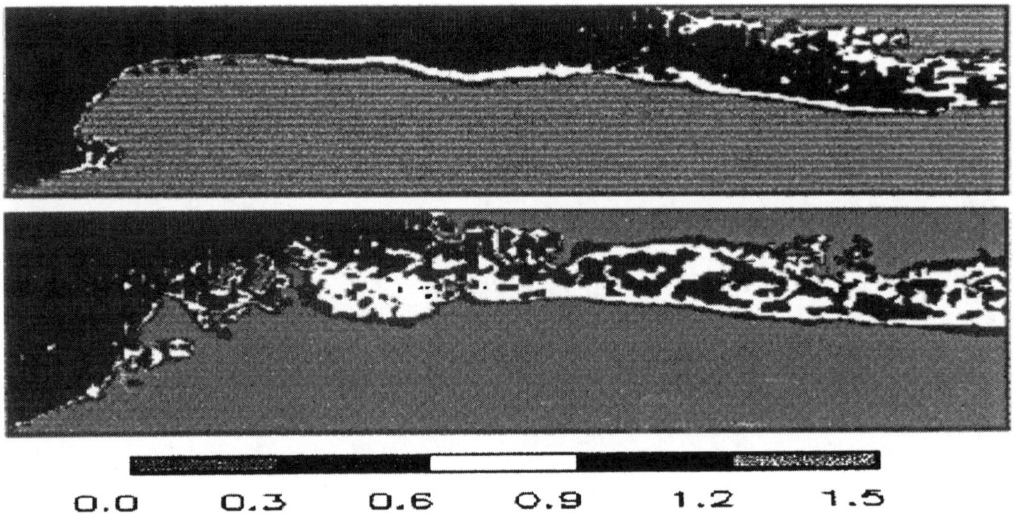

Figure 13-14b. Solvent distribution through cross section using a fractal semivariogram. From Lake and Malik (1993). The mobility ratio is 10. The scale refers to the fractional (at the injected) solvent concentration. Flow is from left to right.

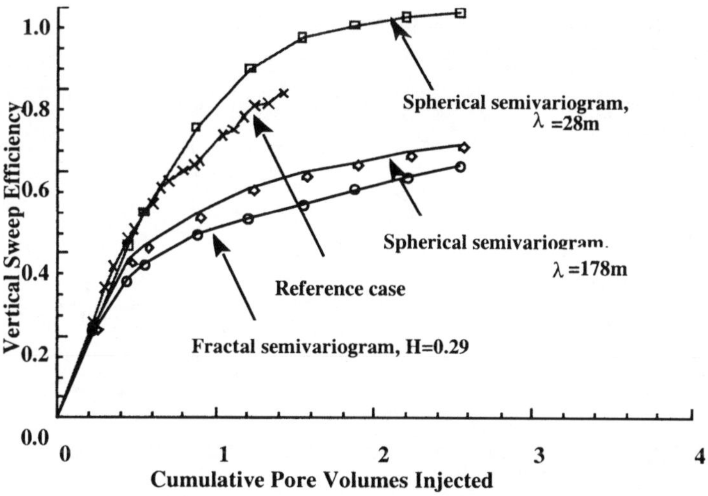

Figure 13-15. Sweep efficiency of the base and CS cases. From Lake and Malik (1993). The deterministic reference case was based entirely on the known geology.

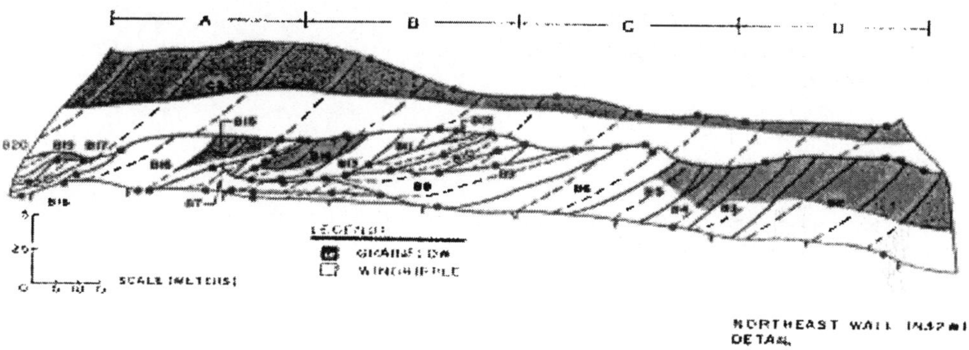

Figure 13-16. Spatial distribution of stratification types on the northeast wall of the Page Sandstone panel. From Kasap and Lake (1993). The thin bounding surfaces are black lines, the high-permeability grain-flow deposits are shaded, and the intermediate-permeability wind-ripples are light regions.

Figure 13-16 shows the actual distribution of stratification types at the Page Sandstone panel based on the northeast wall of the Page Knob. The thin bounding surfaces are readily apparent (black lines), as are the high-permeability grainflow deposits (shaded) and the intermediate-permeability windripples (light). This is the panel for which the simulation in Fig. 13-13 was performed. Even though the entire cross section was from the same eolian environment, the cross section consists of two sands with differing amounts of lateral continuity: a continuous upper sand and a discontinuous lower sand. Both sands require separate statistical treatment because they are so disparate that it is unlikely that the behavior of both could be mimicked with the same population statistics. (Such behavior might be possible with many realizations generated, but it is unlikely that the mean of this ensemble would reproduce the deterministic performance.)

When we divide the sand into a continuous upper portion, in which we use the fractal semivariogram, and a discontinuous lower portion, in which we use the spherical semivariogram, the results improve (Fig. 13-17). Now both the predominant and the secondary flow channels are reproduced. More importantly, the results agree quantitatively as well as qualitatively (Fig. 13-18).

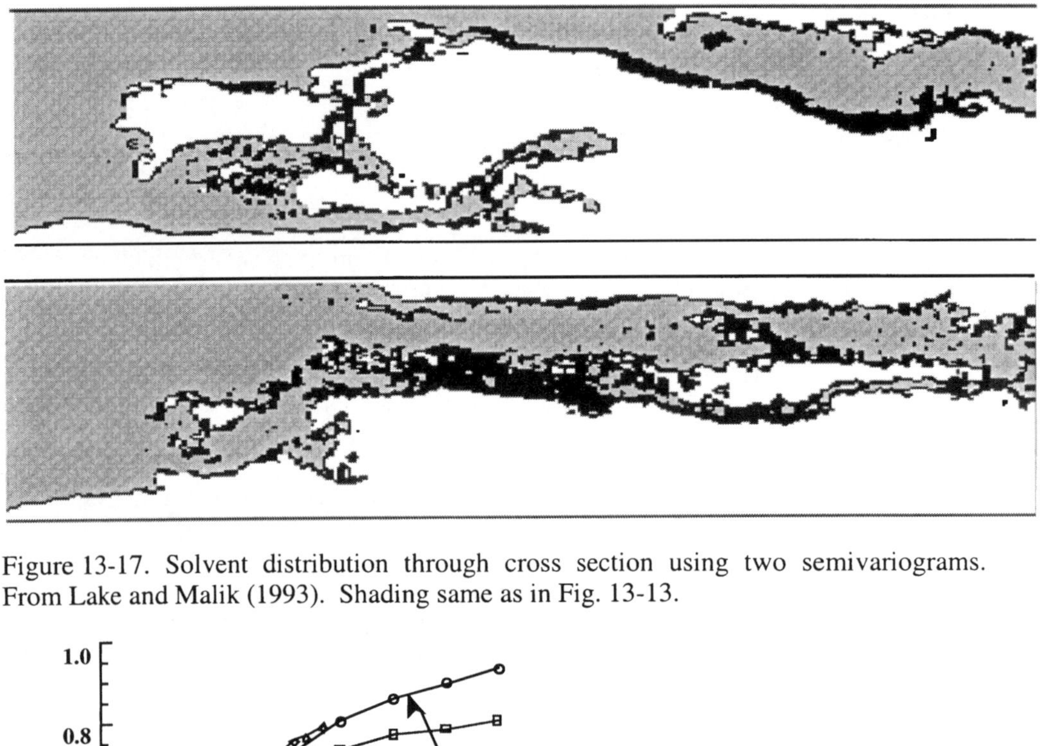

Figure 13-17. Solvent distribution through cross section using two semivariograms. From Lake and Malik (1993). Shading same as in Fig. 13-13.

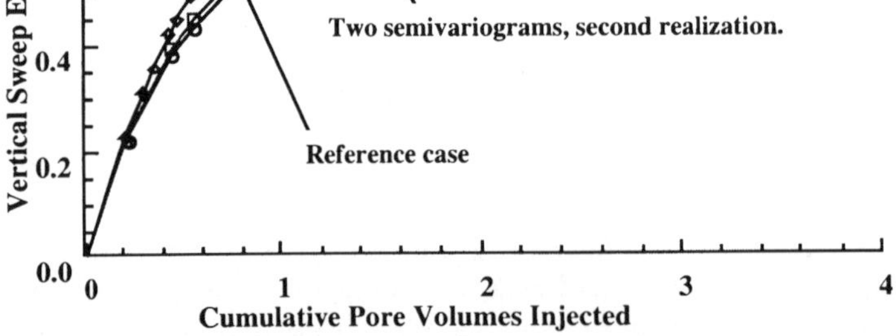

Figure 13-18. Sweep efficiency of the base and dual-population CS cases. From Lake and Malik (1993). The deterministic reference case was based entirely on the known geology.

The existence of these two populations is unlikely to be detected from limited well data with statistics only; hence, we conclude that the best prediction still requires a measure of geological information. The ability to include geology was possible because of the extreme flexibility of CS. The technological success of CS did not diminish the importance of geology; rather, each one showed the need for the other.

13-3 HYBRID MODELLING

If the reservoir lithology can be discriminated into two populations, for example by a bimodal PDF of a particular wireline-log measurement, the autocorrelation can be determined using the resulting discrete lithological indicator.

Rossini et al. (1994) have taken this approach with a sandy dolomitic reservoir where a wireline-log value was used to distinguish sandy facies ("good" reservoir) from dolomitic facies ("poor" reservoir). Sequential indicator simulation (SIS) using facies-indicator semivariograms is used to generate the model facies fields. Porosities within each facies are then assigned as autocorrelated random fields from the appropriate facies-specific semivariograms using sequential Gaussian simulation (SGS). The permeability is determined from the porosity data using a regression relationship and a Monte Carlo procedure. The resulting fields capture both the variability and architecture of the geological structure (i.e., facies architecture)). However, indicator semivariograms are needed, and these often must be imposed from an understanding of the geological structure in the absence of real data.

In this study by Rossini et al. (1994), the quality of the statistical models was assessed by a grading system based on the maximum number of water-cut matches to production data. The high-graded realizations were then used for full-field simulations to develop a management strategy. This interesting case study into model building combines many of the statistical treatments covered in this text—PDF's, transformations, regression, spatial correlation, and statistical modelling—and provides a useful introduction for all students of reservoir geo-modelling.

Fluvial reservoirs, in which the flow units can comprise reservoir units (channel sandstones, crevasse splay sandstones) and nonreservoir units (lagoonal muds, coals), are often modeled with object-based techniques (using the known or inferred geometry of the sands and/or shales). The use of object-based models for the simulation of low net-to-gross reservoirs (e.g., Ness Formation, Upper Brent, North Sea) have been discussed in Williams et al. (1993). These systems have also been modeled using a marked-point process (Tyler et al., 1994), where the "marked points" represent the principal axis of a channel belt to which the modeled channel attributes are attached. In both techniques, discrete architectural elements are distributed in a background matrix.

The realism of the models depends on how well the input parameters (e.g., total reservoir net-to-gross, channel belt azimuth, and channel belt width-to-thickness ratios) can be defined. Net-to-gross can be obtained from well information; width-thickness data rely on appropriate outcrop analog data. Other data sources include "hard" data (e.g., well logs and well test) and "soft" data (e.g., regional channel orientations or maps of seismic attributes) to further condition the models.

The advantage of the statistical techniques is that numerous "equiprobable" realizations can be generated. These realizations are data-hungry and the need for screening prior to use in an engineering application can limit the usefulness of *all* the realizations. Often the connectivity alone is of major concern and this can be determined relatively simply (Williams et al., 1993). The balance between models and realism has to be carefully weighed so that the uncertainty in the engineering decision-making process is adequately quantified within the time frame and budget available.

13-4 MODEL-BUILDING SUMMARY

The few case studies summarized above illustrate several aspects of reservoir modelling using statistics. We can generalize these results, in order to sketch out a method for the selection of an appropriate approach. Remember, the goal is to generate a three-dimensional array of petrophysical properties (facies, porosity, and ultimately permeability) that can be used in numerical simulation.

Geology provides several insights that are useful for model building: categorization of petrophysical dependencies, identification of large-scale trends, identification of reservoir zones or layers, interpretation of statistical measures, and quality control on generated models. The first three serve to bring the statistical analysis closer to satisfying the underlying assumptions. For example, identification of categories, trends and layering and their subsequent removal, will bring data sets closer to being Gaussian and/or to being stationary. The last two are to detect incorrect inferences arising from limited and/or biased sampling. Consequently, we shall see aspects of these in the following procedures.

The first two steps are common for all procedures.

1. Divide the reservoir into structural and/or stratigraphic flow units. Flow units are packages of reservoir rocks with *similar* petrophysical characteristics—not necessarily uniform or of a single rock type—that appear in all or most of the wells. This classification will serve to develop a stratigraphic framework ("stratigraphic grid") for the reservoir model at the interwell (or field) scale. Flow unit boundaries will naturally follow stratigraphic boundaries in this framework.

2. Review the sample petrophysical distributions with respect to geological and statistical stationarity.

What follows next depends on the properties within the flow units, the process to be modeled, and the amount of data available. We presume some degree of heterogeneity within the flow unit. If the fields are very uniform, then analytical engineering solutions might be appropriate.

3. If there are no data present apart from well data and there is no geologic interpretation, the best procedure is to simply use a conditional simulation technique applied directly to the well data (Fig. 13-15). The advantage of this approach is that it requires minimal data: semivariogram models for the three orthogonal directions. (Remember, a grid

block representing a single value of a parameter adds autocorrelation to the field to be simulated.)

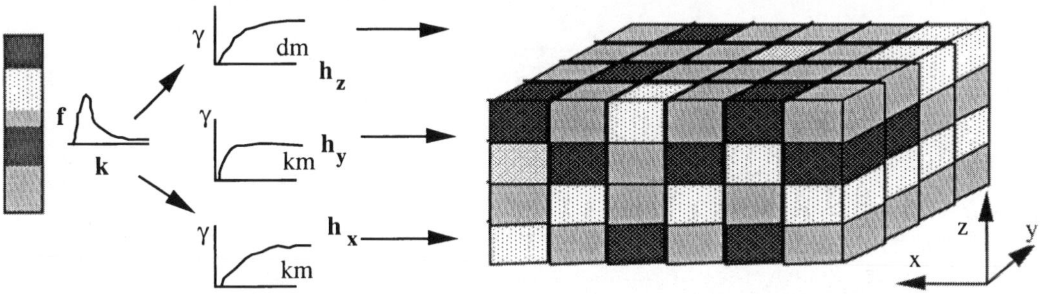

Figure 13-15. Generation of a three-dimensional autocorrelated random field using orthogonal semivariograms. h_x, h_y and h_z are the lags in the coordinate directions. dm is decameters, km kilometers. The column to the left indicates a geological description from core or wireline logs. The petrophysical properties are determined by a PDF and by using semivariograms in three directions and a three-dimensional grid block model (right) can be built.

While parameters for vertical semivariograms are usually available from the well data, the lateral parameters must be estimated from the normally sparse well data and/or reservoir analogs. It is important that uncertainties in these parameters be acknowledged in the modelling schemes. This approach is the most stochastic of those considered here and will lead to the most uncertainty in the predictions.

4. If, from geological analysis, a representative geological structure can be recognized within the flow unit and the flow unit itself is laterally extensive, a deterministic model can be built. The scales of heterogeneity can be handled by geopseudo upscaling (Fig. 13-16). The role of statistics in this case is to assess how representative is the structure. Structure can be identified by nuggets, holes, and/or trends in the semivariograms. Lateral dimensions rely on the quality of the geological interpretation in this model. The Rannoch example followed these steps.

5. If the property distribution is multi-modal, the population in the flow unit can be split into components and indicator conditional simulation can be used to generate the fields. This is useful in fields where the variation between the elements is not clearly defined and distinct objects cannot be measured. This approach (Fig. 13-17) yields fields with jigsaw patterns. This was the approach taken by Rossini et al. (1994), Jordan and Goggin (1995), and Sweet et al. (1996).

6. If the flow unit contains distinctly separate rock types and a PDF can be determined for the dimensions of the objects, the latter can be distributed in a matrix until some conditioning parameter is satisfied (e.g., the ratio of sand to total thickness). This type of object-based modelling lends itself to the representation of labyrinth fluvial reservoirs (Fig. 13-18). More sophisticated rules to follow deterministic stratigraphic trends (e.g., stacking patterns) and interaction between objects (e.g., erosion or

aversion) are available or being developed. A similar model for stochastic shales would place objects representing the shales in a sandstone matrix following the same method.

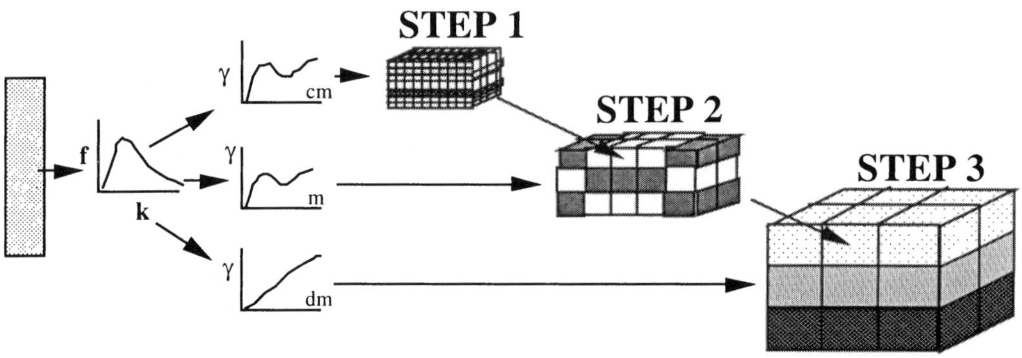

Figure 13-16. Schematic reservoir model at three hierarchical scales capturing well-determined geological scales at each step. The column to the left indicates a geological description from core or wireline logs. A small-scale (cm to m) semivariogram guides selection of representative structure.

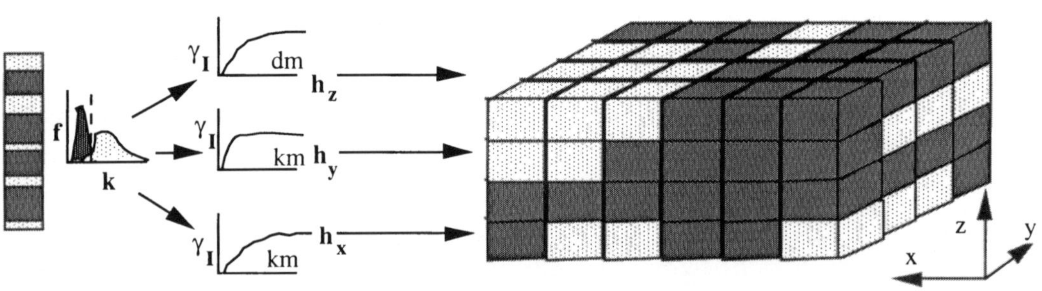

Figure 13-17. Schematic of a sequential indicator simulation for a binary system. Indicator semivariograms are determined from a binary coding of the formation. h_x, h_y and h_z are the lags in the coordinate directions. dm is decameters, km kilometers. The column to the left indicates a geological description from core or wireline logs.

These models are the most realistic of all for handling low net-to-gross fluvial reservoirs; however, they require a good association between petrophysical properties and the CDF's for the geometries. Tyler et al. (1994) give a good example of this application.

If the reservoir contains more than one flow unit, then the procedure must be repeated for the next zone. In the Page example (Section 13-3), the upper layer needed a different correlation model (generated by a fractal semivariogram) from the lower layer (spherical semivariogram).

The modelling of petroleum reservoirs requires an understanding of the heterogeneity, autocorrelation structure (nested or hierarchical) derived from the geological architecture, and the flow process to select the most appropriate technique. As reservoirs have various correlation structures at various scales, flexibility in modelling approach must be maintained. The "right" combination of techniques for a Jurassic reservoir in Indonesia may be entirely inappropriate for a Jurassic North Sea reservoir. The methods discussed here try to emphasize the need for a flexible toolbox.

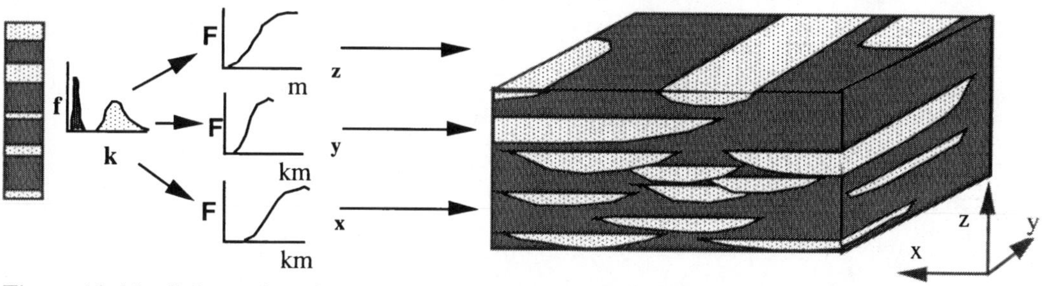

Figure 13-18. Schematic of an object model of a fluvial reservoir. Dimensions of objects are sampled and their locations set by sampling CDF's. The column to the left indicates a geological description from core or wireline logs.

13-5 CONCLUDING REMARKS

We have seen the basics of statistics, how geology and statistics (and other technologies) might be used to complement each other in analysis, and how their combinations produce the best predictions. As we stated in the introduction to this chapter, what constitutes the "best" model depends on the data available, the reservoir type, the processes envisioned, and of course budget and time available. Despite ongoing progress, one thing seems clear: statistical methods (to be sure, along with improved computing power) have vastly improved our ability to predict reservoir performance. We hope that the technologists from the various disciplines will be helped by this volume as the science of generating realistic "numerical" rocks will remain a common ground for geoscientists and statistical research in years to come. The last point to note is that "all models are wrong"; therefore, the aim of the technical experts is to make a judgment about how close the model might be to reality. Capturing reality within the range of predicted outcomes is an honorable technical alternative to promoting a single outcome.

NOMENCLATURE

[=] means "has units of"; L is length, t time, m mass and F force

A, B, C	events
a	arbitrary constant; constant in Archie's law
a,b,c	parameters of a uniform or triangular distribution
a_i	constant weight applied to element i in a summation
b	bias of an estimator [=] parameter units
c	arbitrary constant
df	degrees of freedom
Cov(X,Y)	covariance of two random variables X and Y [=] units of X times units of Y
C_V	coefficient of variation of a random variable
D_p	particle diameter [=] L
D_{pi}	grain diameter of fragment i [=] L
d	total sampling domain length [=] L
d_{max}	maximum vertical separation between two sample CDF's
d_o	sample spacing required for sampling density n_o [=] L
E()	expectation operator
E(X)	expected value of a random variable X [=] units of X
erf(x)	error function of x
exp ()	exponentiation function
F	F-statistic or ratio of variances
F(x;σ,μ)	theoretical normal or Guassian cumulative distribution function of random variable X with mean μ and standard deviation σ
f(x)	probability distribution function of a continuous random variable X [=] (units of X)$^{-1}$

$f(x,y)$	joint probability distribution function for two continuous random varibles X and Y [=] (units of X or Y)$^{-1}$
$f_{Y\|X}$	condition probability distribution function of random variable Y given a fixed value of the random variable $X = x_o$
$f(x;\sigma,\mu)$	theoretical normal of Gaussian probability distribution function with mean μ and standard deviation σ [=] (units of X)$^{-1}$
$F_X(x)$ or $F(x)$	cumulative distribution function (CDF) of a random variable X for a bounding value x
$F(z)$	standarized normal or Gaussian cumulative distribution function with zero mean and unit standard deviation
F, F_a	untruncated and (apparently) truncated or original CDF, respectively
f_A, f_B	fraction of data points taken from populations A and B, respectively
f_b, f_t	fraction of bottom- or top-truncated data points, respectively
f_i	proportion of smaller grain fragment i derived from larger grain, fraction or percent
h_i	thickness of sublayer i in a reservoir zone [=] L
H	Hurst or intermittency coefficient
H_A	alternative hypothesis
H_K	Koval heterogeneity factor
H_o	null hypothesis
h	separation or lag distance [=] L
I	terminal value of index i
${}_I C_s$	total number of ways (combinations) of getting at level s successes in I trials for a series of binomial experiments
I_H	Gelhar-Axness coefficient or heterogeneity index
J	terminal vaueof index j
K	terminal value of index k
k	permeability scalar [=] L^2
k	permeability tensor [=] L^2
k_x, k_y, k_z	permeability in the x,y and z coordinate directions [=] L^2

k_v, k_h	vertical and horizontal permeabilities [=] L^2
k_{rj}	relative permeability of phase j, fraction or percent
L_c	Lorenz coefficient
ln(x)	natural logarithm of x
log(x)	base 10 logarithm of x
M	total number of samples taken from a continuous distribution
$MS\varepsilon$	mean square error or standard error of regression
$^m/_n$	relative frequencey of the positive occurence of m experiments in n trials
$N(\mu,\sigma^2)$	Gaussian normal probability distribution function with mean μ and variance σ^2
P	probability proportion of the occurence of successful outcomes in a series of binomial experiments; tolerance about the sample mean, percent
p	power coefficient in p-normal distribution
Prob(A| B)	probability of the occurence of event A conditioned tothe prior occurence of event B, fraction bounded by [0,1]
Prob(E)	probability of the occurrence of event E, fraction bounded by [0,1]
R^2	coefficient of determination
R'	all real numbers on a line [=] units of mapped variable
R_i	resistivity of formation component i [=] ohm-m
r	multiple factor, constant
r(x,y)	correlation coefficient of two variables x and y
s	number of successful outcomes in a series of binomial experiments
S_j	saturation of phase j, fraction or percent
S_{xy}	sum of squares defined as the sum of the product of random variable X deviations from μ_X times random variable Y deviations from μ_Y [=] units of X times units of Y
$S(\beta_o,\beta_1)$	sum of squared residuals in response data about the regression model line [=] squared units of response variable
$SS\varepsilon$	residual sum of squares

$\hat{S}^2_{\beta_o}$	estimated variablity of sample regression parameter β_o (intercept)
$\hat{S}^2_T$	estimated total variability of all data in several data sets [=] squared units of measured property
$\hat{S}^2_{\bar{X}}$	estimated variability of averages from several data sets [=] squared units of measured property
$\hat{S}_V$	estimated standard error of the Dykstra-Parsons coefficient.
$S_{\hat{\theta}}$	sample standard deviation of an estimated parameter $\hat{\theta}$ [=] paramter units
S^2_x	sample variance of a variable X [=] squared units of X
T_X	sum of ranks of all X_i samples used in a nonparametric Mann-Whitney-Wilcoxon rank sum test
t	estimated value for student's t-distribution
$t\left(\frac{\alpha}{2}, df\right)$	t-value of student's distribution given a confidence level α and df degrees of freedom
t_{β_o}	t-value for regression parameter β_o (intercept)
$t(\alpha,df)$ or t_o	critical value from student's t-distribution for level of conficence α and df degrees of freedom
Var (X)	variance of random variable X, or the second centered moment of X [=] X^2
V_{DP}	Dykstra-Parsons coefficient
$W(X_1, ..., X_I)$	estimation function W or method used to estimate one or more parameter(s) $\hat{\theta}$ from a set of data $\{X_1, X_2 ..., X_I\}$
$X(\omega)$	result of random variable X as a mapping of an outcome ω from the sample space Ω onto the real line (R') [=] units of X random variable
X_i	sample value i from random variable X
x_i	predicted property value at i^{th} recursive iteration in May's equation (ref.) [=] units of predicted property
X, Y, Z	stochastic variables
x, y, z	deterministic analogs of X, Y, and Z; coordinate directions

Acronyms

BLUE	best linear unbiased estimator

CDF	cumulative distibution function, see F(x)
CI	confidence interval
CLT	central limit theorem
GR	gamma ray wireline log response [=] API-GR units
HCS	hummocky cross-stratification
IQR	interquartile range taken as the difference ($X_{0.75}$ - $X_{0.25}$)
JND	joint normal distribution of two correlated random variables
MVUE	minimum variance unbiased estimator
OK	ordinary Kriging
OLS	ordinary least squares
PDF	probability density function, see f(x)
RMA	reduced major axis
RV	random variable
SCS	swaly cross-stratification
SD	sample standard deviation of a random variable [=] units of random variable
SE	standard error
SGS	sequential Gaussian simulation
SK	simple Kriging
WLS	weighted least squares

Subscripts

0	null
0.16, 0.25, 0.50, and 0.75	16th , 25th (1st quartile), median, $75t^{h}$ (3rd quartile) percentile of a random variable
A	alternative; arithmetic mean or average
a	apparent
A,B	populations A or B
G	geometric mean or average
H	harmonic mean or average

I	indicator
i, j, k	indices of a variable with terminal values I, J, K
o	base value of a variable
P	power mean
sand	sandstone sample
silt	siltstone sample
T	true mean, or total
t	true or formation value
w	water phase

Superscripts

-1	inverse function; vector, matrix inverse
c	complementary or complement
m	cementation exponent in Archie's law
n	saturation exponent in Archie's law
o	base value of a variable
*	standard normal variable
p	power exponent
W	estimator W
(1),(2),...(m)	predictor or explanatory variables
T	vector or matrix transpose

Overstrike

^	estimate of a variable or function
-	average (usually arithmetic) of a variable

Greek

α	confidence level, fraction or percent
β_o	intercept parameter for linear regression model [=] units of response variable
β_1	slope parameter for linear regression model [=] ratio of response variable units to predictor variable units
∇	gradient operator
ε	random error, frequently assumed to be N(0,1) distributed

Φ	null set, an event E possessing no outcomes or results
ϕ	porosity of a given rock volume [=] fraction or percent
ϕ_p, ϕ_w	core-plug, wire-line porosities [=] fraction or percent
γ	semivariance
γ_1, γ_2	coefficients of skewness and kurtosis
λ_i	weight applied ot ith data value, typically the Kriging estimator or variance formula
λ	exponential correlogram autocorrelation length (theoretical mean); nonlinearity coefficient in May's equation
λ_I	integral scale
λ_R	spherical semivariogram range
μ	population mean (noncentered moment of order 1)[=] random variable
μ_r	noncentered moment of order r [=] units of random variable raised to the rth power
μ'_r	r^{th} centered moment of a random variable [=] units of random variable raised to the rth power
π	Pythagorean constant = 3.141592.
ρ_g	grain density [=] m/L^3
$\rho(X,Y)$	correlation coefficient of two random variables X and Y
σ	population standard deviation of a random variable [=] units of random variable
σ^2	population variance or second centered moment of a random variable [=] squared units of random variable
σ_ε^2	variance of random error in response about the regression model line [=] squared units of response variable
θ	general parameter
Ω	sample space, a collection of all possible outcomes of an experiment $\Im$, [=] elements units
ω	an outcome or element of the sample space Ω [=] units of Ω
χ^2	PDF for chi-squared distribution
$\Im$	an experiment, the operation of establishing conditions that may produce one of several possible outcomes or results [=] outcome units

Symbols

$\in \{a,b\}$	element of a real-valued set bounded by a and b, inclusively
$\exists$	complementary event, the subset of the sample space Ω that does not contain the elements of event E [=] event elemts units
$\cap$	intersection
$\cup$	union
$\vert$	given that or conditioned upon (i.e. $Y\vert X=x_o$ means Y given that $X = x_o$)

REFERENCES

Abramowitz, M. and I. A. Stegun, *Handbook of Mathematical Functions*, New York: Dover Publications Inc. (1965).

Agterberg, F. P., *Geomathematics*, Amsterdam: Elsevier Scientific Publishing Co. (1974).

Alabert, F. G. and G. J. Massonnat, "Heterogeneity in a Complex Turbiditic Reservoir: Stochastic Modeling of Facies and Petrophysical Variability," SPE 20604, presented at the Society of Petroleum Engineers' 65th Annual Technical Conference and Exhibition, New Orleans., La., Sept. 23-26 1990.

Allen, J. R. L., *Principles of Physical Sedimentology*, London: G. Allen and Unwin (1985).

Anderson, T. W., *An Introduction to Multivariate Statistical Analysis*, New York: John Wiley (1984).

Anguy, Y., Ehrlich, R., Prince, C.M., Riggert, V.L., and Bernard, D., "The sample support problem for permeability assessment in sandstone reservoirs", *in Stochastic Modeling and Geostatistics: Principles, Methods and Case Studies*, J.M. Yarus and R.L.Chambers, AAPG Computer applications in geology, Tulsa, Oklahoma, 3 (1994) 37-54

Archer, J. S. and C. G. Wall, *Petroleum Engineering: Principles and Practice,* London: Graham and Trotman (1986).

Archie, G.E., "The Electrical Resistivity Log as an Aid in Determining Some Reservoir Characteristics," *SPE - AIME Transactions*, 146 (1943) 54 - 62.

Arya, A., T. A. Hewett, R. G. Larson, and L. W. Lake, "Dispersion and Reservoir Heterogeneity," *SPE Reservoir Engineering*, 3 (1988) 140-148.

Aziz, K., and A. Settari, *Petroleum Reservoir Simulation*, Amsterdam: Elsevier Scientific Publishing Co. (1979).

Bard, Y., *Nonlinear Parameter Estimation*, New York: Academic Press (1974).

Barnes, R. J. and Kwangho You, "Adding Bounds to Kriging," *Mathematical Geology*, 24 (1992) 171-177.

Barnes, R. J., "Bounding the Required Sample Size for Geologic Site Characterization," *Mathematical Geology*, 20 (1988) 477-490.

Barnett, V. and T. Lewis, *Outliers in Statistical Data*, New York: John Wiley (1984).

Bashore, W.M., Araktingi, U., Levy, M., and Schweller, W.J., "Importance of a geological framework and seismic data for reservoir modeling and subsequent fluid flow predictions", *in Stochastic Modeling and Geostatistics: Principles, Methods and Case Studies*, J.M. Yarus and R.L.Chambers, AAPG Computer applications in geology, Tulsa, Oklahoma, 3 (1994) 159-176

Bear, J., *Dynamics of Fluids in Porous Media*, Amsterdam: Elsevier Scientific Publishing Co. (1972).

Beard, D. C. and P. K. Weyl, "Influence of Texture on Porosity and Permeability of Unconsolidated Sandstone," *American Association of Petroleum Geologists Bulletin*, 57 (1973) 349-369.

Beveridge, G. S. G. and R.S. Schechter, *Optimization: Theory and Practice*, New York: McGraw-Hill, Inc. (1970).

Bhosale, J. V., "Generation and Analysis of Stochastic Fields by the Source Point Method, " M.S. Thesis, The University of Texas at Austin (1992).

Box, G. E. P. and D. R. Cox, "An Analysis of Transformations," *Jornal of the Royal Statistical Society,* Series B, 26 (1964) 211-252.

Box, G. E. P., W. G. Hunter, and J. S. Hunter, *Statistics for Experimenters*, New York: John Wiley (1978).

Bras, R. L. and I. Rodriguez-Iturbe, *Random Functions and Hydrology,* Reading, Massachusetts: Addison-Wesley Publishing Co. (1985).

Breitenbach, E. A., "Reservoir Simulation: State of the Art," *Journal of Petroleum Technology,* 43 (1991) 1033-1036.

Brown, G.O., Hsieh, H.T., and Lucero, D.A., "Evaluation of laboratory dolomite core sample size using representative elementary volume concepts", *Water Resources Research*, 36 (2000) 1199-1208.

Bruining, J., "Modeling Reservoir Heterogeneity With Fractals," Report of the Enhanced Oil and Gas Recovery Research Program, The University of Texas at Austin (1991).

Burrough, P. A., "Multiscale Sources of Spatial Variation in Soil. I. The Application of Fractal Concepts to Nested Levels of Soil Variation," *Journal of Soil Science,* 34 (1983a) 577-597.

Burrough, P. A., "Multiscale Sources of Spatial Variation in Soil. II. A Non-Brownian Fractal Model and Its Application in Soil Survey," *Journal of Soil Science,* 34 (1983b) 599-620.

Campozana, Fernando P., "Incorporating Dynamic Data Into Geostatistical Reservoir Modeling," Ph.D. Dissertation, University of Texas at Austin (1997).

Carr, J. R., *Numerical Analysis for the Geological Sciences*, Englewood Cliffs, New Jersey: Prentice-Hall (1995).

Chambers, R.L., Zinger, M.A., and Kelly, M.C., "Constraining geostatistical reservoir descriptions with 3-D seismic data to reduce uncertainty", *in Stochastic Modeling and Geostatistics: Principles, Methods and Case Studies*, J.M. Yarus and R.L.Chambers, AAPG Computer applications in geology, Tulsa, Oklahoma, 3 (1994), 143-157.

Chambers, K.T., D. R. DeBaun, L. J. Durlofsky, I. J. Taggart, A. B. Bernath, A. Y. Shen, H. A. Legarre, and D. J. Goggin, "Geologic Modeling, Upscaling, and Simulation of Faulted Reservoirs Using Faulted Stratigraphic Grids," SPE 51889 presented at the *Society of Petroleum Engineers Simulation Symposium, Houston*, Feb. 14-17 (1999), 12 pgs.

Chandler, M. A., G. Kocurek, D. J. Goggin, and L. W. Lake, "Effects of Stratigraphic Heterogeneity on Permeability in Eolian Sandstone Sequences, Page Sandstone, Northern Arizona," *American Association of Petroleum Geologists Bulletin*, 73 (1989) 658-668.

Chang, Y. B., M. T. Chang, M.T. Lim, G. A. Pope, and K. Sepehrnoori, "CO_2 Flow Patterns Under Multiphase Flow: Heterogeneous Field-Scale Conditions," *SPE Reservoir Engineering*, 9 (1994) 208-216.

Cheeney, R. F., *Statistical Methods in Geology*, London: George Allen and Unwin (1983).

Chilès, J.P., and Delfiner, P., *Geostatistics - Modelling spatial uncertainty*, John Wiley and Sons Inc: New York (1999).

Corbett, P. W. M., "Reservoir Characterisation of a Laminated Sediment," PhD Dissertation, Heriot-Watt University, Edinburgh (1993).

Corbett, P.W.M., Anggraeni, S., and Bowen, D., "The use of the probe permeameter in carbonates - addressing the problems of permeability support and stationarity", *Log Analyst*, 40 (1999), 316-326.

Corbett, P. W. M. and J. L. Jensen, "A Comparison of Small-Scale Permeability Measurement Methods for Reservoir Characterisation," presented at the PSTI Advances in Reservoir Technology Conference, Edinburgh, February 21-22 (1991).

Corbett, P. W. M. and J. L. Jensen, "Estimating the Mean Permeability: How Many Measurements Do You Need?," *First Break*, 10 (1992a) 89-94.

Corbett, P. W. M. and J. L. Jensen, "Variation of Reservoir Statistics According to Sample Spacing and Measurement Type for Some Intervals in the Lower Brent Group," *Log Analyst*, 33 (1992b) 22-41.

Corbett, P. W. M. and J. L. Jensen, "Quantification of Variability in Laminated Sediments: A Role for the Probe Permeameter in Improved Reservoir Characterization," in *Characterization of Fluvial and Aeolian Reservoirs,* C. P. North and D. J. Prosser (eds.), London Geological Society Special Publication, 73 (1993a) 433-442.

Corbett, P. W. M. and J. L. Jensen, "Application of Probe Permeametry to the Prediction of Two-Phase Flow Performance in Laminated Sandstones (Lower Brent Group, North Sea)," *Marine and Petroleum Geology*, 10 (1993b) 335-346.

Corbett, P. W. M., P. S. Ringrose, J. L. Jensen,, and K. S. Sorbie, "Laminated Clastic Reservoirs: The Interplay of Capillary Pressure and Sedimentary Architecture" Transactions of 67th Annual Technical Conference and Exhibition of the Society of Petroleum Engineers, Wash. DC, 4-7 Oct. (1992), paper SPE 24699, 11p.

Corbett, P.W.M., T. Good, J. L. Jensen, G. E. Pickup, P. S. Ringrose, K. S. Sorbie, J. J. M. Lewis, "Reservoir Descrption in the 90's: A Perspective from Flow Simulation Through Layercake Parasequence Flow Units", *in AD1995: NW Europe's hydrocarbon industry* , Hurst, A., and Glennie, K., (eds.), The Geological Society, London (1996), 167-176.

Cox, D. R. and D. V. Hinkley, *Theoretical Statistics,* London: Chapman and Hall (1974).

Cramér, H., *Mathematical Methods of Statistics*, Princeton: Princeton University Press (1946).

Cressie, N. A. C., *Statistics for Spatial Data*, New York: John Wiley (1991).

Crutchfield, J. P., J. D. Farmer, N. H. Packard, and R. S. Shaw, "Chaos," *Scientific American*, (Dec. 1986) 46-57.

Cuevas Gozalo, M. C. and A. W. Martinius, "Outcrop Data-Base for the Geological Characterization of Fluvial Reservoirs: An Example From Distal Fluvial Fan Deposits in the Loranca Basin, Spain," in *Characterization of Fluvial and Aeolian Reservoirs,* C. P. North and D. J. Prosser (eds.), London Geological Society Special Publication 73 (1993) pp. 79-94.

Dagan, G., *Flow and Transport in Porous Formations*, New York: Springer-Verlag (1989).

Dake, L. P., *Fundamentals of Reservoir Engineering*, Amsterdam: Elsevier Scientific Publishing Co. (1978.

Dake, L.P., *The Practice of Reservoir Engineering*, Amsterdam: Elsevier Scientific Publishing Co. (1994).

Datta-Gupta, A., G. A. Pope and L. W. Lake, "Heterogeneity and Mixing in Flow Through Porous Media," in *Proceedings, Computational Methods in Water Resources IX, Vol. 2: Mathematical Modeling in Water Resources*, T.F. Russell, R.E. Ewing, C.A. Brebbia, W.G. Gray, G.F. Pinder (eds.), Computational Mechanics Publications and Elsevier Science Publishers, June (1992).

Datta-Gupta, A., L. W. Lake, and G. A. Pope, "Characterizing Heterogeneous Permeable Media with Spatial Statistics and Tracer Data Using Sequential Simulated Annealing," *Mathematical Geology* 27 (1995), 763-787.

Davis, J. C., *Statistics and Data Analysis in Geology*, New York: John Wiley (1973).

Davis, J. C. and T. Chang, "Estimating Potential for Small Fields in Mature Petroleum Province," *American Association of Petroleum Geologists Bulletin*, 73 (1989) 967-976.

de Lima, L. C., "*Large-Scale Conditional Simulation: Domain and Matrix Decomposition and the Moving Template Model,*", Ph.D. Dissertation, The University of Texas at Austin (1995).

de Marsily, G., *Quantitative Hydrogeology*, Orlando: Academic Press, Inc. (1986).

Dehghani, K., P. M. Harris, K. A. Edwards, and W. T. Dees, "Modeling a Vuggy Carbonate Reservoir, McElroy Field, West Texas", *American Association of Petroleum Geologists Bulletin*, 83 (1999) 19-42.

Dershowitz, W. S., and H. H. Einstein, "Characterizing Rock Joint Geometry with Joint System Models," *Rock Mechanics and Rock Engineering*, 21 (1988) 21-51.

Desbarats, A. J., "Numerical Estimation of Effective Permeability in Sand-Shale Formations," *Water Resources Research*, 23 (1987) 273-286.

Desbarats, A. J., "Support Effects and the Spatial Averaging of Transport Properties," *Mathematical Geology*, 21 (1989) 383-389.

Deutsch, C. V. and A. G. Journel, *GSLIB Geostatistical Software Library and User's Guide*, New York: Oxford University Press (1992.

Dreyer, T., A. Scheie, and O. Walderhaug, "Minipermeameter-Based Study of Permeability Trends in Channel Sand Bodies," *American Association of Petroleum Geologists Bulletin*, 74 (1990) 359-374.

Dubrule, O., *Geostatistics in Petroleum Geology*, AAPG Continuing Education Course Note Series, Tulsa, Oklahoma: 38 (1998).

Dykstra, H. and R. L. Parsons, "The Prediction of Oil Recovery by Waterflood," *Secondary Recovery of Oil in the United States*, 2nd ed., New York: American Petroleum Institute (1950, 160-174.

Eisenberg, R.A., P. M. Harris, C. W. Grant, "Modeling Reservoir Heterogeneity Within Outer Ramp Carbonate Facies Using an Outcrop Analog, San Andres Formation of the Permian Basin," *American Association of Petroleum Geologists Bulletin*, 78 (1994), 1337-1359.

Emanuel, A. S., G. K. Alameda, R. A. Behrens, and T. A. Hewett, "Reservoir Performance Prediction Methods Based on Fractal Geostatistics," *SPE Reservoir Engineering,* 4 (1989) 311-317.

Farmer, C. L., "The Mathematical Generation of Reservoir Geology" in *Numerical Rocks* , Joint IMA/SPE European Conference on the Mathematics of Oil Recovery, Robinson College, Cambridge University, July 25-27 (1989.

Feller, W., *An Introduction to Probability Theory and Its Applications, Vol. 1*, New York: John Wiley (1950.

Feller, W., *An Introduction to Probability Theory and Its Applications, Vol. 2*, New York: John Wiley (1971.

Fogg, G. E., F. J. Lucia, and R.K. Senger, "Stochastic Simulation of Interwell-Scale Heterogeneity for Improved Prediction of Sweep Efficiency in a Carbonate Reservoir, in *Reservoir Characterization II*, L. W. Lake, H. B. Carroll, Jr., and T. C. Wesson (eds.), New York: Academic Press, Inc. (1991, 355-381.

Gelhar, L. W. and C. L. Axness, "Three Dimensional Stochastic Analysis of Macrodispersion in Aquifers," *Water Resources Research 19* (1983) 161-180.

Gelhar, L. W., *Stochastic Subsurface Hydrology*, Englewood Cliffs, New Jersey: Prentice-Hall (1993).

Geman, S. and D. Geman, "Stochastic Relaxation, Gibb's Distributions and the Bayesian Restoration of Images," *Proceedings IEEE,* Transactions on Patterns Analysis and Machine Intelligence, PAMI - 6 (1984) 721-741.

Ghori, S. G. and J. P. Heller, "Computed Effect of Heterogeneity on Well-to-Well Tracer Results," New Mexico Petroleum Research Resource Center Report No. 90-33, Socorro, New Mexico, Sept. (1990).

Goggin, D. J., "Probe Permeametry: Is It Worth the Effort?" *Marine and Petroleum Geology,* 10 (1993), pp. 299-308.

Goggin, D. J., "Geologically Sensible Modelling of the Spatial Distribution of Permeability in Eolian Deposits: Page Sandstone (Jurassic) Outcrop, Northern Arizona," PhD dissertation, The University of Texas at Austin (1988.

Goggin, D. J., M. A. Chandler, G. A. Kocurek, and L. W. Lake, "Patterns of Permeability in Eolian Deposits," *SPE Formation Evaluation*, 3 (1988) 297-306.

Goggin, D. J., M. A. Chandler, G. A. Kocurek, and L. W. Lake, "Permeability Transects of Eolian Sands and Their Use in Generating Random Permeability Fields," *SPE Formation Evaluation*, 7 (1992) 7-16.

Goggin, D. J., R. A. Eisenberg, C. W. Grant, and P. M. Harris, "Stratigraphic Controls on Permeability and Fluid Flow Patterns in Ramp Carbonates: An Outcrop Study of the San Andres Formation, New Mexico, USA," in *Subsurface Reservoir Characterization from Outcrop Observations*, R. Eschard, and B. Doligez, (eds.), Editions Technip, 27 rue Ginoux, Paris (1993) pp. 65-95.

Grant, C. W., D. J. Goggin, and P. M. Harris, "Outcrop Analog for Cyclic-Shelf Reservoirs, San Andres Formation of Permian Basin: Stratigraphic Framework, Permeability Distribution, Geostatistics, and Fluid-Flow Modeling," *American Association of Petroleum Geologists Bulletin*, 78 (1994) 23-54.

Graybill, F. A., *An Introduction to Linear Statistical Models*, New York: McGraw-Hill (1961.

Gunter, G. W., J. M. Finneran, D. J. Hartmann, and J. D. Miller, "Early Determination of Reservoir Flow Units Using an Integrated Petrophysical Method", SPE 38679, Proceedings of 1997 Society of Petroleum Engineers Annual Technical Conference and Exhibition, San Antonio, Tx., U.S.A., 5-8 October (1997), 373-380.

Hahn, G. J. and W. Q. Meeker, *Statistical Intervals*, New York: John Wiley (1991).

Hald, A., *Statistical Theory with Engineering Applications,* New York: John Wiley (1952).

Haldorsen, H. H., and L. W. Lake, "A New Approach to Shale Management in Field-Scale Models," *Society of Petroleum Engineers Journal*, 24 (1984) p. 447-457.

Halvorsen, C. and A. Hurst, "Principles, Practice and Applications of Laboratory Minipermeametry," in *Advances in Core Evaluation and Accuracy and Prediction*, P. F. Worthington (ed.), London: Gordon and Breach Science Publishers, (1990), pp. 521-549.

Hardy, H. H. and R. A. Beier, *Fractals in Reservoir Engineering*, Singapore: World Scientific Publishing Co. (1994).

Hearn, C.L., W.J. Ebanks, Jr., R.S. Tye and V. Ranaganathan, "Geological factors influencing reservoir performance of the Hartzog Draw field," *Journal of Petroleum Technology*, 36 (1984), p.1335-1344.

Hearn, G. J. A., O. S. Krakstad, D. T. Rian, and M. Skaug, "The Application of New Approaches for Shale management in a Three-Dimensional Simulation Study of the Frigg Field," *SPE Formation Evaluation*, 3 (1988) 493-502.

Heller, J. P., "Observations of Mixing and Diffusion in Porous Media," Proceedings of the Second Symposium on the Fundamentals of Transport Phenomena in Porous Media, IAHR-ISS, Ontario, August 7-11 (1972).

Hewett, T. A., "Fractal Distributions of Reservoir Heterogeneity and Their Influence on Fluid Transport," SPE 15386, presented at the 61st Annual Technical Conference and Exhibition of the Society of Petroleum Engineers, New Orleans, Louisiana, October 5-8 (1986).

Hoaglin, D. C., F. Mosteller, and J. W. Tukey, *Understanding Robust and Exploratory Data Analysis*, New York: John Wiley (1983.

Hoffmann, B., *Albert Einstein, Creator and Rebel*, with H. Dukas, New York: Viking Press (1972).

Hohn, M. E., *Geostatistics and Petroleum Geology*, Boston: Kluwer Academic Pubs. (1998).

Hosking, J. R. M., and J. R. Wallis, *Regional Frequency Analysis*, New York: Cambridge U. Press (1997).

House, M. R., "Orbital Forcing Timescales: An Introduction," in *Orbital Forcing Timescales and Cyclostratigraphy*, M. R. House and A. S. Gale (eds.), London: The Geological Society, 85 (1995) 1-18.

Hoye, T., E. Damsleth, and K. Hollund, "Stochastic structural modelling of Troll West, with special emphasis on the thin oil zone", *in Stochastic Modeling and Geostatistics: Principles, Methods and Case Studies*, J.M. Yarus and R.L.Chambers, AAPG Computer applications in geology, Tulsa, Oklahoma, 3 (1994), 217-239.

Hurst, A. and K. J. Rosvoll, "Permeability Variations in Sandstones and Their Relationship to Sedimentary Structures" in *Reservoir Characterization II*, L. W. Lake, H. B. Carroll Jr., and T. C. Wesson (eds.), Orlando: Academic Press, Inc. (1991), p. 166-196.

Hurst, A. and D. J. Goggin, "Probe Permeametry: An Overview and Bibliography," *American Association of Petroleum Geologists Bulletin*, 79 (1995) 463-473.

Isaaks, E.H. and R.M. Srivastava, *Applied Geostatistics*, New York: Oxford University Press (1989).

Jacobsen, T. and H. Rendall, "Permeability Patterns in Some Fluvial Sandstones: An Outcrop Study From Yorkshire, North-East England," in *Reservoir Characterization II*, L. W. Lake, H. B. Carroll Jr., and T. C. Wesson (eds.), Orlando: Academic Press, Inc. (1991), pp. 315-338.

Jenkins, G. M. and D. G. Watts, *Spectral Analysis and Its Applications*, San Francisco: Holden-Day (1968).

Jennings, J. W., Ruppel, S.C., and Ward, B., "Geostatistical analysis of petrophysical data and modeling of fluid flow effects in carbonate outcrops", SPE 49025, proceedings 1998 Annual Technical Conference, New Orleans (1988), 1-23.

Jensen, J. L. and I. D. Currie, "A New Method for Estimating the Dykstra-Parsons Coefficient To Characterize Reservoir Heterogeneity," *SPE Reservoir Engineering*, 5 (1990) 369-374.

Jensen, J. L. and L. W. Lake, "Optimization of Regression-Based Porosity-Permeability Predictions," Paper R, Tenth Annual Symposium of the Canadian Well Logging Society, Calgary, 29 September-2 October (1985).

Jensen, J. L. and L. W. Lake, "The Influence of Sample Size and Permeability Distribution upon Heterogeneity Measures," *SPE Reservoir Engineering*, 3 (1988) 629-637.

Jensen, J. L., "A Statistical Study of Reservoir Permeability Distributions," Ph.D. Dissertation, The University of Texas at Austin (1986).

Jensen, J. L., "Use of the Geometric Average for Effective Permeability Estimation," *Mathematical Geology*, 23 (1991) 833-840.

Jensen, J. L., "Some Statistical Properties of Power Averages for Lognormal Samples," *Water Resources Research*, 34 (1998) 2415-2418.

Jensen, J. L., C. A. Glasbey, and P. W. M. Corbett, "On the Interaction of Geology, Measurement, and Statistical Analysis of Small-Scale Permeability Measurements," *Terra Nova*, 6 (1994) 397-403.

Jensen, J. L., D. V. Hinkley, and L. W. Lake, "A Statistical Study of Reservoir Permeability: Distributions, Correlations and Averages," *SPE Formation Evaluation*, 2 (1987) 461-468.

Jensen, J. L., P. W. M. Corbett, G. E. Pickup, and P. S. Ringrose, "Permeability Semivariograms, Geological Structure, and Flow Performance," *Mathematical Geology*, 28 (1996) 419-436.

Johnson, N. L. and S. Kotz, *Continuois Univariate Distributions - 1*, New York: John Wiley (1970).

Jordan, D. L., and D. J. Goggin, "An Application of Categorical Indicator Geostatistics for Facies Modeling in Sand-Rich Turbidite Systems," SPE 30603 presented at the SPE Annual Technical Conference and Exhibition of the Society of Petroleum Engineers, Dallas, TX., Oct. 22-25 (1995).

Journel, A. G. and C. J. Huijbregts, *Mining Geostatistics*, Orlando: Academic Press (1978.

Journel, A. G., "Constrained Interpolation and Qualitative Information - the Soft Kriging Approach," *Mathematical Geology*, 18 (1992) 269-286.

Kasap, E. and L. W. Lake, "Calculating the Effective Permeability Tensor of a Grid Block," *SPE Formation Evaluation* 5 (1990), 192-200.

Kasap, E., "Analytic Methods to Calculate an Effective Permeability Tensor and Effective Relative Permeabilities for Cross-Bedded Flow Units," Ph.D. dissertation, The University of Texas at Austin (1990).

Kempers, L. J. T. M., "Dispersive Mixing in Unstable Displacements," in *Proceedings of the Second European Conference on The Mathematics of Oil Recovery*, D. Guerillot and O. Guillon (eds.), Paris: Editions Technip (1990), pp. 197-204.

Kendall, M. and A. Stuart, *The Advanced Theory of Statistics, Vol. 1: Distribution Theory,* 4th ed., New York: MacMillian Publishing Co. (1977).

Kenney, B. C., "Beware of Spurious Self-Correlations!," *Water Resources Research*, 18 (1982) 1041-1048.

Kerans, C., F. J. Lucia, and R. K. Senger, "Integrated Characterization of Carbonate Ramp Reservoirs Using Permian San Andres Formation Outcrop Analogs," *American Association of Petroleum Geologists Bulletin*, 78 (1994) 181-216.

Kermack, K. A., and J. B. S. Haldane, "Organic Correlation and Allometry," *Biometrika*, 37 (1950) 30-41.

Kitanidis, P.K., *Introduction to Geostatistics*, Cambridge: Cambridge University Press (1997).

Kittridge, M. G., L. W. Lake, F. J. Lucia, and G. E. Fogg, "Outcrop/Subsurface Comparisons of Heterogeneity in the San Andres Formation," *SPE Formation Evaluation*, 5 (1990) 233-240.

Klir, G. J. and T. A. Folger, *Fuzzy Sets, Uncertainty, and Information*, Englewood Cliffs, New Jersey: Prentice-Hall (1988).

Knutsen, H.P. and Y.C. Kim, *A Short Course on Geostatistical Ore Reserve Estimation,* Dept. of Mining and Geological Engineering, College of Mines, The University of Arizona, May (1978).

Koopmans, L. H., D. B. Owen, and J. I. Rosenblatt, "Confidence Intervals for the Coefficient of Variation for the Normal and Log Normal Distributions," *Biometrika,* 51 (1964) 25-32.

Koval, E. J., "A Method for Predicting the Performance of Unstable Miscible Displacement in Heterogeneous Media," *Soceity of Petroleum EngineersJournal*, 3 (1963) 145-154.

Krige, D.G., "A Statistical Approach to Some Basic Mine Valuation Problems on the Witwatersrand," *J.ournal of the Chemical, Metallurgical and Mining Soceity of South Africa*, 52 (1951) 119-139.

Lake, L. W. and J. L. Jensen, "A Review of Heterogeneity Measures Used in Reservoir Characterization," *In Situ*, 15 (1991) 409-440.

Lake, L. W. and M. A. Malik, "Modelling Fluid Flow Through Geologically Realistic Media," in *Characterization of Fluvial and Aeolian Reservoirs*, C. P. North, and D. J. Prosser (eds.), Geological Society Special Publication 73, (1993), pp. 367-375.

Lake, L. W., *Enhanced Oil Recovery*, Englewood Cliffs, N.J: Prentice Hall (1989.

Lambert, M. E., "A Statistical Study of Reservoir Heterogeneity," M.S. Thesis, The University of Texas at Austin, Aug. (1981).

Larsen, R. J. and M. L. Marx, *An Introduction to Mathematical Statistics and its Applications*, Englewood Cliffs, New Jersey: Prentice Hall (1986).

Lewis, J. J. M. and B. Lowden, "A Quantification of the Scales and Magnitudes of Permeability Heterogeneity in a Tidally-Dominated, Shallow Marine Sand Body," Geological Society Advances in Reservoir Geology Conference, London (1990.

Lewis, P. A. W. and E. J. Orav, *Simulation Methodology for Statisticians, Operations Analysts, and Engineers, Vol. 1*, Belmont, California: Wadsworth and Brooks/Cole (1989.

Li, D. and L. W. Lake, "A Moving Window Semivariance Estimator, *Water Resources Research*, 30 (1994) 1479-1489.

Li, D. and L. W. Lake, "Scaling Flow Through Heterogeneous Permeable Media", SPE 26648 presented at the 68th Annual Technical Conference and Exhibition of the Society of Petroleum Engineers, Houston, Texas, October 3-6 (1993.

Limbrunner, J. F., R. M. Vogel, and L. C. Brown, "Estimation of Harmonic Mean of a Lognormal Variable," *Journal of Hydrologic Engineering*, 5 (2000) 59-66.

Liu, K., P. Boult, S. Painter, and L. Paterson, "Outcrop analog for sandy braided stream reservoirs: Permeability patterns in the Triassic Hawkesbury Sandstone, Sydney Basin, Australia", *American Association of Petroleum Geologists Bulletin* 80 (1996), 1850-1866.

Lucia, F. J. and G. E. Fogg, "Geologic/Stochastic Mapping of Heterogeneity in a Carbonate Reservoir," *Journal of Petroleum Technology*, 42 (1990) 1298-1303.

Lucia, F. J., *Carbonate Reservoir Characterization*, Berlin: Springer-Verlag (1999).

MacCleod, M., R. A. Behrens, and T. T. Tran, "Incorporating Seismic Attribute Maps in 3D Reservoir Rocks," 36499, presented at the 71st Annual Technical Conference and Exhibition of the Society of Petroleum Engineers, Denver, CO: Oct. 6-9 (1996).

MacKean, K., "The Orderly Pursuit of Disorder," *Discover*, (Jan. 1987), 72-81.

Mann, C. J., "Misuses of Linear Regression in Earth Sciences," in *Use and Abuse of Statistical Methods in the Earth Sciences*, W. B. Size (ed.), New York: Oxford University Press (1987).

Mantoglou, A. and J. L. Wilson, "The Turning Bands Method for Simulation of Random Fields Using Line Generation by a Spectral Method," *Water Resources Research*, 18 (1982) 1379-1394.

Martinius, A.W. and R. A. Nieuwenhuijs, "Geological Description of Flow Units in Channel Sandstones in a Fluvial Reservoir Analogue (Loranc Basin, Spain)," *Petroleum Geoscience*, 1 (1995), 237-252.

Martino, R. L. and D. D. Sanderson, "Fourier and Autocorrelation Analysis of Estuarine Tidal Rhythmites, Lower Breathitt Formation (Pennsylvanian), Eastern Kentucky, USA," *Journal of Sedimentary Petrology*, 63 (1993) 105-119.

Matheron, G., *Éléments Pour une Théorie des Milieux Poreux*, Paris: Masson and Cie (1967).

Matheron, G., *Les Variables Régionalisées et leur Estimation*, Paris: Masson and Cie (1965).

McCray, A. W., *Petroleum Evaluations and Economic Decisions*, Englewood Cliffs, New Jersey: Prentice-Hall, Inc., (1975).

Metropolis, N., A. Rosenbluth, M. Rosenbluth, A. Teller, and E. Teller, "Equation of State Calculations by Fast Computing Machinges," *Journal of Chemical Physics*, 21 (1953) 1087-1092.

Meyer, P. L., *Introductory Probability and Statistical Applications*, Reading, Massachusetts: Addison-Wesley Publishing Co. (1966.

Miall, A. D., "Reservoir Heterogeneities in Fluvial Sandstones: Lessons from Outcrop Studies," *American Association of Petroleum Geologists Bulletin*, 72 (1988) 682-697.

Middleton, G. V., *Nonlinear Dynamics, Chaos, and Fractals*, St. John's, Newfoundland: Geological Association of Canada (1991).

Miller, R. G., *Beyond ANOVA, Basics of Applied Statistics*, New York: John Wiley (1986).

Miller, R. G., *Simultaneous Statistical Inference*, New York: Springer-Verlag (1981).

Moissis, D. E. and M. F. Wheeler, "Effect of the Structure of the Porous Medium on Unstable Miscible Displacement," in *Dynamics of Fluids in Hierarchical Porous Media*, J. H. Cushman (ed.), Orlando: Academic Press (1990).

Montgomery, D. C. and E. A. Peck, *Introduction to Linear Regression Analysis*, New York: John Wiley (1982.

Morgan, B. J. T., *Elements of Simulation*, London: Chapman and Hall Ltd. (1984).

Muskat, M., *Flow of Homogeneous Fluids*, New York: McGraw-Hill (1938).

Newendorp, P. D., "Bayesian Analysis--A Method for Updating Risk Estimates," *Journal of Petroleum Technology*, 24 (1972) 193-198.

Noetinger, B., "The Effective Permeability of a Heterogeneous Porous Medium," *Transport in Porous Media*, 15 (1994) 99-127.

Olea, R. A., *Geostatistics for Engineers and Earth Scientists*, Boston: Kluwer Academic Pubs. (1999).

Panda, Manmath, "A Forward Modeling of the Interwell Property Assignment in Eolian Depositions," Ph.D. Dissertation, The University of Texas at Austin, August (1994).

Panda, M. N. and L. W. Lake, "Parallel Simulated Annealing for Stochastic Reservoir Modeling," SPE 26418, presented at the 68th Annual Technical Conference and Exhibition of the Society of Petroleum Engineers, Houston, Texas, October 3-6 (1993).

Panda, M.N. and L. W. Lake, "A Physical Model of Cementation and Its Effects on Single-Phase Permeability, *American Association of Petroleum Geolositsts Bulletin*, 79, 3, (1995), 431-443.

Papoulis, A., *Probability, Random Variables, and Stochastic Processes*, New York: McGraw-Hill (1965).

Paul, G. W., L. W. Lake, G. A. Pope, and G. B. Young, "A Simplified Predictive Model for Micellar-Polymer Flooding," SPE 10733, presented at the 1982 California Regional Meeting of the Society of Petroleum Engineers, San Francisco, California, March 24-26 (1982).

Peaceman, D.W., *Fundamentals of Reservoir Simulation*, Amsterdam: Elsevier Scientific Publishing Co. (1978).

Pettijohn, F.J., P. E. Potter,, and R. Siever, *Sand and Sandstone*, New York: Springer-Verlag (1987).

Priestly, M. B., *Spectral Analysis and Time Series*, Orlando: Academic Press (1981).

Quigley, K. A., "A Technical Survey of Immiscible Gas Injection Projects," M.S. Report, The University of Texas at Austin (1984).

Rachmawati, S. B., R. J. Sustakoski, T. P. Whitacre, D. J. Goggin, M. Levy, A. B. Bernath, K. Anbarci, G. Wu, "Pattern Waterflood Development in a Giant, Mature Oil Field: Minas NW Segment Reservoir Characterization, Scale-Up, and Flow Modeling," SPE 38705, presented at the 1997 SPE Annual Technical Conference and Exhibition, San Antonio, Texas, October 5-8, Reservoir Engr., p. 535-547.

Ravnaas, R. D., R. F. Strickland, A. P. Yang, M. A. Malik, D. R. Prezbindowski, T. Mairs and L. W. Lake, "Three-Dimensional Conditional Simulation of Schneider (Buda) Field, Wood County, Texas," SPE 23970 presented at the 1992 SPE Permian Basin Oil and Gas Recovery Conference, Midland, Texas, March 18-20.

Rice, J. A., *Mathematical Statistics and Data Analysis*, Belmont, California: Wadsworth and Brooks/Cole (1988).

Ringrose, P. S., G. E. Pickup, J. L. Jensen, and K. S. Sorbie. "The Use of Correlation Statistics for Modelling Immiscible Displacements in Petroleum Reservoirs," Delft Conference on Mathematics of Oil Recovery, June 17-19 (1992).

Ringrose, P. S., K. S. Sorbie, P. W. M. Corbett, and J. L. Jensen, "Immiscible Flow Behaviour in Laminated and Cross-Bedded Sandstones," *Journal of Petroleum Science and Engineering*, 9 (1993) 103-124.

Rock, N. M. S., *Numerical Geology*, Berlin: Springer-Verlag (1988).

Rossini, C., F. Brega, L. Piro, M. Rovellini, and G. Spotti, Combined Geostatistical and Dynamic Simulations for Developing a Reservoir Management Strategy: A Case History, *Journal of Petroleum Technology*, 46 (1994) 979-985.

Satterthwaite, F. E., "An Approximate Distribution of Estimates of Variance Components," *Biometrics Bulletin*, 2 (1946) 110-114.

Schiffelbein, P., "Calculation of Confidence Limits for Geologic Measurements," in *Use and Abuse of Statistical Methods in the Earth Sciences*, W. B. Size (ed.), New York: Oxford University Press (1987).

Schwarzacher, W., *Sedimentation Models and Quantitative Stratigraphy,* Amsterdam: Elsevier Scientific Publishing Co. (1975).

Seber, G. A. F., *Linear Regresion Analysis*, New York: John Wiley (1977).

Seifert, D., J. J. M. Lewis, C. Y. Hern, and N. C. T. Steel, "Well Placement Optimisation and Risking Using 3-D Stochastic Reservoir Modelling Techniques," SPE 35520 presented at the 1996 SPE/NPF European 3-D Reservoir Modelling Conference, Stavanger, 16-17 April, 289-300.

Shafer, J. M. and M. D. Varljen, "Approximation of Confidence Limits on Sample Semivariograms From Single Realizations of Spatially Correlated Fields," *Water Resources Research*, 26 (1990) 1787-1802.

Shinozuka, M. and C. M. Jan, "Digital Simulation of the Random Processes and its Applications," *Journal of Sound and Vibration*, 25 (1972) 111-128.

Sinclair, A. J., "Applications of Probability Graphs in Mineral Exploration," *Association of Exploration Geologists Special Volume No. 1*, Richmond, British Columbia (1976).

Smith, P. J., D. J. Hendry, and A. R. Crowther, "The Quantification and Management of Uncertainty in Reserves," SPE 26056, presented at the Western Regional Meeting, Anchorage, 26-28 May (1993), 275-284.

Snedecor, G. W. and W. G. Cochran, *Statistical Methods,* 7th ed., Ames, Iowa: Iowa State University Press (1980).

Sokal, R. R. and F. J. Rohlf, *Biometry*, 2nd ed., New York: W. H. Freeman and Co. (1981).

Sorbie, K. S., F. Feghi, G. E. Pickup, P. S. Ringrose, and J. L. Jensen, "Flow Regimes in Miscible Displacements in Heterogeneous Correlated Random Fields," *SPE Advanced Technology Series*, 2 (1994) 78-87.

Sweet, M.L., Blewden, C.H., Carter, A.M. and Mills, C.A., "Modeling Heterogeneity in a Low-Permeability Gas Reservoir using Geostatistical Techniques, Hyde Field, Southern North Sea," *American Association of Petroleum Geologists Bulletin* 80 (1996), 1719-1735.

Tarantola, A., *Inverse Problem Theory*, Amsterdam: Elsevier Scientific Publishing Co. (1987).

Tehrani, D. H., "An Analysis of a Volumetric Balance Equation for Calculation of Oil in Place and Water Influx," *Journal of Petroleum Technology,* 37 (1985), 1664-1670.

Teissier, G., "La Relation d'Allométrie: Sa Signification Statistique et Biologique," *Journal of Biometrics*, 4 (1948), 14-53.

Thomas, D. C. and V. J. Pugh, "A Statistical Analysis of the Accuracy and Reproducibility of Standard Core Analysis," *Log Analyst,* 30 (1989), 71-77.

Thomas, S.D., P. W. M., Corbett, and J. L. Jensen, "Upscaling Vertical Permeability within a Fluvio-aeolian Reservoir", Proc. 4th US DOE et al., Reservoir Characterization Int. Tech. Conf., Houston, March (1997), 583-601.

Tjolsen, C.B., Johnsen, G., Halvorsen, G., Ryseth, A., and Damsleth, E., Seismic Data can Improve Stochastic Facies Modelling," *SPE Formation Evaluation*, 11 (1996), 141-146.

Toro-Rivera, M. L. E., P. W. M. Corbett, and G. Stewart, "Well Test Interpretation in a Heterogeneous Braided Fluvial Reservoir," SPE 28828, presented at the European Petroleum Conference, London, October 25-27 (1994), 223-238.

Troutman, B. M. and G. P. Williams, "Fitting Straight Lines in the Earth Sciences," in *Use and Abuse of Statistical Methods in the Earth Sciences*, W. B. Size (ed.), New York: Oxford University Press (1987), pp. 107-128.

Tyler, K. J., T. Svances, and A. Henriquez, "Heterogeneity Modeling Used for Production Simulation of a Fluvial Reservoir," *SPE Formation Evaluation,* 9 (1994) 85-92.

van Wagoner, J. C., C. R. Mitchum, K. M. Campion, and V. D Rahmanian, "Siliclastic Sequence Stratigraphy in Well Logs, Cores and Outcrops," *American Association of Petroleum Geologists Methods in Exploration Series,* 7 (1990) 55.

Vanmarcke, E., *Random Fields: Analysis and Synthesis,* Cambridge, Massachusetts: The MIT Press (1984).

Waggoner, J. R., J. L. Castillo, and L. W. Lake, "Simulation of EOR Processes in Stochastically Generated Permeable Media," *SPE Formation Evaluation*, 7 (1992) 173-180.

Warren, J. E. and H. S. Price, "Flow in Heterogeneous Porous Media," *Society of Petroleum EngineersJournal*, 1 (1961) 153-169.

Warren, J. E., F. F. Skiba, and H. S. Price, "An Evaluation of the Significance of Permeability Measurements," *Journal of Petroleum Technology*, 13 (1961), 739-744.

Weber, K. J. and L. C. van Geuns, "Framework for Constructing Clastic Reservoir Simulation Models," *Journal of Petroleum Technology,* 42 (1990) 1248-53 and 1296-1297.

Weber, K.J., "Influence of Common Sedimentary Structures on Fluid Flow in Reservoir Models, *Journal of Petroleum Technology*, 34 (1982), 665-672.

White, C. D. and B. J. Willis, "A Method to Estimate Length Distributions from Outcrop Data," *Mathematical Geology*, 32 (2000), 389-419.

Willhite, G. P., *Waterflooding*, SPE Textbook Series, Dallas, Texas: Society of Petroleum Engineers (1986.

Williams, J. K., A. J. Pearce, and G. Geehan, "Modelling Complex Channel Environments, Assessing Connectivity and Recovery," presented at 7th European IOR Symposium, Moscow, Oct. 27-29 (1993.

Willis, B.J., "Permeability structure of a compound valley fill in the Cretaceous Fall River Formation of South Dakota", *American Association of Petroleum Geologists Bulletin,* 82 (1998), 206-227.

Wolf, D.J., K. D. Withers, K.D., and M. D. Burnaman, "Integration of Well and Seismic Data Using Geostatistics", *in Stochastic Modeling and Geostatistics: Principles, Methods and Case Studies*, J.M. Yarus and R.L.Chambers, AAPG Computer applications in geology, Tulsa, Oklahoma, 3 (1994), 177-200.

Yang, A. P. and L. W. Lake, "The Accuracy of Autocorrelation Estimates," *In Situ,* 12 (1988) 227-274.

Yang, A. P., "Stochastic Heterogeneity and Dispersion," Ph.D. dissertation, The University of Texas at Austin (1990).

Zeito, G. A., "Interbedding of Shale Breaks and Reservoir Heterogeneities," *Journal of Petroleum Technology*, 17 (1965) 1223-1228.

Zheng, S., P. W. M. Corbett, A. Ryseth, and G. Stewart, "Uncertainty in Well Test and Core Permeability Analysis: A Case Study in Fluvial Channel Reservoirs, Northern North Sea, Norway", *American Association of Petroleum Geologists Bulletin* 84 (2000), in press.

INDEX